Advanced Level Mathematics
Pure Mathematics 2 & 3
Hugh Neill and Douglas Quadling

CAMBRIDGE
UNIVERSITY PRESS

The publishers would like to acknowledge the contributions of the following people to this series of books: Tim Cross, Richard Davies, Maurice Godfrey, Chris Hockley, Lawrence Jarrett, David A. Lee, Jean Matthews, Norman Morris, Charles Parker, Geoff Staley, Rex Stephens, Peter Thomas and Owen Toller.

CAMBRIDGE UNIVERSITY PRESS
Cambridge, New York, Melbourne, Madrid, Cape Town, Singapore, São Paulo, Delhi

Cambridge University Press
The Edinburgh Building, Cambridge CB2 8RU, UK

www.cambridge.org
Information on this title: www.cambridge.org/9780521530125

First published 2002
8th printing 2009

Printed in the United Kingdom at the University Press, Cambridge

A catalogue record for this publication is available from the British Library

ISBN 978-0-521-53012-5 paperback

Contents

Introduction

Cambridge International Examinations (CIE) Advanced Level Mathematics has been created especially for the new CIE mathematics syllabus. There is one book corresponding to each syllabus unit, except for this book which covers two units, the second and third Pure Mathematics units, P2 and P3.

The syllabus content is arranged by chapters which are ordered so as to provide a viable teaching course. The first eleven chapters are required for unit P2; all the chapters are required for unit P3. This is indicated by the vertical grey bars on the contents page.

A few sections include important results that are difficult to prove or outside the syllabus. These sections are marked with an asterisk (*) in the section heading, and there is usually a sentence early on explaining precisely what it is that the student needs to know.

Some paragraphs within the text appear in *this type style*. These paragraphs are usually outside the main stream of the mathematical argument, but may help to give insight, or suggest extra work or different approaches.

Graphic calculators are not permitted in the examination, but they are useful aids in learning mathematics. In the book the authors have noted where access to a graphic calculator would be especially helpful but have not assumed that they are available to all students.

Numerical work is presented in a form intended to discourage premature approximation. In ongoing calculations inexact numbers appear in decimal form like 3.456..., signifying that the number is held in a calculator to more places than are given. Numbers are not rounded at this stage; the full display could be, for example, 3.456 123 or 3.456 789. Final answers are then stated with some indication that they are approximate, for example '1.23 correct to 3 significant figures'.

There are plenty of exercises, and each chapter ends with a Miscellaneous exercise which includes some questions of examination standard. There are two Revision exercises for the material common to units P2 and P3, and a further Revision exercise for unit P3. There are also two Practice examination papers for unit P2 at the end of P2&3, and two Practice examination papers for unit P3 at the end of P3.

Some exercises include questions that go beyond the likely requirements of the examinations, either in difficulty or in length or both. In the P2&3 chapters some questions may be more appropriate for P3 than for P2 students. Questions marked with an asterisk require knowledge of results or techniques outside the syllabus.

Cambridge University Press would like to thank OCR (Oxford, Cambridge and RSA Examinations), part of the University of Cambridge Local Examinations Syndicate (UCLES) group, for permission to use past examination questions set in the United Kingdom.

The authors thank UCLES and Cambridge University Press, in particular Diana Gillooly, for their help in producing this book. However, the responsibility for the text, and for any errors, remains with the authors.

Unit P2 and Unit P3

The subject content of unit P2 is a subset of the subject content of unit P3. This part of the book (pages 1–158) comprises the subject content of unit P2, and is required for both units P2 and P3. The additional material required to complete unit P3 is contained in the second part of the book.

1 Polynomials

This chapter is about polynomials, which include linear and quadratic expressions. When you have completed it, you should

- be able to add, subtract, multiply and divide polynomials
- understand the words 'quotient' and 'remainder' used in dividing polynomials
- be able to use the method of equating coefficients
- be able to use the remainder theorem and the factor theorem.

1.1 Polynomials

You already know a good deal about polynomials from your work on quadratics in Chapter 4 of Pure Mathematics 1 (unit P1), because a quadratic is a special case of a polynomial. Here are some examples of polynomials.

$$3x^3 - 2x^2 + 1 \qquad 3 \qquad 4 - 2x \qquad x^2 \qquad 1$$
$$2x^4 \qquad 1 - 2x + 3x^5 \qquad \sqrt{2}x^2 \qquad \tfrac{1}{2}x^{17} \qquad x$$

> A (non-zero) **polynomial**, $p(x)$, is an expression in x of the form
>
> $$ax^n + bx^{n-1} + \ldots + jx + k$$
>
> where $a, b, c, \ldots, k$ are real numbers, $a \neq 0$, and n is a non-negative integer.

The number n is called the **degree** of the polynomial. The expressions ax^n, bx^{n-1}, $\ldots$, jx and k which make up the polynomial are called **terms**. The numbers $a, b, c, \ldots, j$ and k are called **coefficients**; a is the **leading coefficient**. The coefficient k is the **constant term**.

Thus, in the quadratic polynomial $4x^2 - 3x + 1$, the degree is 2; the coefficients of x^2 and x, and the constant term, are 4, -3 and 1 respectively.

Polynomials with low degree have special names: if the polynomial has
- degree 0 it is called a **constant polynomial**, or a **constant**
- degree 1 it is called a **linear polynomial**
- degree 2 it is called a **quadratic polynomial**, or a **quadratic**
- degree 3 it is called a **cubic polynomial**, or a **cubic**
- degree 4 it is called a **quartic polynomial**, or a **quartic**.

When a polynomial is written as $ax^n + bx^{n-1} + \ldots + jx + k$, with the term of highest degree first and the other terms in descending degree order finishing with the constant term, the terms are said to be in **descending order**. If the terms are written in the reverse order, they are said to be in **ascending order** (or ascending powers of x). For example, $3x^4 + x^2 - 7x + 5$ is in descending order; in ascending order it is $5 - 7x + x^2 + 3x^4$. It is the same polynomial whatever order the terms are written in.

The functions $\dfrac{1}{x} = x^{-1}$ and $\sqrt{x} = x^{\frac{1}{2}}$ are not polynomials, because the powers of x are not positive integers or zero.

Polynomials have much in common with integers. You can add them, subtract them and multiply them together and the result is another polynomial. You can even divide a polynomial by another polynomial, as you will see in Section 1.4.

1.2 Addition, subtraction and multiplication of polynomials

To add or subtract two polynomials, you simply add or subtract the coefficients of corresponding powers; in other words, you collect like terms. Suppose that you want to add $2x^3 + 3x^2 - 4$ to $x^2 - x - 2$. Then you can set out the working like this:

$$
\begin{array}{rrrrrrr}
2x^3 & + & 3x^2 & & & - & 4 \\
     &   & x^2 & - & x & - & 2 \\
\hline
2x^3 & + & 4x^2 & - & x & - & 6
\end{array}
$$

Notice that you must leave gaps in places where the coefficient is zero. You need to do addition so often that it is worth getting used to setting out the work in a line, thus:

$$\left(2x^3 + 3x^2 - 4\right) + \left(x^2 - x - 2\right) = (2+0)x^3 + (3+1)x^2 + (0+(-1))x + ((-4)+(-2))$$
$$= 2x^3 + 4x^2 - x - 6.$$

You will soon find that you can miss out the middle step and go straight to the answer.

The result of the polynomial calculation $\left(2x^3 + 3x^2 - 4\right) - \left(2x^3 + 3x^2 - 4\right)$ is 0. This is a special case, and it is called the **zero polynomial**. It has no degree.

Look back at the definition of a polynomial, and see why the zero polynomial was not included there.

Multiplying polynomials is harder. It relies on the rules for multiplying out brackets,

$$a(b + c + \ldots + k) = ab + ac + \ldots + ak \qquad \text{and} \qquad (b + c + \ldots + k)a = ba + ca + \ldots + ka.$$

To apply these rules to multiplying the two polynomials $5x + 3$ and $2x^2 - 5x + 1$, replace $2x^2 - 5x + 1$ for the time being by z. Then

$$(5x + 3)\left(2x^2 - 5x + 1\right) = (5x + 3)z$$
$$= 5xz + 3z$$
$$= 5x\left(2x^2 - 5x + 1\right) + 3\left(2x^2 - 5x + 1\right)$$
$$= \left(10x^3 - 25x^2 + 5x\right) + \left(6x^2 - 15x + 3\right)$$
$$= 10x^3 - 19x^2 - 10x + 3.$$

In practice, it is easier to note that every term in the left bracket multiplies every term in the right bracket. You can show this by setting out the steps in the following way.

	$2x^2$	$-$		$5x$	$+$	1	$\times$
$10x^3$	$-$		$25x^2$	$+$		$5x$	$5x$
	$+$		$6x^2$	$-$		$15x + 3$	$+3$
$10x^3$	$+$	$(-25+6)x^2$	$+$	$(5-15)x$	$+$	3	

giving the result $10x^3 - 19x^2 - 10x + 3$.

It is worth learning to work horizontally. The arrows below show the term $5x$ from the first bracket multiplied by $-5x$ from the second bracket to get $-25x^2$.

$$(5x+3)(2x^2 - 5x + 1) = 5x(2x^2 - 5x + 1) + 3(2x^2 - 5x + 1)$$
$$= (10x^3 - 25x^2 + 5x) + (6x^2 - 15x + 3)$$
$$= 10x^3 - 19x^2 - 10x + 3.$$

You could shorten the process and write

$$(5x+3)(2x^2 - 5x + 1) = 10x^3 - 25x^2 + 5x + 6x^2 - 15x + 3$$
$$= 10x^3 - 19x^2 - 10x + 3.$$

If you multiply a polynomial of degree m by a polynomial of degree n, you have a calculation of the type

$$\left(ax^m + bx^{m-1} + \ldots\right)\left(Ax^n + Bx^{n-1} + \ldots\right) = aAx^{m+n} + \ldots$$

in which the largest power of the product is $m+n$. Also the coefficient aA is not zero because neither of a and A is zero. This shows that:

> When you multiply two polynomials, the degree of the product polynomial is the sum of the degrees of the two polynomials.

Exercise 1A

1 State the degree of each of the following polynomials.

(a) $x^3 - 3x^2 + 2x - 7$ (b) $5x+1$ (c) $8 + 5x - 3x^2 + 7x + 6x^4$

(d) 3 (e) $3 - 5x$ (f) x^0

2 In each part find $p(x) + q(x)$, and give your answer in descending order.

(a) $p(x) = 3x^2 + 4x - 1$, $q(x) = x^2 + 3x + 7$

(b) $p(x) = 4x^3 + 5x^2 - 7x + 3$, $q(x) = x^3 - 2x^2 + x - 6$

(c) $p(x) = 3x^4 - 2x^3 + 7x^2 - 1$, $q(x) = -3x - x^3 + 5x^4 + 2$

(d) $p(x) = 2 - 3x^3 + 2x^5$, $q(x) = 2x^4 + 3x^3 - 5x^2 + 1$

(e) $p(x) = 3 + 2x - 4x^2 - x^3$, $q(x) = 1 - 7x + 2x^2$

3 For each of the pairs of polynomials given in Question 2 find $p(x) - q(x)$.

4 Note that $p(x) + p(x)$ may be shortened to $2p(x)$. Let $p(x) = x^3 - 2x^2 + 5x - 3$ and $q(x) = x^2 - x + 4$. Express each of the following as a single polynomial.

    (a) $2p(x) + q(x)$ (b) $3p(x) - q(x)$ (c) $p(x) - 2q(x)$ (d) $3p(x) - 2q(x)$

5 Find the following polynomial products.

    (a) $(2x - 3)(3x + 1)$ (b) $(x^2 + 3x - 1)(x - 2)$

    (c) $(x^2 + x - 3)(2x + 3)$ (d) $(3x - 1)(4x^2 - 3x + 2)$

    (e) $(x^2 + 2x - 3)(x^2 + 1)$ (f) $(2x^2 - 3x + 1)(4x^2 + 3x - 5)$

    (g) $(x^3 + 2x^2 - x + 6)(x + 3)$ (h) $(x^3 - 3x^2 + 2x - 1)(x^2 - 2x - 5)$

    (i) $(1 + 3x - x^2 + 2x^3)(3 - x + 2x^2)$ (j) $(2 - 3x + x^2)(4 - 5x + x^3)$

    (k) $(2x + 1)(3x - 2)(x + 5)$ (l) $(x^2 + 1)(x - 3)(2x^2 - x + 1)$

6 In each of the following products find the coefficient of x and the coefficient of x^2.

    (a) $(x + 2)(x^2 - 3x + 6)$ (b) $(x - 3)(x^2 + 2x - 5)$

    (c) $(2x + 1)(x^2 - 5x + 1)$ (d) $(3x - 2)(x^2 - 2x + 7)$

    (e) $(2x - 3)(3x^2 - 6x + 1)$ (f) $(2x - 5)(3x^3 - x^2 + 4x + 2)$

    (g) $(x^2 + 2x - 3)(x^2 + 3x - 4)$ (h) $(3x^2 + 1)(2x^2 - 5x + 3)$

    (i) $(x^2 + 3x - 1)(x^3 + x^2 - 2x + 1)$ (j) $(3x^2 - x + 2)(4x^3 - 5x + 1)$

7 In each of the following the product of $Ax + B$ with another polynomial is given. Using the fact that A and B are constants, find A and B.

    (a) $(Ax + B)(x - 3) = 4x^2 - 11x - 3$ (b) $(Ax + B)(x + 5) = 2x^2 + 7x - 15$

    (c) $(Ax + B)(3x - 2) = 6x^2 - x - 2$ (d) $(Ax + B)(2x + 5) = 6x^2 + 11x - 10$

    (e) $(Ax + B)(x^2 - 1) = x^3 + 2x^2 - x - 2$ (f) $(Ax + B)(x^2 + 4) = 2x^3 - 3x^2 + 8x - 12$

    (g) $(Ax + B)(2x^2 - 3x + 4) = 4x^3 - x + 12$ (h) $(Ax + B)(3x^2 - 2x - 1) = 6x^3 - 7x^2 + 1$

1.3 Equations and identities

In this chapter so far you have learned how to add, subtract and multiply polynomials, and you can now carry out calculations such as

$$(2x + 3) + (x - 2) = 3x + 1,$$

$$\left(x^2 - 3x - 4\right) - (2x + 1) = x^2 - 5x - 5 \quad \text{and}$$

$$(1 - x)\left(1 + x + x^2\right) = 1 - x^3$$

fairly automatically.

However, you should realise that these are not equations in the normal sense, because they are true for all values of x.

In P1 Section 10.6, you saw that when two expressions take the same values for every value of the variable, they are said to be **identically equal**, and a statement such as

$$(1-x)\left(1+x+x^2\right)=1-x^3$$

is called an **identity**.

To emphasise that an equation is an identity, the symbol $\equiv$ is used. The statement $(1-x)\left(1+x+x^2\right)\equiv 1-x^3$ means that $(1-x)\left(1+x+x^2\right)$ and $1-x^3$ are equal for all values of x.

But now suppose that $Ax+B\equiv 2x+3$. What can you say about A and B? As $Ax+B\equiv 2x+3$ is an identity, it is true for all values of x. In particular, it is true for $x=0$. Therefore $A\times 0+B=2\times 0+3$, giving $B=3$. But the identity is also true when $x=1$, so $A\times 1+3=2\times 1+3$, giving $A=2$. Therefore:

> If $\quad Ax+B\equiv 2x+3,\quad$ then $\quad A=2$ and $B=3$..

This is an example of the process called **equating coefficients**. The full result is·

> If $\quad ax^n+bx^{n-1}+\ldots+k\equiv Ax^n+Bx^{n-1}+\ldots+K,$
>
> then $\quad a=A,\ b=B,\ \ldots,\ k=K.$

The statement in the box says that, if two polynomials are equal for all values of x, then all the coefficients of corresponding powers of x are equal.

This result may not surprise you, but you should be aware that you are using it. Indeed, it is very likely that you have used it before now without being aware of it.

Example 1.3.1
One factor of $3x^2-5x-2$ is $x-2$. Find the other factor.

There is nothing wrong in writing down the answer by inspection as $3x+1$. But the process behind this quick solution is as follows.

Suppose that the other factor is $Ax+B$. Then $(Ax+B)(x-2)\equiv 3x^2-5x-2$, and, multiplying out, you get

$$Ax^2+(-2A+B)x-2B\equiv 3x^2-5x-2.$$

By equating coefficients of x^2, you get $A=3$. Equating coefficients of x^0, the constant term, you get $-2B=-2$, giving $B=1$. Therefore the other factor is $3x+1$.

You can also check that the middle term, $-2A+B=-6+1=-5$, is correct.

You should continue to write down the other factor by inspection if you can. However, in some cases, it is not easy to see what the answer will be without intermediate working.

Example 1.3.2

If $4x^3 + 2x^2 + 3 \equiv (x - 2)(Ax^2 + Bx + C) + R$, find A, B, C and R.

Multiplying out the right side gives

$$4x^3 + 2x^2 + 3 \equiv Ax^3 + (-2A + B)x^2 + (-2B + C)x + (-2C + R).$$

Equating coefficients of x^3: $4 = A$.
Equating coefficients of x^2: $2 = -2A + B = -2 \times 4 + B = -8 + B$, so $B = 10$.
Equating coefficients of x: $0 = -2B + C = -20 + C$, so $C = 20$.
Equating coefficients of x^0: $3 = -2C + R = -40 + R$, giving $R = 43$.

Therefore $A = 4$, $B = 10$, $C = 20$ and $R = 43$, so
$$4x^3 + 2x^2 + 3 \equiv (x - 2)(4x^2 + 10x + 20) + 43.$$

In practice, people often use the symbol for equality, $=$, when they really mean the symbol for identity, $\equiv$. The context usually suggests which meaning is intended.

Exercise 1B

1 In each of the following quadratic polynomials one factor is given. Find the other factor.

(a) $x^2 + x - 12 \equiv (x + 4)(\quad)$ (b) $x^2 + 14x - 51 \equiv (x - 3)(\quad)$

(c) $3x^2 + 5x - 22 \equiv (x - 2)(\quad)$ (d) $35x^2 + 48x - 27 \equiv (5x + 9)(\quad)$

(e) $2x^2 - x - 15 \equiv (2x + 5)(\quad)$ (f) $14x^2 + 31x - 10 \equiv (2x + 5)(\quad)$

2 In each of the following identities find the values of A, B and R.

(a) $x^2 - 2x + 7 \equiv (x + 3)(Ax + B) + R$ (b) $x^2 + 9x - 3 \equiv (x + 1)(Ax + B) + R$

(c) $15x^2 - 14x - 8 \equiv (5x + 2)(Ax + B) + R$ (d) $6x^2 + x - 5 \equiv (2x + 1)(Ax + B) + R$

(e) $12x^2 - 5x + 2 \equiv (3x - 2)(Ax + B) + R$ (f) $21x^2 - 11x + 6 \equiv (3x - 2)(Ax + B) + R$

3 In each of the following identities find the values of A, B, C and R.

(a) $x^3 - x^2 - x + 12 \equiv (x + 2)(Ax^2 + Bx + C) + R$

(b) $x^3 - 5x^2 + 10x + 10 \equiv (x - 3)(Ax^2 + Bx + C) + R$

(c) $2x^3 + x^2 - 3x + 4 \equiv (2x - 1)(Ax^2 + Bx + C) + R$

(d) $12x^3 + 11x^2 - 7x + 5 \equiv (3x + 2)(Ax^2 + Bx + C) + R$

(e) $4x^3 + 4x^2 - 37x + 5 \equiv (2x - 5)(Ax^2 + Bx + C) + R$

(f) $9x^3 + 12x^2 - 15x - 10 \equiv (3x + 4)(Ax^2 + Bx + C) + R$

4 In each of the following identities find the values of A, B, C, D and R.

(a) $2x^4 + 3x^3 - 5x^2 + 11x - 5 \equiv (x + 3)(Ax^3 + Bx^2 + Cx + D) + R$

(b) $4x^4 - 7x^3 - 2x^2 - 2x + 7 \equiv (x - 2)(Ax^3 + Bx^2 + Cx + D) + R$

(c) $6x^4 + 5x^3 - x^2 + 3x + 2 \equiv (2x + 1)(Ax^3 + Bx^2 + Cx + D) + R$

(d) $3x^4 - 7x^3 + 17x^2 - 14x + 5 \equiv (3x - 1)(Ax^3 + Bx^2 + Cx + D) + R$

1.4 Division of polynomials

You can, if you wish, carry out division of polynomials using a layout like the one for long division of integers. You may already have seen and used such a process. However, you can also use the method of equating coefficients for division.

When you divide 112 by 9, you get an answer of 12 with 4 over. The number 9 is called the divisor, 12 is the quotient and 4 the remainder. You can express this as an equation in integers, $112 = 9 \times 12 + 4$. The remainder r has to satisfy the inequality $0 \leqslant r < 9$.

Now look back at Example 1.3.2. You will see that it is an identity of just the same shape, but with polynomials instead of integers. So you can say that, when $4x^3 + 2x^2 + 3$ is divided by the divisor $x - 2$, the quotient is $4x^2 + 10x + 20$ and the remainder is 43. The degree of the remainder (in this case 0) has to be less than the degree of the divisor. The degree of the quotient $4x^2 + 10x + 20$, which is 2, is equal to the difference between the degree of the polynomial $4x^3 + 2x^2 + 3$, which is 3, and the degree of the divisor $x - 2$, which is 1.

> When a polynomial, $a(x)$, is divided by a non-constant **divisor**, $b(x)$, the **quotient** $q(x)$ and the **remainder** $r(x)$ are defined by the identity
>
> $$a(x) \equiv b(x)q(x) + r(x),$$
>
> where the degree of the remainder is less than the degree of the divisor.
>
> The degree of the quotient is equal to the degree of $a(x)$ − the degree of $b(x)$.

Example 1.4.1
Find the quotient and remainder when $x^4 + x + 2$ is divided by $x + 1$.

Using the result in the box, as the degree of $x^4 + x + 2$ is 4 and the degree of $x + 1$ is 1, the degree of the quotient is $4 - 1 = 3$. And as the degree of the remainder is less than 1, the remainder is a constant.

Let the quotient be $Ax^3 + Bx^2 + Cx + D$, and let the remainder be R. Then

$$x^4 + x + 2 \equiv (x + 1)\left(Ax^3 + Bx^2 + Cx + D\right) + R,$$

so $\quad x^4 + x + 2 \equiv Ax^4 + (A + B)x^3 + (B + C)x^2 + (C + D)x + D + R.$

Equating coefficients of x^4: $\quad 1 = A.$
Equating coefficients of x^3: $\quad 0 = A + B$, so $B = -A$, giving $B = -1$.
Equating coefficients of x^2: $\quad 0 = B + C$, so $C = -B$, giving $C = 1$.
Equating coefficients of x: $\quad 1 = C + D$, so $D = 1 - C$, giving $D = 0$.
Equating coefficients of x^0: $\quad 2 = D + R$, so $R = 2 - D$, giving $R = 2$.

The quotient is $x^3 - x^2 + x$ and the remainder is 2.

Example 1.4.2

Find the quotient and remainder when $x^4 + 3x^2 - 2$ is divided by $x^2 - 2x + 2$.

The result in the box states that the degree of the remainder is less than 2, so assume that it is a linear polynomial. Let the quotient be $Ax^2 + Bx + C$, and the remainder be $Rx + S$. Then

$$x^4 + 3x^2 - 2 \equiv \left(x^2 - 2x + 2\right)\left(Ax^2 + Bx + C\right) + Rx + S,$$

so $x^4 + 3x^2 - 2 \equiv Ax^4 + (-2A + B)x^3 + (2A - 2B + C)x^2$
$$+ (2B - 2C + R)x + 2C + S.$$

Equating coefficients of x^4: $1 = A$.
Equating coefficients of x^3: $0 = -2A + B$, so $B = 2A$, giving $B = 2$.
Equating coefficients of x^2: $3 = 2A - 2B + C$, so $C = 3 - 2A + 2B$, giving $C = 5$.
Equating coefficients of x: $0 = 2B - 2C + R$, so $R = -2B + 2C$, giving $R = 6$.
Equating coefficients of x^0: $-2 = 2C + S$, so $S = -2 - 2C$, giving $S = -12$.

The quotient is $x^2 + 2x + 5$ and the remainder is $6x - 12$.

When you are dividing by a *linear* polynomial, there is a quick way of finding the remainder. For example, in Example 1.4.1, when $x^4 + x + 2$ was divided by $x + 1$, the first line of the solution was:

$$x^4 + x + 2 \equiv (x + 1)\left(Ax^3 + Bx^2 + Cx + D\right) + R.$$

Since this is an identity, it is true for all values of x and, in particular, it is true for $x = -1$. Putting $x = -1$ in the left side, you get $(-1)^4 + (-1) + 2 = 2$; putting $x = -1$ in the right side, you get $0 \times \left(A(-1)^3 + B(-1)^2 + C(-1) + D\right) + R$, which is simply R. Therefore $R = 2$.

Similar reasoning leads to the remainder theorem.

> **Remainder theorem**
> When a polynomial $p(x)$ is divided by $x - t$,
> the remainder is the constant $p(t)$.

Proof   When $p(x)$ is divided by $x - t$, let the quotient be $q(x)$ and the remainder be R. Then

$$p(x) \equiv (x - t)q(x) + R.$$

Putting $x = t$ in this identity gives $p(t) = 0 \times q(t) + R = R$, so $R = p(t)$.

Example 1.4.3

Find the remainder when $x^3 - 3x + 4$ is divided by $x + 3$.

Let $p(x) \equiv x^3 - 3x + 4$. Then $p(-3) = (-3)^3 - 3 \times (-3) + 4 = -27 + 9 + 4 = -14$.
By the remainder theorem, the remainder is -14.

Example 1.4.4

When the polynomial $p(x) \equiv x^3 - 3x^2 + ax + b$ is divided by $x - 1$ the remainder is -4. When $p(x)$ is divided by $x - 2$ the remainder is also -4. Find the remainder when $p(x)$ is divided by $x - 3$.

By the remainder theorem, when $p(x)$ is divided by $x - 1$, the remainder is $p(1) = 1^3 - 3 \times 1^2 + a + b = a + b - 2$. Therefore $a + b - 2 = -4$, so $a + b = -2$.

Similarly, $p(2) = 2^3 - 3 \times 2^2 + 2a + b = 2a + b - 4$, so $2a + b - 4 = -4$ and $2a + b = 0$.

Solving the equations $a + b = -2$ and $2a + b = 0$ simultaneously gives $a = 2$ and $b = -4$, making the polynomial $p(x) \equiv x^3 - 3x^2 + 2x - 4$.

The remainder on division by $x - 3$ is $p(3) = 3^3 - 3 \times 3^2 + 2 \times 3 - 4 = 2$.

The remainder theorem is useful for finding the remainder when you divide a polynomial by a linear polynomial such as $x - 2$, but it doesn't tell you how to find the remainder when you divide by a linear polynomial such as $3x - 2$. To do this, you need the extended form of the remainder theorem.

Remainder theorem: extended form

When a polynomial $p(x)$ is divided by $sx - t$,

the remainder is the constant $p\left(\dfrac{t}{s}\right)$.

Proof When $p(x)$ is divided by $sx - t$, let the quotient be $q(x)$ and the remainder be R. Then $p(x) \equiv (sx - t)q(x) + R$.

Putting $x = \dfrac{t}{s}$ in this identity,

$$p\left(\frac{t}{s}\right) = \left(s \times \frac{t}{s} - t\right) \times q\left(\frac{t}{s}\right) + R = 0 \times q\left(\frac{t}{s}\right) + R = R, \quad \text{so} \quad R = p\left(\frac{t}{s}\right).$$

This proves that the remainder is the constant $p\left(\dfrac{t}{s}\right)$.

Example 1.4.5

Find the remainder when $x^3 - 3x + 4$ is divided by $2x + 3$.

Let $p(x) \equiv x^3 - 3x + 4$. Then $p\left(-\frac{3}{2}\right) = \left(-\frac{3}{2}\right)^3 - 3 \times \left(-\frac{3}{2}\right) + 4 = -\frac{27}{8} + \frac{9}{2} + 4 = 5\frac{1}{8}$.

By the remainder theorem in its extended form, the remainder is $5\frac{1}{8}$.

Exercise 1C

1 Find the quotient and the remainder when

(a) $x^2 - 5x + 2$ is divided by $x - 3$,

(b) $x^2 + 2x - 6$ is divided by $x + 1$,

(c) $2x^2 + 3x - 1$ is divided by $x - 2$,

(d) $2x^2 + 3x + 1$ is divided by $2x - 1$,

(e) $6x^2 - x - 2$ is divided by $3x + 1$,

(f) x^4 is divided by x^3.

2 Find the quotient and the remainder when the first polynomial is divided by the second.

(a) $x^3 + 2x^2 - 3x + 1$, $x + 2$

(b) $x^3 - 3x^2 + 5x - 4$, $x - 5$

(c) $2x^3 + 4x - 5$, $x + 3$

(d) $5x^3 - 3x + 7$, $x - 4$

(e) $2x^3 - x^2 - 3x - 7$, $2x + 1$

(f) $6x^3 + 17x^2 - 17x + 5$, $3x - 2$

3 Find the quotient and the remainder when

(a) $x^4 - 2x^3 - 7x^2 + 7x + 5$ is divided by $x^2 + 2x - 1$,

(b) $x^4 - x^3 + 7x + 2$ is divided by $x^2 + x - 1$,

(c) $2x^4 - 4x^3 + 3x^2 + 6x + 5$ is divided by $x^3 + x^2 + 1$,

(d) $6x^4 + x^3 + 13x + 10$ is divided by $2x^2 - x + 4$.

4 Find the remainder when the first polynomial is divided by the second.

(a) $x^3 - 5x^2 + 2x - 3$, $x - 1$

(b) $x^3 + x^2 - 6x + 5$, $x + 2$

(c) $2x^3 - 3x + 5$, $x - 3$

(d) $4x^3 - 5x^2 + 3x - 7$, $x + 4$

(e) $x^3 + 3x^2 - 2x + 1$, $2x - 1$

(f) $2x^3 + 5x^2 - 3x + 6$, $3x + 1$

(g) $x^4 - x^3 + 2x^2 - 7x - 2$, $x - 2$

(h) $3x^4 + x^2 - 7x + 6$, $x + 3$

5 When $x^3 + 2x^2 - px + 1$ is divided by $x - 1$ the remainder is 5. Find the value of p.

6 When $2x^3 + x^2 - 3x + q$ is divided by $x - 2$ the remainder is 12. Find the value of q.

7 When $x^3 + 2x^2 + px - 3$ is divided by $x + 1$ the remainder is the same as when it is divided by $x - 2$. Find the value of p.

8 When $x^3 + px^2 - x - 4$ is divided by $x - 1$ the remainder is the same as when it is divided by $x + 3$. Find the value of p.

9 When $3x^3 - 2x^2 + ax + b$ is divided by $x - 1$ the remainder is 3. When divided by $x + 1$ the remainder is -13. Find the values of a and b.

10 When $x^3 + ax^2 + bx + 5$ is divided by $x - 2$ the remainder is 23. When divided by $x + 1$ the remainder is 11. Find the values of a and b.

11 When $x^3 + ax^2 + bx - 5$ is divided by $x - 1$ the remainder is -1. When divided by $x + 1$ the remainder is -5. Find the values of a and b.

12 When $2x^3 - x^2 + ax + b$ is divided by $x - 2$ the remainder is 25. When divided by $x + 1$ the remainder is -5. Find the values of a and b.

1.5 The factor theorem

When you solve an equation $p(x) = 0$ by factors, writing $p(x) \equiv (x - t)(x - u)(x - v)\dots$, you deduce that $x = t$ or $x = u$ or $x = v$ or $\dots$. So when you substitute $x = t$ in $p(x)$, you find that $p(t) = 0$. The converse is not so obvious: that if $p(t) = 0$, then $x - t$ is a factor of $p(x)$. This result, a special case of the remainder theorem, is called the factor theorem.

> Let $p(x)$ be a polynomial. Then
> (a) if $x - t$ is a factor of $p(x)$, then $p(t) = 0$;
> (b) if $p(t) = 0$, then $x - t$ is a factor of $p(x)$.
>
> The second of these results is called the **factor theorem**.

Proof

(a) If $x - t$ is a factor of $p(x)$, then $p(x) \equiv (x - t)q(x)$, where $q(x)$ is a polynomial. Putting $x = t$ into this identity shows that $p(t) = (t - t)q(t) = 0$.

(b) When $p(x)$ is divided by $x - t$, let the quotient be $q(x)$ and the remainder be R. Then $p(x) \equiv (x - t)q(x) + R$.

Putting $x = t$ into this identity gives $p(t) = R$ (this is the remainder theorem again). Thus if $p(t) = 0$, $R = 0$, so $x - t$ is a factor of the polynomial $p(x)$.

You can use the factor theorem to search for factors of a polynomial when its coefficients are small.

When you search for factors of a polynomial such as $x^3 - x^2 - 5x - 3$, you need only try factors of the form $x - t$ where t divides the constant coefficient, in this case 3. Thus you need only try $x - 1$, $x + 1$, $x - 3$ and $x + 3$.

Example 1.5.1

Find the factors of $x^3 - x^2 - 5x - 3$, and hence solve the equation $x^3 - x^2 - 5x - 3 = 0$.

Denote $x^3 - x^2 - 5x - 3$ by $p(x)$.

Could $x - 1$ be a factor? $p(1) = 1^3 - 1^2 - 5 \times 1 - 3 = -8 \neq 0$, so $x - 1$ is not a factor.

Try $x + 1$ as a factor. $p(-1) = (-1)^3 - (-1)^2 - 5 \times (-1) - 3 = 0$, so $x + 1$ is a factor.

Dividing $x^3 - x^2 - 5x - 3$ by $x + 1$ in the usual way, you find

$$x^3 - x^2 - 5x - 3 \equiv (x + 1)(x^2 - 2x - 3).$$

Since $x^2 - 2x - 3 \equiv (x + 1)(x - 3)$, you can now factorise $x^3 - x^2 - 5x - 3$ completely to get

$$x^3 - x^2 - 5x - 3 \equiv (x + 1)(x + 1)(x - 3) \equiv (x + 1)^2(x - 3).$$

The solution of the equation $x^3 - x^2 - 5x - 3 = 0$ is $x = -1$ (repeated) and $x = 3$.

Example 1.5.2

Find the factors of $x^4 + x^3 - x - 1$ and solve the equation $x^4 + x^3 - x - 1 = 0$.

Let $p(x) \equiv x^4 + x^3 - x - 1$.

Since $p(1) = 1 + 1 - 1 - 1 = 0$, $x - 1$ is a factor of $p(x)$.

Writing $x^4 + x^3 - x - 1 \equiv (x - 1)(Ax^3 + Bx^2 + Cx + D)$ and multiplying out the right side shows that

$$x^4 + x^3 - x - 1 \equiv Ax^4 + (B - A)x^3 + (C - B)x^2 + (D - C)x - D.$$

Equating coefficients of x^4 and the constant terms gives $A = 1$ and $D = 1$, and you can see by inspection that the other coefficients are $B = 2$ and $C = 2$. So

$$p(x) \equiv (x - 1)(x^3 + 2x^2 + 2x + 1).$$

Let $q(x) \equiv x^3 + 2x^2 + 2x + 1$. Then $q(1) \neq 0$, so $x - 1$ is not a factor of $q(x)$, but $q(-1) = -1 + 2 - 2 + 1 = 0$, so $x + 1$ is a factor of $q(x)$.

Writing $x^3 + 2x^2 + 2x + 1 \equiv (x + 1)(Ex^2 + Fx + G)$ and equating coefficients shows that $E = 1$, $G = 1$ and $F = 1$.

Therefore $x^4 + x^3 - x - 1 \equiv (x - 1)(x + 1)(x^2 + x + 1)$.

As the discriminant of $x^2 + x + 1$ is $1^2 - 4 \times 1 \times 1 = -3 < 0$, $x^2 + x + 1$ does not split into linear factors, so $(x - 1)(x + 1)(x^2 + x + 1)$ cannot be factorised further.

Also the equation $x^2 + x + 1 = 0$ doesn't have real roots. So the solution of the equation $x^4 + x^3 - x - 1 = 0$ is $x = 1$ or $x = -1$.

Like the remainder theorem, the factor theorem has an extended form.

> Let $p(x)$ be a polynomial. Then
>
> (a) if $sx - t$ is a factor of $p(x)$, then $p\left(\dfrac{t}{s}\right) = 0$;
>
> (b) if $p\left(\dfrac{t}{s}\right) = 0$, then $sx - t$ is a factor of $p(x)$.
>
> The second result is the **extended form of the factor theorem.**

To prove this, modify the proof of the factor theorem on page 13 in the same way as the proof of the remainder theorem was modified in Section 1.4. Simply replace $p(x) \equiv (x - t)q(x)$ by $p(x) \equiv (sx - t)q(x)$, and put $x = \dfrac{t}{s}$ in the identity.

You can save a lot of effort when you apply this form of the factor theorem by using the fact that, if the coefficients of $p(x) \equiv ax^n + bx^{n-1} + \ldots + k$ are all integers, and if $sx - t$ is a factor of $p(x)$, then s divides a and t divides k. (This can be proved by using properties of prime factors in arithmetic, but the proof is not included in this course.)

Example 1.5.3

Find the factors of $p(x) \equiv 3x^3 + 4x^2 + 5x - 6$.

Begin by noting that, if $sx - t$ is a factor, s divides 3 and t divides 6. So s can only be ± 1 or ± 3, and t can only be ± 1, ± 2, ± 3 or ± 6.

You can further reduce the number of possibilities in two ways.
- $sx - t$ is not really a different factor from $-sx + t$. So you need consider only positive values of s.
- The factors can't be $3x \pm 3$ or $3x \pm 6$ since then 3 would be a common factor of the coefficients of $p(x)$, which it isn't.

So there are only twelve possible factors: $x \mp 1$, $x \mp 2$, $x \mp 3$, $x \mp 6$, $3x \mp 1$ and $3x \mp 2$. You can test these by evaluating $p(x)$ for $x = \pm 1$, ± 2, ± 3, ± 6, $\pm \frac{1}{3}$ and $\pm \frac{2}{3}$ until you get a zero.

Working through these in turn, you will eventually find that

$$p\left(\tfrac{2}{3}\right) = 3 \times \left(\tfrac{2}{3}\right)^3 + 4 \times \left(\tfrac{2}{3}\right)^2 + 5 \times \tfrac{2}{3} - 6 = \tfrac{8}{9} + \tfrac{16}{9} + \tfrac{10}{3} - 6 = 0.$$

So $3x - 2$ is a factor, and by division $p(x) \equiv (3x - 2)(x^2 + 2x + 3)$.

Since $x^2 + 2x + 3 \equiv (x + 1)^2 + 2$, which has no factors, $p(x)$ doesn't factorise further.

Exercise 1D

1. Use the factor theorem to factorise the following cubic polynomials $p(x)$. In each case write down the real roots of the equation $p(x) = 0$.
   (a) $x^3 + 2x^2 - 5x - 6$ (b) $x^3 - 3x^2 - x + 3$ (c) $x^3 - 3x^2 - 13x + 15$
   (d) $x^3 - 3x^2 - 9x - 5$ (e) $x^3 + 3x^2 - 4x - 12$ (f) $2x^3 + 7x^2 - 5x - 4$
   (g) $3x^3 - x^2 - 12x + 4$ (h) $6x^3 + 7x^2 - x - 2$ (i) $x^3 + 2x^2 - 4x + 1$

2. Use the factor theorem to factorise the following quartic polynomials $p(x)$. In each case write down the real roots of the equation $p(x) = 0$.
   (a) $x^4 - x^3 - 7x^2 + x + 6$ (b) $x^4 + 4x^3 - x^2 - 16x - 12$
   (c) $2x^4 - 3x^3 - 12x^2 + 7x + 6$ (d) $6x^4 + x^3 - 17x^2 - 16x - 4$
   (e) $x^4 - 2x^3 + 2x - 1$ (f) $4x^4 - 12x^3 + x^2 + 12x + 4$

3. Factorise the following.
   (a) $x^3 - 8$ (b) $x^3 + 8$ (c) $x^3 - a^3$
   (d) $x^3 + a^3$ (e) $x^4 - a^4$ (f) $x^5 + a^5$

4. (a) Show that $x - a$ is a factor of $x^n - a^n$.
   (b) Under what conditions is $x + a$ a factor of $x^n + a^n$? Under these conditions, find the other factor.

Miscellaneous exercise 1

1 It is given that

$$(x+a)\left(x^2+bx+2\right) \equiv x^3 - 2x^2 - x - 6$$

where a and b are constants. Find the value of a and the value of b.   (OCR)

2 Find the remainder when $(1+x)^4$ is divided by $x+2$.

3 Show that $(x-1)$ is a factor of $6x^3 + 11x^2 - 5x - 12$, and find the other two linear factors of this expression.   (OCR)

4 The cubic polynomial $x^3 + ax^2 + bx - 8$, where a and b are constants, has factors $(x+1)$ and $(x+2)$. Find the values of a and b.   (OCR)

5 Find the value of a for which $(x-2)$ is a factor of $3x^3 + ax^2 + x - 2$.

Show that, for this value of a, the cubic equation $3x^3 + ax^2 + x - 2 = 0$ has only one real root.   (OCR)

6 Solve the equation $4x^3 + 8x^2 + x - 3 = 0$ given that one of the roots is an integer.   (OCR)

7 The cubic polynomial $x^3 - 2x^2 - 2x + 4$ has a factor $(x-a)$, where a is an integer.

(a) Use the factor theorem to find the value of a.

(b) Hence find exactly all three roots of the cubic equation $x^3 - 2x^2 - 2x + 4 = 0$.   (OCR)

8 The cubic polynomial $x^3 - 2x^2 - x - 6$ is denoted by $f(x)$. Show that $(x-3)$ is a factor of $f(x)$. Factorise $f(x)$. Hence find the number of real roots of the equation $f(x) = 0$, justifying your answer.

Hence write down the number of points of intersection of the graphs with equations

$$y = x^2 - 2x - 1 \quad \text{and} \quad y = \frac{6}{x},$$

justifying your answer.   (OCR)

9 Given that $(2x+1)$ is a factor of $2x^3 + ax^2 + 16x + 6$, show that $a = 9$.

Find the real quadratic factor of $2x^3 + 9x^2 + 16x + 6$. By completing the square, or otherwise, show that this quadratic factor is positive for all real values of x.   (OCR)

10 Show that both $\left(x - \sqrt{3}\right)$ and $\left(x + \sqrt{3}\right)$ are factors of $x^4 + x^3 - x^2 - 3x - 6$.

Hence write down one quadratic factor of $x^4 + x^3 - x^2 - 3x - 6$, and find a second quadratic factor of this polynomial.   (OCR)

11 The diagram shows the curve

$$y = -x^3 + 2x^2 + ax - 10.$$

The curve crosses the x-axis at $x = p$, $x = 2$ and $x = q$.

(a) Show that $a = 5$.

(b) Find the exact values of p and q.   (OCR)

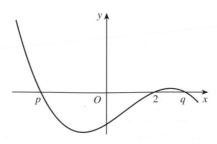

12 The polynomial $x^3 + 3x^2 + ax + b$ leaves a remainder of 3 when it is divided by $x + 1$ and a remainder of 15 when it is divided by $x - 2$. Find the remainder when it is divided by $(x - 2)(x + 1)$.

13 Find the quotient and the remainder when $x^4 + 4$ is divided by $x^2 - 2x + 2$.

14 Let $p(x) = 4x^3 + 12x^2 + 5x - 6$.

    (a) Calculate $p(2)$ and $p(-2)$, and state what you can deduce from your answers.

    (b) Solve the equation $4x^3 + 12x^2 + 5x - 6 = 0$.

15 It is given that $f(x) = x^4 - 3x^3 + ax^2 + 15x + 50$, where a is a constant, and that $x + 2$ is a factor of $f(x)$.

    (a) Find the value of a.

    (b) Show that $f(5) = 0$ and factorise $f(x)$ completely into exact linear factors.

    (c) Find the set of values of x for which $f(x) > 0$.   (OCR)

16 The diagram shows the graph of $y = x^2 - 3$ and the part of the graph of $y = \dfrac{2}{x}$ for $x > 0$.

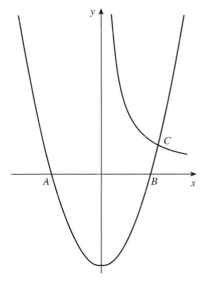

The two graphs intersect at C, and A and B are the points of intersection of $y = x^2 - 3$ with the x-axis. Write down the exact coordinates of A and B.

Show that the x-coordinate of C is given by the equation $x^3 - 3x - 2 = 0$.

Factorise $x^3 - 3x - 2$ completely.

Hence

    (a) write down the x-coordinate of C,

    (b) describe briefly the geometrical relationship between the graph of $y = x^2 - 3$ and the part of the graph of $y = \dfrac{2}{x}$ for which $x < 0$.   (OCR)

17 The polynomial $x^5 - 3x^4 + 2x^3 - 2x^2 + 3x + 1$ is denoted by $f(x)$.

    (a) Show that neither $(x - 1)$ nor $(x + 1)$ is a factor of $f(x)$.

    (b) By substituting $x = 1$ and $x = -1$ in the identity

$$f(x) \equiv \left(x^2 - 1\right)q(x) + ax + b,$$

    where $q(x)$ is a polynomial and a and b are constants, or otherwise, find the remainder when $f(x)$ is divided by $\left(x^2 - 1\right)$.

    (c) Show, by carrying out the division, or otherwise, that when $f(x)$ is divided by $\left(x^2 + 1\right)$, the remainder is $2x$.

    (d) Find all the real roots of the equation $f(x) = 2x$.   (OCR)

2 The modulus function

This chapter introduces the modulus function, written as $|x|$. When you have completed it, you should

- know the definition of modulus, and recognise $|x|$ as a function
- know how to draw graphs of functions involving modulus
- know how to use modulus algebraically and geometrically
- be able to solve simple equations and inequalities involving modulus.

2.1 The modulus function and its graph

You met the modulus notation briefly in P1 Section 3.4, and have used it from time to time since then. Since $|x|$ is defined for all real numbers x, it is another example of a function of x. Its domain is the set of real numbers, $\mathbb{R}$ (see P1 Section 11.3), and its range is $\mathbb{R}, y \geqslant 0$.

> The **modulus** of x, denoted by $|x|$, is defined by
>
> $|x| = x \qquad$ if $x \geqslant 0$,
>
> $|x| = -x \qquad$ if $x < 0$.

On some calculators the modulus function is [mod]; on others it is [abs], short for 'the absolute value of x'. This book always uses the notation $|x|$.

Fig. 2.1 shows the graph of $y = |x|$. The graph has a 'V' shape, with both branches making an angle of 45° with the x-axis, provided that the scales are the same on both axes.

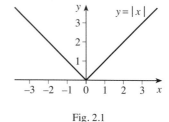

Fig. 2.1

2.2 Graphs of functions involving modulus

Suppose that you want to draw the graph of $y = |x-2|$. You can do this directly from the definition of modulus. When $x \geqslant 2$, $x - 2 \geqslant 0$, so $|x-2| = x-2$. For these values of x, the graphs of $y = |x-2|$ and $y = x - 2$ are the same.

When $x < 2$, $x - 2 < 0$, so $|x-2| = -(x-2) = 2 - x$. So for these values of x, the graph of $y = |x-2|$ is the same as the graph of $y = 2 - x$.

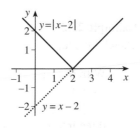

Fig. 2.2

Another way of dealing with the case $x < 2$ is to note that the graph of $y = -(x-2)$ is the reflection of $y = x - 2$ in the x-axis. So you can draw the graph of $y = |x-2|$ by first drawing the graph of $y = x - 2$ and then reflecting in the x-axis that part of the line which is below the x-axis. This is illustrated in Fig. 2.2.

This method can always be used to get the graph of $y = |f(x)|$ from the graph of $y = f(x)$. In the definition of $|x|$ in the box on page 18, you can write any expression in place of x. So, replacing x by $f(x)$,

$$|f(x)| = f(x) \text{ if } f(x) \geq 0, \quad \text{and} \quad |f(x)| = -f(x) \text{ if } f(x) < 0.$$

It follows that, for the parts of the graph $y = f(x)$ which are on or above the x-axis, the graphs of $y = f(x)$ and $y = |f(x)|$ are the same. But for the parts of $y = f(x)$ below the x-axis, $y = |f(x)| = -f(x)$ is obtained from $y = f(x)$ by reflection in the x-axis.

A nice way of showing this is to draw the graph of $y = f(x)$ on a transparent sheet. You can then get the graph of $y = |f(x)|$ by folding the sheet along the x-axis so that the negative part of the sheet lies on top of the positive part.

Example 2.2.1
Sketch the graphs of   (a) $y = |2x - 3|$,   (b) $y = |(x - 1)(x - 3)|$.

Figs. 2.3 and 2.4 show the graphs of (a) $y = 2x - 3$ and (b) $y = (x - 1)(x - 3)$ with the part below the x-axis (drawn dotted) reflected in the x-axis to give the graphs required.

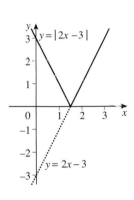

Fig. 2.3

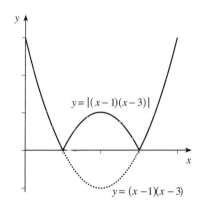

Fig. 2.4

Graphs which involve the modulus function are likely to have sharp corners. If you have access to a graphic calculator, show the graphs in Example 2.2.1 on it.

Example 2.2.2
Sketch the graph of $y = |x - 2| + |1 - x|$.

With two moduli involved it is usually best to go back to the definition of modulus. For $|x - 2|$ you have to consider $x - 2 \geq 0$ and $x - 2 < 0$ separately, and for $|1 - x|$ you have to consider $1 - x \geq 0$ and $1 - x < 0$. So altogether there are three intervals to investigate: $x \leq 1$, $1 < x < 2$ and $x \geq 2$.

When $x \leq 1$, $|x - 2| = -(x - 2)$ and $|1 - x| = 1 - x$, so $y = -x + 2 + 1 - x = 3 - 2x$.
When $1 < x < 2$, $|x - 2| = -(x - 2)$ and $|1 - x| = -(1 - x)$, so $y = -x + 2 - 1 + x = 1$.
When $x \geq 2$, $|x - 2| = x - 2$ and $|1 - x| = -(1 - x)$, so $y = x - 2 - 1 + x = 2x - 3$.

The graph is therefore in three parts, as shown in Fig. 2.5.

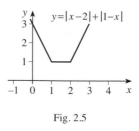

Fig. 2.5

You may sometimes also want to get the graph of $y = f(|x|)$ from the graph of $y = f(x)$. From the definition, $f(|x|)$ is the same as $f(x)$ when $x \geqslant 0$, but $f(|x|) = f(-x)$ when $x < 0$. So the graph of $y = f(|x|)$ is the same as the graph of $y = f(x)$ to the right of the y-axis, but to the left of the y-axis it is the reflection in the y-axis of $y = f(x)$ for $x > 0$.

Example 2.2.3

Sketch the graph of $y = \sin|x|$.

To the right of the y-axis, where $x > 0$, the graph is the same as the graph of $y = \sin x$. The graph is completed to the left of the y-axis, where $x < 0$, by reflecting in the y-axis the graph of $y = \sin x$ for $x > 0$. Fig. 2.6 shows the result.

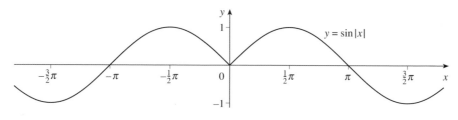

Fig. 2.6

2.3   Some algebraic properties

Let a and b be two real numbers. Since $|a|$ is always equal to either $-a$ or a, it follows that a is always equal to $-|a|$ or $|a|$. Similarly, b is always equal to $-|b|$ or $|b|$. So $a \times b$ is always equal to $|a| \times |b|$ or $-|a| \times |b|$. And since $|a| \times |b|$ is positive or zero, you can deduce that $|a \times b| = |a| \times |b|$.

A similar argument holds for division.

> If a and b are real numbers,
>
> $$|a \times b| = |a| \times |b| \quad \text{and} \quad \left|\frac{a}{b}\right| = \frac{|a|}{|b|} \qquad \text{(provided that } b \neq 0\text{)}.$$

Example 2.3.1

Show that   (a) $|4x + 6| = 2 \times |2x + 3|$,   (b) $|3 - x| = |x - 3|$.

(a) $|4x + 6| = |2(2x + 3)| = |2| \times |2x + 3| = 2 \times |2x + 3|$.

(b) $|3 - x| = |(-1) \times (x - 3)| = |-1| \times |x - 3| = 1 \times |x - 3| = |x - 3|$.

But beware! Similar rules don't hold for addition and subtraction. For example, if $a = 2$ and $b = -3$, $|a + b| = |2 + (-3)| = |-1| = 1$, but $|a| + |b| = 2 + 3 = 5$. So, for these values of a and b, $|a + b|$ does not equal $|a| + |b|$. See Exercise 2A Question 5.

2.4 Modulus on the number line

Some results about modulus can be illustrated by the distance between points on a number line. Let A and B be two points on a line with coordinates a and b (which can be positive, negative or zero) relative to an origin O, as in Fig. 2.7. Then the distance AB is given by $b - a$ if $b \geqslant a$, or $b - a \geqslant 0$; and by $a - b$, which is $-(b - a)$, if $b < a$, or $b - a < 0$. You will recognise this as the definition of $|b - a|$.

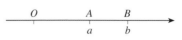

Fig. 2.7

> The distance between points on the number line
> with coordinates a and b is $|b - a|$.

As a special case, if a point X has coordinate x, then $|x|$ is the distance of X from the origin. This is used in the next example.

Example 2.4.1

What can you deduce about x if you know that   (a) $|x| = 3$,   (b) $|x| \leqslant 3$?

(a) If $|x| = 3$, X is a point 3 units from O. But the only two points 3 units from O are $x = 3$ or $x = -3$, so if $|x| = 3$, then $x = 3$ or $x = -3$.

The converse is also true. For if $x = 3$ or $x = -3$, then $|x| = 3$.

Therefore

$$|x| = 3 \text{ is equivalent to } x = 3 \text{ or } x = -3.$$

(b) If $|x| \leqslant 3$, X is a point 3 units or less from O. So x is between -3 and 3 (inclusive). It follows that if $|x| \leqslant 3$, then $-3 \leqslant x \leqslant 3$.

If $-3 \leqslant x \leqslant 3$, then X is 3 units or less from O, so $|x| \leqslant 3$.

Therefore

$$|x| \leqslant 3 \text{ is equivalent to } -3 \leqslant x \leqslant 3.$$

You can prove the result in Example 2.4.1(b) more formally from the definition of $|x|$. If $|x| \leqslant 3$, then either $x \geqslant 0$ and $x = |x| \leqslant 3$, so $0 \leqslant x \leqslant 3$; or $x < 0$ and $x = -|x| \geqslant -3$, so $-3 \leqslant x < 0$. In either case, $-3 \leqslant x \leqslant 3$.

The converse is also true. For if you know that $-3 \leqslant x \leqslant 3$, you have $-3 \leqslant x$ and $x \leqslant 3$. This is the same as $-x \leqslant 3$ and $x \leqslant 3$. Since $|x|$ is equal to either $-x$ or x, it follows that $|x| \leqslant 3$.

Putting the two results together gives

$|x| \leqslant 3$ is equivalent to $-3 \leqslant x \leqslant 3$.

The phrase '**is equivalent to**' connecting two statements means that each can be deduced from the other. In any mathematical argument you can then replace the first statement by the second, or the second by the first.

You can also say that two statements are equivalent by saying that one statement is true 'if and only if' the other is true.

You can use the argument in Example 2.4.1(b) to show that

if $a > 0$,   then $|x| \leqslant a$ is equivalent to $-a \leqslant x \leqslant a$.

What happens if $a = 0$? In that case $|x| \leqslant a$ means that $|x| \leqslant 0$, so $x = 0$, and $-a \leqslant x \leqslant a$ means that $-0 \leqslant x \leqslant 0$, so $x = 0$. Combining this result with the previous one gives:

> If $a \geqslant 0$, then $|x| \leqslant a$ is equivalent to $-a \leqslant x \leqslant a$.

Taking this a little further, you can deduce a useful generalisation about the inequality $|x - k| \leqslant a$. Let $y = x - k$, so that $|y| \leqslant a$. Then $-a \leqslant y \leqslant a$, so $-a \leqslant x - k \leqslant a$ and $k - a \leqslant x \leqslant k + a$.

Working in reverse, if $k - a \leqslant x \leqslant k + a$, then $-a \leqslant x - k \leqslant a$ and $-a \leqslant y \leqslant a$, so $|y| \leqslant a$, that is $|x - k| \leqslant a$.

This has proved that:

> If $a \geqslant 0$, then $|x - k| \leqslant a$ is equivalent to $k - a \leqslant x \leqslant k + a$.

This kind of inequality is involved when you give a number correct to a certain number of decimal places. For example, to say that $x = 3.87$ 'correct to 2 decimal places' is in effect saying that $|x - 3.87| \leqslant 0.005$. The statement $|x - 3.87| \leqslant 0.005$ is equivalent to

$$3.87 - 0.005 \leqslant x \leqslant 3.87 + 0.005,$$

or $3.865 \leqslant x \leqslant 3.875$.

Fig. 2.8

This is illustrated in Fig. 2.8.

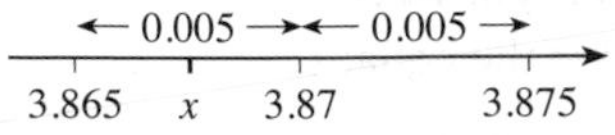

Exercise 2A

1 Sketch the following graphs.

(a) $y = |x + 3|$

(b) $y = |3x - 1|$

(c) $y = |x - 5|$

(d) $y = |3 - 2x|$

(e) $y = 2|x + 1|$

(f) $y = 3|x - 2|$

(g) $y = -2|2x - 1|$

(h) $y = 3|2 - 3x|$

(i) $y = |x + 4| + |3 - x|$

(j) $y = |6 - x| + |1 + x|$

(k) $y = |x - 2| + |2x - 1|$

(l) $y = 2|x - 1| - |2x + 3|$

2 Sketch each of the following sets of graphs.

(a) $y = x^2 - 2$ and $y = |x^2 - 2|$

(b) $y = \sin x$ and $y = |\sin x|$

(c) $y = (x - 1)(x - 2)(x - 3)$ and $y = |(x - 1)(x - 2)(x - 3)|$

(d) $y = \cos 2x$ and $y = |\cos 2x|$ and $y = \cos|2x|$

(e) $y = |x - 2|$ and $y = ||x| - 2|$

3 Write the given inequalities in equivalent forms of the type $a < x < b$ or $a \leq x \leq b$.

(a) $|x - 3| < 1$

(b) $|x + 2| \leq 0.1$

(c) $|2x - 3| \leq 0.001$

(d) $|4x - 3| \leq 8$

4 Rewrite the given inequalities using modulus notation.

(a) $1 \leq x \leq 2$

(b) $-1 < x < 3$

(c) $-3.8 \leq x \leq -3.5$

(d) $2.3 < x < 3.4$

5 Investigate the value of $|a + b|$ for various positive and negative choices for the real numbers a and b, and make a conjecture about the largest possible value for $|a + b|$.

See also if you can make a conjecture about the smallest possible value of $|a + b|$.

6 Construct an argument like that on page 20 to show that $\left|\dfrac{a}{b}\right| = \dfrac{|a|}{|b|}$, provided that $b \neq 0$.

2.5 Equations involving modulus

You can now use the results of the preceding sections to solve equations which involve the modulus function.

The examples which follow use several methods of solution.

- Method 1 is graphical.
- Method 2 uses the definition of modulus.
- Method 3 uses the idea that $|x - a|$ is the distance of x from a.

Not all the methods are used for each example.

Example 2.5.1

Solve the equation $|x - 2| = 3$.

Method 1 From the graphs of $y = |x - 2|$ and $y = 3$ in Fig. 2.9, the solution is $x = -1$ or $x = 5$.

Method 2 $|x - 2| = 3$ means that $x - 2 = 3$ or $-(x - 2) = 3$. Thus the solution is $x = 5$ or $x = -1$.

Method 3 $|x - 2|$ is the distance of x from 2. If this distance is 3, then, thinking geometrically, $x = 2 + 3 = 5$ or $x = 2 - 3 = -1$.

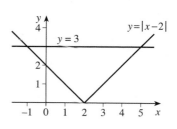

Fig. 2.9

Example 2.5.2

Solve the equation $|x-2|=|2x-1|$.

>   **Method 2**   Either $x-2=2x-1,$ giving $x=-1$
>
>   or $x-2=-(2x-1),$ giving $x=1.$

The solution is $x=-1$ or $x=1$.

>   **Method 3**   Since $|2x-1|=\left|2\left(x-\frac{1}{2}\right)\right|=|2|\times\left|x-\frac{1}{2}\right|$, the equation can be
>   written as $|x-2|=2\times\left|x-\frac{1}{2}\right|$. This means that you want the points x on the
>   number line such that the distance of x from 2 is
>   twice the distance of x from $\frac{1}{2}$ (see Fig. 2.10). It is
>   easy to see that, if x is between $\frac{1}{2}$ and 2 then
> $x=1$; and if x is to the left of $\frac{1}{2}$, then $x=-1$.

Fig. 2.10

2.6   Inequalities involving modulus

The examples which follow use a variety of methods for solving inequalities.

- Method 1 is graphical.
- Method 2 uses the definition of modulus.
- Method 3 uses the result 'If $a\geqslant 0$, then $|x|\leqslant a$ is equivalent to $-a\leqslant x\leqslant a$.'

Not all the methods are used for each example.

Example 2.6.1

Solve the inequality $|x-2|<3$.

>   **Method 2**   From the definition of modulus, you have to separate the cases $x<2$
>   and $x\geqslant 2$.
>
>   When $x<2$, $|x-2|=-(x-2)<3$; this gives $x>-1$, which together with $x<2$
>   gives $-1<x<2$. When $x\geqslant 2$, $|x-2|=x-2<3$; this gives $x<5$, which
>   together with $x\geqslant 2$ gives $2\leqslant x<5$.
>
>   Therefore $-1<x<5$.
>
>   **Method 3**   From the result 'If $a\geqslant 0$, the inequalities $|x-k|\leqslant a$ and
> $k-a\leqslant x\leqslant k+a$ are equivalent', the solution is $2-3<x<2+3$, which is
> $-1<x<5$.

Example 2.6.2

Solve the inequality $|x-2|\geqslant|2x-3|$.

>   **Method 1**   Consider the graphs of $y=|x-2|$
>   and $y=|2x-3|$. These were drawn in Figs. 2.2 and
>   2.3. They are reproduced together in Fig. 2.11; the
>   graph of $y=|2x-3|$ is shown with a dashed line.
>
>   The solid line is above or coincides with the dashed
>   line when $1\leqslant x\leqslant 1\frac{2}{3}$.

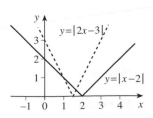

Fig. 2.11

Method 2   In $|x-2|$ you have to separate the cases $x < 2$ and $x \geqslant 2$; and in $|2x-3|$ you have to separate $x < 1\frac{1}{2}$ and $x \geqslant 1\frac{1}{2}$. So it is necessary to consider the cases $x < 1\frac{1}{2}$, $1\frac{1}{2} \leqslant x < 2$ and $x \geqslant 2$.

When $x < 1\frac{1}{2}$, $|x-2| = -(x-2)$ and $|2x-3| = -(2x-3)$, so $-x+2 \geqslant -2x+3$, giving $x \geqslant 1$. So the inequality is satisfied when $1 \leqslant x < 1\frac{1}{2}$.

When $1\frac{1}{2} \leqslant x < 2$, $|x-2| = -(x-2)$ and $|2x-3| = 2x-3$, so $-x+2 \geqslant 2x-3$, giving $x \leqslant 1\frac{2}{3}$. So the inequality is satisfied when $1\frac{1}{2} \leqslant x \leqslant 1\frac{2}{3}$.

When $x \geqslant 2$, $|x-2| = x-2$ and $|2x-3| = 2x-3$, so $x-2 \geqslant 2x-3$, giving $x \leqslant 1$. This is inconsistent with $x \geqslant 2$.

Since the inequality is satisfied when $1 \leqslant x < 1\frac{1}{2}$ and when $1\frac{1}{2} \leqslant x \leqslant 1\frac{2}{3}$, the complete solution is $1 \leqslant x \leqslant 1\frac{2}{3}$.

Example 2.6.3
Solve the inequality $|x-2| \geqslant 2x+1$.

Method 1   Consider the graphs of $y = |x-2|$ and $y = 2x+1$, shown in Fig. 2.12.

The solid line is above or on the dashed line when $x \leqslant \frac{1}{3}$.

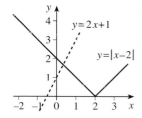

Fig. 2.12

Method 2   If $x < 2$, then $|x-2| = 2-x$, so $2-x \geqslant 2x+1$, giving $x \leqslant \frac{1}{3}$. So the inequality is satisfied when $x \leqslant \frac{1}{3}$.

If $x \geqslant 2$, then $|x-2| = x-2$, so $x-2 \geqslant 2x+1$, giving $x \leqslant -3$. This is inconsistent with $x \geqslant 2$.

So the complete solution is $x \leqslant \frac{1}{3}$.

2.7   Squares, square roots and moduli

You know that, if x is any real number, then $x^2 \geqslant 0$. It follows that $|x^2| = x^2$. Also, from the rule $|a \times b| = |a| \times |b|$, it follows that $|x^2| = |x| \times |x| = |x|^2$.

> If x is any real number, $|x^2| \equiv |x|^2 \equiv x^2$.

Now since $|x|^2 = x^2$, and $|x|$ is positive or zero, it follows that $|x|$ is the square root of x^2. You can show this by evaluating the composite function

$$x \;\to\; [\text{ square }] \;\to\; x^2 \;\to\; [\sqrt{}] \;\to\; \sqrt{x^2}$$

on your calculator with various inputs for x, positive or negative. If you put $x = 3$, say, then you will get the display sequence $3, 9, 3$. But if you put $x = -3$, you will get $-3, 9, 3$, because $\sqrt{}$ always gives the positive square root. That is, $\sqrt{x^2}$ is equal to x when $x \geqslant 0$, but equal to $-x$ when $x < 0$. This is just the definition of $|x|$. It follows that:

> If x is any real number, $\sqrt{x^2} \equiv |x|$.

If you have access to a graphic calculator, verify this identity by displaying the graphs of $y = \sqrt{x^2}$ *and* $y = |x|$ *on it. In fact, if your calculator does not have a key for the modulus function, use* $y = \sqrt{x^2}$.

Example 2.7.1
Find the distance between the points with coordinates (a, k) and (b, k).

Method 1   Both points have the same y-coordinate, so the distance is the same as the distance between points with coordinates a and b on the number line, which is $|b - a|$.

Method 2   By the formula for the distance between two points in P1 Section 1.1, the distance is

$$\sqrt{(b-a)^2 + (k-k)^2} = \sqrt{(b-a)^2 + 0} = \sqrt{(b-a)^2} = |b-a|,$$

using $\sqrt{x^2} = |x|$ with $x = b - a$.

Useful results connecting squares with moduli can be got from the identity

$$x^2 - a^2 \equiv |x|^2 - |a|^2 \equiv \big(|x| - |a|\big)\big(|x| + |a|\big).$$

Suppose first that $a \neq 0$. Then $|a| > 0$, so $|x| + |a| > 0$. It then follows that

$x^2 - a^2 = 0$ is equivalent to $|x| - |a| = 0$,

$x^2 - a^2 > 0$ is equivalent to $|x| - |a| > 0$,

$x^2 - a^2 < 0$ is equivalent to $|x| - |a| < 0$.

You can easily check that the first two of these are also true when $a = 0$; but the third is impossible if $a = 0$, since it gives $x^2 < 0$, which can never occur for any real number x.

It is useful to introduce the symbol $\Leftrightarrow$ for 'is equivalent to'.

> $|x| = |a| \quad \Leftrightarrow \quad x^2 = a^2,$
>
> $|x| > |a| \quad \Leftrightarrow \quad x^2 > a^2;$
>
> if $a \neq 0, |x| < |a| \quad \Leftrightarrow \quad x^2 < a^2.$

These relations are sometimes useful in solving equations and inequalities. They are effective because, although squaring is involved, the two sides are logically equivalent.

The usual warning that squaring may introduce extra roots which don't satisfy the original equation (see P1 Example 4.7.2) doesn't apply.

Example 2.7.2 (see Example 2.5.2)
Solve the equation $\left| x - 2 \right| = \left| 2x - 1 \right|$.

$$\left| x - 2 \right| = \left| 2x - 1 \right| \quad \Leftrightarrow \quad (x - 2)^2 = (2x - 1)^2$$
$$\Leftrightarrow \quad x^2 - 4x + 4 = 4x^2 - 4x + 1$$
$$\Leftrightarrow \quad 3x^2 - 3 = 0$$
$$\Leftrightarrow \quad 3(x + 1)(x - 1) = 0$$
$$\Leftrightarrow \quad x = -1 \quad \text{or} \quad x = 1.$$

Example 2.7.3 (see Example 2.6.2)
Solve the inequality $\left| x - 2 \right| \geq \left| 2x - 3 \right|$.

$$\left| x - 2 \right| \geq \left| 2x - 3 \right| \quad \Leftrightarrow \quad (x - 2)^2 \geq (2x - 3)^2$$
$$\Leftrightarrow \quad x^2 - 4x + 4 \geq 4x^2 - 12x + 9$$
$$\Leftrightarrow \quad 3x^2 - 8x + 5 \leq 0$$
$$\Leftrightarrow \quad (x - 1)(3x - 5) \leq 0$$
$$\Leftrightarrow \quad 1 \leq x \leq 1\tfrac{2}{3}.$$

This method is very quick when it works, but there is a drawback. It can only be used for a very specific type of equation or inequality. It is easy to fall into the trap of assuming it can be applied to equations and inequalities of forms other than $\left| f(x) \right| = \left| g(x) \right|$ or $\left| f(x) \right| < \left| g(x) \right|$, and this can have disastrous consequences.

Example 2.7.4
Solve the equation $\left| x - 2 \right| + \left| 1 - x \right| = 0$.

It is obvious from the answer to Example 2.2.2 that this equation has no solution.

False solution

$$\left| x - 2 \right| + \left| 1 - x \right| = 0 \quad \Leftrightarrow \quad \left| x - 2 \right| = -\left| 1 - x \right|$$
$$\Leftrightarrow (!) \quad (x - 2)^2 = (1 - x)^2$$
$$\Leftrightarrow \quad x^2 - 4x + 4 = 1 - 2x + x^2$$
$$\Leftrightarrow \quad 2x = 3$$
$$\Leftrightarrow \quad x = 1\tfrac{1}{2}.$$

There is no justification for the step marked (!). The previous line has the form $\left| x \right| = -\left| a \right|$, not $\left| x \right| = \left| a \right|$, so the result in the box can't be used.

Exercise 2B

1 Solve the following equations, using at least two methods for each case.

(a) $|x+2|=5$ (b) $|x-1|=7$

(c) $|2x-3|=3$ (d) $|3x+1|=10$

(e) $|x+1|=|2x-3|$ (f) $|x-3|=|3x+1|$

(g) $|2x+1|=|3x+9|$ (h) $|5x+1|=|11-2x|$

2 Solve the following inequalities, using at least two methods for each case.

(a) $|x+2|<1$ (b) $|x-3|>5$

(c) $|2x+7|\leqslant 3$ (d) $|3x+2|\geqslant 8$

(e) $|x+2|<|3x+1|$ (f) $|2x+5|>|x+2|$

(g) $|x|>|2x-3|$ (h) $|4x+1|\leqslant|4x-1|$

3 Solve the equations

(a) $|x+1|+|1-x|=2$,   (b) $|x+1|-|1-x|=2$,   (c) $-|x+1|+|1-x|=2$.

4 Solve the equations

(a) $|x|=|1-x|+1$,   (b) $|x-1|=|x|+1$,   (c) $|x-1|+|x|=1$.

5 Are the following statements true or false? Give a counterexample where appropriate.

(a) The graph of $y=|f(x)|$ never has negative values for y.

(b) The graph of $y=f(|x|)$ never has negative values for y.

Miscellaneous exercise 2

1 Solve the inequality $|x+1|<|x-2|$.   (OCR)

2 Find the greatest and least values of x satisfying the inequality $|2x-1|\leqslant 5$.   (OCR)

3 Sketch, on a single diagram, the graphs of $x+2y=6$ and $y=|x+2|$. Hence, or otherwise, solve the inequality $|x+2|<\frac{1}{2}(6-x)$.   (OCR)

4 Solve the equation $|x|=|2x+1|$.   (OCR)

5 Sketch the graph of $y=|x+2|$ and hence, or otherwise, solve the inequality $|x+2|>2x+1$.   (OCR)

6 Solve the equation $4|x| = |x-1|$.

On the same diagram sketch the graphs of $y = 4|x|$ and $y = |x-1|$, and hence, or otherwise, solve the inequality $4|x| > |x-1|$.

7 Sketch, on separate diagrams, the graphs of $y = |x|$, $y = |x-3|$ and $y = |x-3| + |x+3|$. Find the solution set of the equation $|x-3| + |x+3| = 6$. (OCR)

8 The functions f and g are defined on the set of real numbers as follows:

$$f: x \mapsto |2\sin x°|, \qquad g: x \mapsto |\sin 2x°|.$$

(a) (i) Make clearly labelled sketches of the graphs of $y = f(x)$ and $y = g(x)$ in the interval $-270 \leqslant x \leqslant 270$.

(ii) State the range of each function.

(b) Decide whether or not each function is periodic and, if so, state its period. (OCR)

9 Solve the inequality $|x| < 4|x-3|$.

10 Rewrite the function $k(x)$ defined by $k(x) = |x+3| + |4-x|$ for the following three cases, without using the modulus in your answer.

(a) $x > 4$ (b) $-3 \leqslant x \leqslant 4$ (c) $x < -3$

11 Solve the equations   (a) $x + |2x-1| = 3$,   (b) $3 + |2x-1| = x$.

12 Sketch the graph of $y = |2x-3| + |5-x|$.

(a) Calculate the y coordinate of the point where the graph cuts the y-axis.

(b) Determine the gradient of the graph where $x < -5$. (OCR, adapted)

13 A graph has equation $y = x + |2x-1|$. Express y as a linear function of x (that is, in the form $y = mx + c$ for constants m and c) in each of the following intervals for x.

(a) $x > \frac{1}{2}$ (b) $x < \frac{1}{2}$ (OCR)

14 Sketch the graphs of the following functions.

(a) $y = \sin 3x°$ (b) $y = |\sin 3x°|$ (c) $y = \sin|3x|°$

15* Solve the following inequalities.

(a) $\dfrac{x+1}{x-1} < 4$ (b) $\dfrac{|x|+1}{|x|-1} < 4$ (c) $\left|\dfrac{x+1}{x-1}\right| < 4$ (OCR)

3 Exponential and logarithmic functions

This chapter investigates the function b^x which appears in the equation for exponential growth, and its inverse $\log_b x$. When you have completed it, you should

- understand the idea of continuous exponential growth and decay
- know the principal features of exponential functions and their graphs
- know the definition and properties of logarithmic functions
- be able to switch between the exponential and logarithmic forms of an equation
- understand the idea and possible uses of a logarithmic scale
- be familiar with logarithms to the special bases e and 10
- be able to solve equations and inequalities with the unknown in the index
- be able to use logarithms to identify models of the forms $y = ab^x$ and $y = ax^n$.

3.1 Continuous exponential growth

In P1 Section 14.4 you met the idea of exponential growth and decay, defined by an equation of the form $u_i = ar^i$. In this equation a is the initial value; r is the rate of growth if $r > 1$, or the rate of decay if $r < 1$; and i is the number of time-units after the start. The equation defines a geometric sequence with common ratio r, but with the first term denoted by u_0 instead of u_1.

Exponential growth doesn't only occur in situations which increase by discrete steps. Rampant inflation, a nuclear chain reaction, the spread of an epidemic and the growth of cells are phenomena which take place in continuous time, and they need to be described by functions having the real numbers rather than the natural numbers for their domain.

For continuous exponential growth, the equation $u_i = ar^i$, where $i \in \mathbb{N}$, is replaced by

$$f(x) = ab^x, \quad \text{where } x \in \mathbb{R} \text{ and } x > 0.$$

In this equation a stands for the initial value when $x = 0$, and b is a constant which indicates how fast the quantity is growing. (The idea of a 'common ratio' no longer applies in the continuous case, so a different letter is used.) In many applications the variable x represents time. The graph of $f(x)$ is shown in Fig. 3.1.

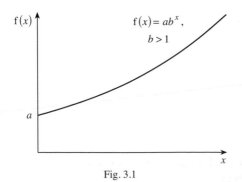

Fig. 3.1

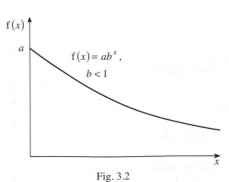

Fig. 3.2

For exponential growth b has to be greater than 1. If $0 < b < 1$ the graph takes the form
shown in Fig. 3.2; for large values of x the graph gets closer to the x-axis but never reaches
it. This then represents exponential decay. Examples of this are the level of radioactivity in a
lump of uranium ore, and the concentration of an antibiotic in the blood stream.

Example 3.1.1

The population of the USA grew exponentially from the end of the War of Independence
until the Civil War. It increased from 3.9 million at the 1790 census to 31.4 million in 1860.
What would the population have been in 1990 if it had continued to grow at this rate?

If the population x years after 1790 is P million, and if the growth were exactly
exponential, then P and x would be related by an equation of the form

$$P = 3.9b^x,$$

where $P = 31.4$ when $x = 70$. The constant b therefore satisfies the equation

$$31.4 = 3.9b^{70}, \text{ so } b = \left(\frac{31.4}{3.9}\right)^{\frac{1}{70}} = 1.030\ldots.$$

At this rate the population in 1990 would have grown to about
$3.9 \times 1.030\ldots^{200}$ million, which is between 1.5 and 1.6 billion.

You can shorten this calculation as follows. In 70 years, the population multiplied by $\dfrac{31.4}{3.9}$.

In 200 years, it would therefore multiply by $\left(\dfrac{31.4}{3.9}\right)^{\frac{200}{70}}$. The 1990 population can then
be calculated as $3.9 \times \left(\dfrac{31.4}{3.9}\right)^{\frac{200}{70}}$ million, without working out b as an intermediate step.

Example 3.1.2

Carbon dating in archaeology is based on the decay of the isotope carbon-14, which has
a half-life of 5715 years. By what percentage does carbon-14 decay in 100 years?

The half-life of a radioactive isotope is the time it would take for half of any sample of
the isotope to decay. After t years one unit of carbon-14 is reduced to b^t units, where

$$b^{5715} = 0.5 \qquad \text{(since 0.5 units are left after 5715 years)}$$

so $\quad b = 0.5^{\frac{1}{5715}} = 0.999\,878\,721.$

When $t = 100$ the quantity left is $b^{100} \approx 0.988$ units, a reduction of 0.012 units, or 1.2%.

3.2   Exponential functions

In the equation $y = ab^x$ for exponential growth the constant a simply sets a scale on the
y-axis. The essential features of the relationship can be studied in the function

$$f(x) = b^x, \quad \text{where } x \in \mathbb{R}.$$

A function of this form is called an **exponential function**, because the variable x
appears in the exponent (another word for the index).

This definition needs some points of explanation. First, it makes sense only if b is positive. To see this, note that, for some values of x, b^x has no meaning for negative b; for example, $b^{\frac{1}{2}} = \sqrt{b}$. Secondly, if $b = 1$, b^x has the constant value 1. So the definition of an exponential function applies only if $b > 0$, $b \neq 1$. With this restriction, the values of b^x are always positive.

However, there is no need to restrict x to positive values. Since $b^0 = 1$, the graphs of all exponential functions contain the point $(0,1)$. Notice also that

$$b^{-x} = \frac{1}{b^x} = \left(\frac{1}{b}\right)^x.$$

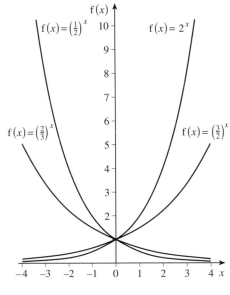

Therefore if b^x is greater than 1 then b^{-x} lies between 0 and 1. A further consequence of this relationship is that the reflection in the y-axis of the graph of $y = b^x$ is $y = \left(\dfrac{1}{b}\right)^x$.

These points are illustrated in Fig. 3.3, which shows the graph of exponential functions for several values of b. Note that the functions are increasing if $b > 1$, and decreasing if $0 < b < 1$.

Fig. 3.3

Lastly, you should notice that up to now the expression b^x has only been defined when x is a positive or negative fraction (that is, x is rational). So the definition has to be extended to all of $\mathbb{R}$ by filling in the gaps where x is irrational.

As an example, suppose that you want to give a meaning to 2^π. Now π is an irrational number ($3.141\,592\,65\ldots$), but you can find pairs of rational numbers very close together such that π lies between them. For example, since $3.141\,592\,6 < \pi < 3.141\,592\,7$, 2^π ought to lie between $2^{\frac{31\,415\,926}{10\,000\,000}}$ and $2^{\frac{31\,415\,927}{10\,000\,000}}$, that is between $8.824\,977\,499\ldots$ and $8.824\,978\,11\ldots$; so $2^\pi = 8.824\,98$ correct to 5 decimal places. If you want to find 2^π to a greater degree of accuracy, you can sandwich π between a pair of rational numbers which are even closer together.

You could, if you wished, define 2^π as the limit, as n tends to infinity, of a sequence 2^{u_r}, where u_r is a sequence of numbers which tends to π. It can be proved that this definition gives a unique answer, and that values of 2^x defined in this way obey the rules for working with indices given in P1 Section 2.3.

3.3   Logarithmic functions

The graphs in Fig. 3.3 show that the exponential function $x \mapsto b^x$ has for its natural domain the set of all real numbers, and the corresponding range is the positive real numbers. The function is increasing if $b > 1$, and decreasing if $b < 1$; in either case it is one–one.

It follows that this function has an inverse whose domain is the set of positive real numbers and whose range is all real numbers. (See P1 Section 11.6.) This inverse function is called the **logarithm to base b**, and is denoted by $\log_b$.

$$y = b^x \Leftrightarrow x = \log_b y, \quad \text{where } x \in \mathbb{R}, y \in \mathbb{R}, y > 0.$$

To draw the graph of $y = \log_b x$ you can use the general result proved in P1 Section 11.8, that the graphs of $y = f(x)$ and $y = f^{-1}(x)$ are reflections of each other in the line $y = x$. This is illustrated in Fig. 3.4, which shows graphs of $y = b^x$ and $y = \log_b x$ using the same axes.

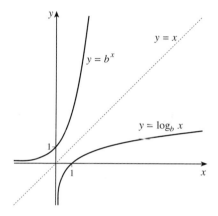

Fig. 3.4

The figure is drawn for $b = 3$, and it is typical of the graphs for any base $b > 1$. The definition of $\log_b$ is still valid if $0 < b < 1$, in which case the graphs have a different form; but this is not important, since in practice logarithms are rarely used with bases less than 1.

Notice that, since the point $(0,1)$ lies on $y = b^x$, its reflection $(1,0)$ lies on $y = \log_b x$ for every base b. That is:

$$\log_b 1 = 0.$$

Other points on $y = b^x$ are $(1,b)$, $\left(2,b^2\right)$ and $\left(-1,\dfrac{1}{b}\right)$, so that other points on $y = \log_b x$ are $(b,1)$, $\left(b^2,2\right)$ and $\left(\dfrac{1}{b},-1\right)$. That is,

$$\log_b(b) = 1, \quad \log_b\left(b^2\right) = 2 \quad \text{and} \quad \log_b\left(\frac{1}{b}\right) = -1.$$

These are important special cases of the following statement:

$$\text{For any } n, \ \log_b b^n = n.$$

This is simply an application of the general result given in P1 Section 11.6, that $f^{-1}f$ is the identity function. With $f : x \mapsto b^x$ and $f^{-1} : x \mapsto \log_b x$, it follows that

$$f^{-1}f : x \mapsto \log_b b^x \text{ is the identity function.}$$

Example 3.3.1

Find   (a) $\log_3 81$,   (b) $\log_{81} 3$,   (c) $\log_3\left(\dfrac{1}{81}\right)$,   (d) $\log_{\frac{1}{3}} 81$.

(a) Since $81 = 3^4$, $\log_3 81 = 4$.   (b) $3 = 81^{\frac{1}{4}}$, so $\log_{81} 3 = \frac{1}{4}$.

(c) $\frac{1}{81} = 3^{-4}$, so $\log_3\left(\dfrac{1}{81}\right) = -4$.   (d) $81 = \dfrac{1}{(1/3)^4} = \left(\frac{1}{3}\right)^{-4}$, so $\log_{\frac{1}{3}} 81 = -4$.

Exercise 3A

1   A rumour spreads exponentially through a college. 100 people have heard it by noon, and 200 by 1 p.m. How many people have heard it

   (a) by 3 p.m.,   (b) by 12.30 p.m.,   (c) by 1.45 p.m.?

2   An orchestra tunes to a frequency of 440, which sounds the A above middle C. Each octave higher doubles the frequency, and each of the 12 semitones in the octave increases the frequency in the same ratio.

   (a) What is this ratio?   (b) Find the frequency of middle C.

   (c) Where on the scale is a note with a frequency of 600?

3   A cup of coffee at 85 °C is placed in a freezer at 0 °C. The temperature of the coffee decreases exponentially, so that after 5 minutes it is 30 °C.

   (a) What is its temperature after 3 minutes?

   (b) Find, by trial, how long it will take for the temperature to drop to 5 °C.

4   A radioactive substance decays at a rate of 12% per hour.

   (a) Find, by trial, after how many hours half of the radioactive material will be left.

   (b) How many hours earlier did it have twice the current amount of radioactive material?

5   With the same axes, sketch the graphs of

   (a) $y = 1.25^x$,   (b) $y = 0.8^x$,   (c) $y = 0.8^{-x}$.

6   Write each of the following in the form $y = b^x$.

   (a) $\log_2 8 = 3$ (b) $\log_3 81 = 4$ (c) $\log_5 0.04 = -2$

   (d) $\log_7 x = 4$ (e) $\log_x 5 = t$ (f) $\log_p q = r$

7   Write each of the following in the form $x = \log_b y$.

   (a) $2^3 = 8$ (b) $3^6 = 729$ (c) $4^{-3} = \frac{1}{64}$

   (d) $a^8 = 20$ (e) $h^9 = g$ (f) $m^n = p$

8   Evaluate the following.

   (a) $\log_2 16$ (b) $\log_4 16$ (c) $\log_7 \frac{1}{49}$

   (d) $\log_4 1$ (e) $\log_5 5$ (f) $\log_{27} \frac{1}{3}$

   (g) $\log_{16} 8$ (h) $\log_2 2\sqrt{2}$ (i) $\log_{\sqrt{2}} 8\sqrt{2}$

9 Find the value of y in each of the following.

(a) $\log_y 49 = 2$

(b) $\log_4 y = -3$

(c) $\log_3 81 = y$

(d) $\log_{10} y = -1$

(e) $\log_2 y = 2.5$

(f) $\log_y 1296 = 4$

(g) $\log_{\frac{1}{2}} y = 8$

(h) $\log_{\frac{1}{2}} 1024 = y$

(i) $\log_y 27 = -6$

3.4 Properties of logarithms

It was shown in P1 Section 2.3 that expressions involving indices can be simplified by applying a number of rules, including the multiplication and division rules and the power-on-power rule. There are corresponding rules for logarithms, which can be deduced from the index rules by using the equivalence

$$\log_b x = y \iff x = b^y.$$

These rules hold for logarithms to any base b, so the notation $\log_b x$ has been simplified to $\log x$.

Power rule:	$\log x^n = n \log x$
nth root rule:	$\log \sqrt[n]{x} = \dfrac{1}{n} \log x$
Multiplication rule:	$\log(pq) = \log p + \log q$
Division rule:	$\log\left(\dfrac{p}{q}\right) = \log p - \log q$

Here are proofs of these rules.

Power rule

If $\log x = r$, then $x = b^r$; so $x^n = \left(b^r\right)^n = b^{rn}$.

In logarithmic form this is $\log x^n = rn = n \log x$.

In this proof n can be any real number, although the rule is most often used with integer values of n.

nth root rule

This is the same as the power rule, since the nth root of x is $x^{\frac{1}{n}}$.

Multiplication rule

If $\log p = r$ and $\log q = s$, then $p = b^r$ and $q = b^s$, so $pq = b^r b^s = b^{r+s}$.

In logarithmic form this is $\log(pq) = r + s = \log p + \log q$.

Division rule

The proof is the same as for the multiplication rule, but with division in place of multiplication and subtraction in place of addition.

Example 3.4.1

If $\log 2 = r$ and $\log 3 = s$, express in terms of r and s (a) $\log 16$,   (b) $\log 18$,   (c) $\log 13.5$.

(a) $\log 16 = \log 2^4 = 4 \log 2 = 4r$.

(b) $\log 18 = \log\left(2 \times 3^2\right) = \log 2 + \log 3^2 = \log 2 + 2 \log 3 = r + 2s$.

(c) $\log 13.5 = \log \dfrac{3^3}{2} = \log 3^3 - \log 2 = 3 \log 3 - \log 2 = 3s - r$.

Example 3.4.2

Find the connection between $\log_b c$ and $\log_c b$.

$$\log_b c = x \iff c = b^x \iff c^{\frac{1}{x}} = \left(b^x\right)^{\frac{1}{x}} = b^1 \iff b = c^{\frac{1}{x}} \iff \log_c b = \frac{1}{x}.$$

Therefore $\log_c b = \dfrac{1}{\log_b c}$.

Historically logarithms were important because for many years, before calculators and computers were available, they provided the most useful form of calculating aid. With a table of logarithms students would, for example, find the cube root of 100 by looking up the value of $\log 100$ and dividing it by 3. By the nth root rule, this gave $\log \sqrt[3]{100}$, and the cube root could then be obtained from a table of the inverse function.

You could simulate this process on your calculator by keying in $[100, \log, \div, 3, =, 10^x]$, giving successive displays 100, 2, $0.666\,666\,6\ldots$ and the answer $4.641\,588\,83\ldots$. But of course you don't need to do this, since your calculator has a special key for working out roots directly.

Exercise 3B

1   Write each of the following in terms of $\log p$, $\log q$ and $\log r$. The logarithms have base 10.

(a) $\log pqr$

(b) $\log pq^2 r^3$

(c) $\log 100 pr^5$

(d) $\log \sqrt{\dfrac{p}{q^2 r}}$

(e) $\log \dfrac{pq}{r^2}$

(f) $\log \dfrac{1}{pqr}$

(g) $\log \dfrac{p}{\sqrt{r}}$

(h) $\log \dfrac{qr^7 p}{10}$

(i) $\log \sqrt{\dfrac{10p^{10} r}{q}}$

2   Express as a single logarithm, simplifying where possible. (All the logarithms have base 10, so, for example, an answer of $\log 100$ simplifies to 2.)

(a) $2 \log 5 + \log 4$

(b) $2 \log 2 + \log 150 - \log 6000$

(c) $3 \log 5 + 5 \log 3$

(d) $2 \log 4 - 4 \log 2$

(e) $\log 24 - \frac{1}{2} \log 9 + \log 125$

(f) $3 \log 2 + 3 \log 5 - \log 10^6$

(g) $\frac{1}{2} \log 16 + \frac{1}{3} \log 8$

(h) $\log 64 - 2 \log 4 + 5 \log 2 - \log 2^7$

3 If $\log 3 = p$, $\log 5 = q$ and $\log 10 = r$, express the following in terms of p, q and r. (All the logarithms have the same unspecified base.)

   (a) $\log 2$ (b) $\log 45$ (c) $\log \sqrt{90}$

   (d) $\log 0.2$ (e) $\log 750$ (f) $\log 60$

   (g) $\log \frac{1}{6}$ (h) $\log 4.05$ (i) $\log 0.15$

3.5 Special bases

Although the base of the logarithm function can be any real positive number except 1, only two bases are in common use. One is a number denoted by e, for which the logarithm function has a number of special properties; these are explored in the next chapter. Logarithms to base e are denoted by 'ln', and can be found using the [LN] key on your calculator.

The other base is 10, which is important because our system of writing numbers is based on powers of 10. On your calculator the key labelled [LOG] gives logarithms to base 10. In Sections 3.5 to 3.9, if no base is specified, the symbol ' $\log$' will stand for $\log_{10}$.

When logarithms were used to do calculations, students used tables which gave $\log x$ only for values of x between 1 and 10. So to find $\log 3456$, you would use the rules in Section 3.4 to write

$$\log 3456 = \log\left(3.456 \times 10^3\right) = \log 3.456 + \log 10^3 = \log 3.456 + 3.$$

The tables gave $\log 3.456$ as 0.5386 (correct to 4 decimal places), so $\log 3456$ is 3.5386. Notice that the number 3 before the decimal point is the same as the index when 3456 is written in standard form.

Logarithms to base 10 are sometimes useful in constructing logarithmic scales. As an example, suppose that you want to make a diagram to show the populations of countries which belong to the United Nations. In 1999 the largest of these was China, with about 1.2 billion people, and the smallest was San Marino, with $25\,000$. If you represented the population of China by a line of length 12 cm, then Nigeria would have length 1.1 cm, Malaysia just over 2 mm, and the line for San Marino would be only 0.0025 mm long!

Fig. 3.5 is an alternative way of showing the data.

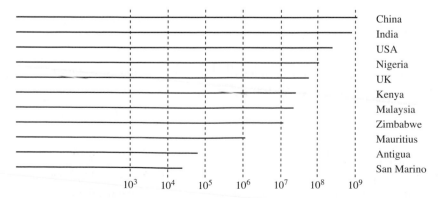

Fig. 3.5

Fig. 3.5 uses a logarithmic scale, in which a country with population P is shown by a line of length $\log P$ cm. China now has a length of just over 9 cm, and San Marino a length of between 4 and 5 cm. You have to understand the diagram in a different way; an extra cm in length implies a population 10 times as large, rather than 100 million larger. But the countries are still placed in the correct order, and the population of any country can be found as 10^x where x is the length of its line in centimetres.

3.6 Equations and inequalities

You know that $\log_2 2 = 1$ and $\log_2 4 = 2$, but how can you find $\log_2 3$?

Suppose that $\log_2 3 = x$. Then from the definition,

$$2^x = 3.$$

So the problem is to solve an equation where the unknown appears in the index. The trick is to use logarithms and to write the equation as

$$\log 2^x = \log 3.$$

This is often described as 'taking logarithms of both sides of the equation'. You can now use the power rule to write this as

$$x \log 2 = \log 3.$$

Using the [LOG] key on the calculator, this is

$$x \times 0.301\ldots \approx 0.477\ldots,$$

which gives $x = \log_2 3 = \dfrac{0.477\ldots}{0.301\ldots} = 1.58$, correct to 3 significant figures.

This type of equation arises in various applications.

Example 3.6.1

Iodine-131 is a radioactive isotope used in treatment of the thyroid gland. It decays so that, after t days, 1 unit of the isotope is reduced to 0.9174^t units. How many days does it take for the amount to fall to less than 0.1 units?

This requires solution of the inequality $0.9174^t < 0.1$. Since $\log$ is an increasing function, taking logarithms gives

$$\log(0.9174^t) < \log 0.1 \quad \Leftrightarrow \quad t \log 0.9174 < \log 0.1.$$

Now beware! The value of $\log 0.9174$ is negative, so when you divide both sides by $\log 0.9174$ you must change the direction of the inequality:

$$t > \frac{\log 0.1}{\log 0.9174} = 26.708\ldots.$$

The amount of iodine-131 will fall to less than 0.1 units after about 26.7 days.

Example 3.6.2

How many terms of the geometric series $1 + 1.01 + 1.01^2 + 1.01^3 + \ldots$ must be taken to give a sum greater than 1 million?

The sum of n terms of the series is given by the formula (see P1 Section 14.2)

$$\frac{1.01^n - 1}{1.01 - 1} = 100(1.01^n - 1).$$

The problem is to find the smallest value of n for which

$$\frac{1.01^n - 1}{1.01 - 1} = 100(1.01^n - 1) > 1\,000\,000, \text{ which gives } 1.01^n > 10\,001.$$

Taking logarithms of both sides,

$$\log 1.01^n > \log 10\,001, \text{ so } n \log 1.01 > \log 10\,001.$$

Since $\log 1.01$ is positive,

$$n > \frac{\log 10\,001}{\log 1.01} = 925.6\ldots.$$

The smallest integer n satisfying this inequality is 926.

3.7* A relation between logarithmic functions

The equation $2^x = 3$ in Section 3.6 was solved using logarithms to base 10, but the steps leading to

$$x = \frac{\log_b 3}{\log_b 2}$$

could have been made with any base b. For example, you could choose base e, using the [LN] key on your calculator to give

$$x = \frac{\ln 3}{\ln 2} = \frac{1.098\ldots}{0.693\ldots} = 1.58, \text{ correct to 3 significant figures.}$$

The answer is the same, because logarithms to different bases are proportional to each other.

Suppose that your calculator had a [LN] key but no [LOG] key, and that you wanted to calculate a value for $\log x$. Then you could argue as follows.

In exponential form, $y = \log x$ becomes $10^y = x$.

The equation $10^y = x$ can be solved by taking logarithms to base e of both sides, giving

$$\ln(10^y) = \ln x;$$

that is,

$$y \ln 10 = \ln x.$$

So $\log x = y = \dfrac{\ln x}{\ln 10}$.

Since $\ln 10 = 2.302\ldots$, $\dfrac{1}{\ln 10} = 0.434\ldots$,

so $\log x = 0.434\ldots \times \ln x$.

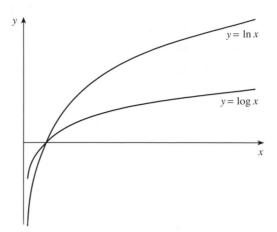

This is illustrated by the graphs in Fig. 3.6. The y-values for $\log x$ are just $0.434\ldots$ times the those for $\ln x$. That is, you can get the $\log x$ graph by scaling down the graph of $\ln x$ in the y-direction by a factor of $0.434\ldots$.

This relation is true more generally. If b and c are any two different bases, then

Fig. 3.6

$\log_c x$ is a constant multiple of $\log_b x$ as x varies.

Exercise 3C

1 Solve the following equations, giving inexact answers correct to 3 significant figures.

(a) $3^x = 5$ (b) $7^x = 21$ (c) $6^{2x} = 60$

(d) $5^{2x-1} = 10$ (e) $4^{\frac{1}{2}x} = 12$ (f) $2^{x+1} = 3^x$

(g) $\left(\frac{1}{2}\right)^{3x+2} = 25$ (h) $2^x \times 2^{x+1} = 128$ (i) $\left(\frac{1}{4}\right)^{2x-1} = 7$

2 Solve the following inequalities, giving your answers correct to 3 significant figures.

(a) $3^x > 8$ (b) $5^x < 10$ (c) $7^{2x+5} \leqslant 24$

(d) $0.5^x < 0.001$ (e) $0.4^x < 0.0004$ (f) $0.2^x > 25$

(g) $4^x \times 4^{3-2x} \leqslant 1024$ (h) $0.8^{2x+5} \geqslant 4$ (i) $0.8^{1-3x} \geqslant 10$

3 How many terms of the geometric series $1 + 2 + 4 + 8 + \ldots$ must be taken for the sum to exceed 10^{11}?

4 How many terms of the geometric series $2 + 6 + 18 + 54 + \ldots$ must be taken for the sum to exceed 3 million?

5 How many terms of the geometric series $1 + \frac{1}{2} + \frac{1}{4} + \frac{1}{8} + \ldots$ must be taken for its sum to differ from 2 by less than 10^{-8}?

6 How many terms of the geometric series $2 + \frac{1}{3} + \frac{1}{18} + \frac{1}{108} + \ldots$ must be taken for its sum to differ from its sum to infinity by less than 10^{-5}?

7 A radioactive isotope decays so that after t days an amount 0.82^t units remains. How many days does it take for the amount to fall to less than 0.15 units?

8 Jacques is saving for a new car which will cost $29 000. He saves by putting $400 a month into a savings account which gives 0.1% interest per month. After how many months will he be able to buy his car? Assume it does not increase in price!

9 To say that a radioactive isotope has a half-life of 6 days means that 1 unit of isotope is reduced to $\frac{1}{2}$ unit in 6 days. So if the daily decay rate is given by r, then $r^6 = 0.5$.

(a) For this isotope, find r.

(b) How long will it take for the amount to fall to 0.25 units?

(c) How long will it take for the amount to fall to 0.1 units?

10 A biological culture contains 500 000 bacteria at 12 noon on Monday. The culture increases by 10% every hour. At what time will the culture exceed 4 million bacteria?

11 A dangerous radioactive substance has a half-life of 90 years. It will be deemed safe when its activity is down to 0.05 of its initial value. How long will it be before it is deemed safe?

12 Finding $\log_3 10$ is equivalent to solving the equation $x = \log_3 10$, which itself is equivalent to solving $3^x = 10$. Find the following logarithms by forming and solving the appropriate equations. Give your answers correct to 3 significant figures.

(a) $\log_4 12$

(b) $\log_7 100$

(c) $\log_8 2.75$

(d) $\log_{\frac{1}{2}} 250$

(e) $\log_3 \pi$

(f) $\log_{\frac{1}{4}} 0.04$

3.8 Graphs of exponential growth

The technique of taking logarithms is often useful when you are dealing with economic, social or scientific data which you think might exhibit exponential growth or decay.

Suppose that a quantity y is growing exponentially, so that its value at time t is given by

$$y = ab^t,$$

where a and b are constants. Taking logarithms of both sides of this equation, to any base,

$$\log y = \log(ab^t) = \log a + \log b^t = \log a + t \log b.$$

The expression on the right increases linearly with t. So if $\log y$ is plotted against t, the graph would be a straight line with gradient $\log b$ and intercept $\log a$.

Example 3.8.1

If $\log y = 0.322 - 0.531t$, where $\log y$ denotes $\log_{10} y$, express y in terms of t.

Equating the right side to $\log a + t \log b$, $\log a = 0.322$ and $\log b = -0.531$. So, since the logarithms are to base 10, $a = 10^{0.322} = 2.10$ and $b = 10^{-0.531} = 0.294$ (both to 3 significant figures). In exponential form the equation for y is therefore

$$y = 2.10 \times 0.294^t.$$

An alternative way of writing this calculation is based on the property that if $\log y = x$ then $y = 10^x$, so $y = 10^{\log y}$. Therefore

$$y = 10^{\log y} = 10^{0.322 - 0.531t} = 10^{0.322} \times (10^{-0.531})^t = 2.10 \times 0.294^t.$$

Example 3.8.2

An investment company claims that the price of its shares has grown exponentially over the past six years, and supports its claim with Fig. 3.7. Is this claim justified?

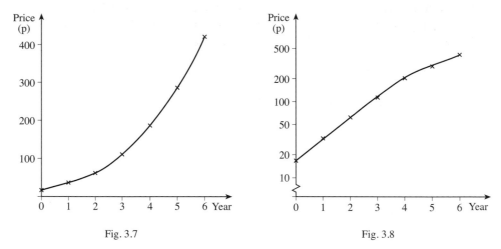

Fig. 3.7   Fig. 3.8

If the graph is drawn with the price shown on a logarithmic scale, you get Fig. 3.8. If the claim were true, this graph would be a straight line. This seems approximately true for the first three years, but more recently the graph has begun to bend downwards, suggesting that the early promise of exponential growth has not been sustained.

The ideas of the last two examples can be combined, not just to investigate whether there is an exponential relationship, but also to find the numerical constants in the equation.

Example 3.8.3

Use the following census data for the USA to justify the statement in Example 3.1.1, that the population grew exponentially from 1790 to 1860.

Year	1790	1800	1810	1820	1830	1840	1850	1860
Population (millions)	3.9	5.3	7.2	9.6	12.9	17.0	23.2	31.4

If you plot these figures on a graph, as in Fig. 3.9, it is clear that the points lie on a smooth curve with a steadily increasing gradient, but this doesn't by itself show that the growth is exponential.

To approach the question scientifically, the first step is to choose appropriate notation. For the population, you may as well work in millions of people, as in the table;

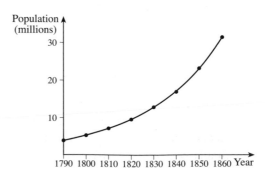

Fig. 3.9

there is no point in cluttering the data with lots of zeros, which would in any case give a false illusion of accuracy. So let P stand for the number of millions of people in the population. As for the date, since you are only interested in the period from 1790 to 1860, it is better to choose a variable t to stand for the number of years after 1790 rather than the actual year number. The theory then being investigated is that P and t are related by an equation of the form

$$P = ab^t \qquad \text{for } 0 \leqslant t \leqslant 70.$$

To convert this into a linear equation, take logarithms of both sides of the equation. You can use logarithms to any base you like; if you choose e, the equation becomes

$$\ln P = \ln a + t \ln b,$$

in which the independent variable is t and the dependent variable is $\ln P$. So make a new table of values in terms of these variables.

t	0	10	20	30	40	50	60	70
$\ln P$	1.36	1.67	1.97	2.26	2.56	2.83	3.14	3.45

These values are used to plot the graph in Fig. 3.10. You can see that the points very nearly lie on a straight line, though not exactly so; you wouldn't expect a population to follow a precise mathematical relationship. However, it is quite close enough to justify the claim that the growth of the population was exponential.

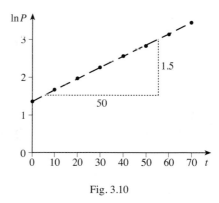

Fig. 3.10

The dashed line in Fig. 3.10 is an attempt to draw by eye a line that best fits the plotted points. By measurement, it seems that the intercept on the vertical axis is about 1.37; and, by using a suitable gradient triangle (shown with dotted lines), you can find that the gradient is about $\dfrac{1.5}{50} = 0.03$.

So the line has equation

$$\ln P = 1.37 + 0.03t,$$

which is of the desired form $\ln P = \ln a + t \ln b$ with $\ln a \approx 1.37$ and $\ln b \approx 0.03$. To find a, remember that $\ln a$ is $\log_e a$, and $\log_e a \approx 1.37 \Leftrightarrow a \approx e^{1.37}$. You can calculate this using the $\left[e^x \right]$ key on your calculator, which gives $a \approx 3.94$. Similarly $b \approx e^{0.03} \approx 1.03$.

It follows that, over the period from 1790 to 1860, the growth of the population could be described to a good degree of accuracy by the law

$$P = 3.94 \times 1.03^t.$$

An equation like $P = 3.94 \times 1.03^t$ is called a **mathematical model.** It is not an exact equation giving the precise size of the population, but it is an equation of a simple form which describes the growth of the population to a very good degree of accuracy. For example, if you wanted to know the population in 1836, when $t = 46$, you could calculate $3.94 \times 1.03^{46} = 15.3\ldots$, and assert with confidence that in that year the population of the USA was between 15 and $15\frac{1}{2}$ million.

Example 3.8.4

A thousand people waiting at a medical centre were asked to record how long they had to wait before they saw a doctor. Their results are summarised as follows.

Waiting time (minutes)	0 to 5	5 to 10	10 to 15	15 to 20	20 to 30	30 to 60	more than 60
Number of people	335	218	155	90	111	85	6

Show that the proportion p of people who had to wait at least t minutes can be modelled by an equation of the form $p = e^{-kt}$, and find the value of k.

Obviously all of the people had to wait at least 0 minutes; all but 335, that is $1000 - 335 = 665$, had to wait at least 5 minutes; of these, $665 - 218 = 447$ had to wait at least 10 minutes; and so on. So you can make a table of p, the proportion that had to wait at least t minutes, for various values of t.

t	0	5	10	15	20	30	60
p	1	0.665	0.447	0.292	0.202	0.091	0.006

If you plot these values for yourself, you will see that they appear to fit an exponential decay graph; but to show this conclusively it is necessary to rewrite the equation so that it can be represented by a straight line.

Now if $p = e^{-kt}$ as suggested in the question, $\ln p = -kt$, so a graph of $\ln p$ against t would be a straight line through the origin with gradient $-k$. So make a table of values of $\ln p$:

t	0	5	10	15	20	30	60
$\ln p$	0	−0.41	−0.81	−1.23	−1.60	−2.40	−5.12

These values are plotted in Fig. 3.11.

This example differs from Example 3.8.3 in that you know that the graph must pass through the origin. So draw the best line that you can through the origin to fit the plotted points. From Fig. 3.11 the gradient of this line is about −0.082.

So the proportion who had to wait more than t minutes is modelled by the equation $p = e^{-0.082t}$.

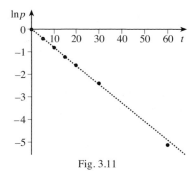

Fig. 3.11

3.9 Power law models

Another type of model which can be investigated using logarithms is the power law, where one variable is related to another by an equation of the form $y = ax^n$. In this case, when you take logarithms (to any base) of both sides, you get

$$\log y = \log(ax^n) = \log a + \log x^n = \log a + n \log x.$$

With such a law, if you plot values of $\log y$ against $\log x$, you get a straight line with gradient n and intercept $\log a$.

You know quite a few examples of power laws in geometry. For example, the surface area of a cube of side x is given by $A = 6x^2$, and the volume of a sphere of radius r is $V = \frac{4}{3}\pi r^3$; laws of this kind also occur frequently in experimental science, and in modelling situations in geography and economics.

Example 3.9.1

These figures have been given for the typical daily metabolic activity of various species of mammal.

	Weight (kg)	Energy expended (calories per kg)
Rabbit	2	58
Man	70	33
Horse	600	22
Elephant	4000	13

Investigate the relation between the energy expenditure (E calories per kg) and the weight (W kg) of the various animals.

This is the kind of situation where a power law model, of the form $E = aW^n$, may be appropriate, so try plotting $\log E$ against $\log W$. Using logarithms to base 10, the corresponding values for the four animals are:

$\log W$	0.30	1.85	2.78	3.60
$\log E$	1.76	1.52	1.34	1.11

These are plotted in Fig. 3.12. There are four points, one for each animal.

Since all the figures are statistical averages, and energy expenditure can't be very precisely measured, you wouldn't expect the points to lie exactly on a straight line. However, they do suggest a trend that might be generalised to apply to other mammals in a similar environment. This is expressed by the equation of the line, which is approximately

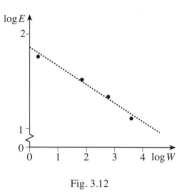

Fig. 3.12

$$\log E = 1.84 - 0.19 \log W .$$

This is of the form

$$\log E = \log a + n \log W$$

obtained by taking logarithms in the power equation $E = aW^n$, with $\log a = 1.84$ and $n = -0.19$. This gives $a = 10^{1.84} \approx 69$, so the power equation is approximately $E = 69W^{-0.19}$.

The uncertainty of the data, and the approximations shown up by the graph, mean that this model can do little more than suggest an order of magnitude for the dependent variable E. It would therefore be unwise to give the coefficients in the model to more than 1 significant figure, since that would suggest a degree of accuracy that couldn't be justified. The best you can assert is that, on the evidence of the data, the daily energy expenditure of a range of mammals of widely differing sizes can be modelled approximately by the formula $70W^{-0.2}$ calories per kilogram.

Exercise 3D

1 (a) If $\log_{10} y = 0.4 + 0.6x$, express y in terms of x.

   (b) If $\log_{10} y = 12 - 3x$, express y in terms of x.

   (c) If $\log_{10} y = 0.7 + 1.7x$, express y in terms of x.

   (d) If $\log_{10} y = 0.7 + 2\log_{10} x$, express y in terms of x.

   (e) If $\log_{10} y = -0.5 - 5\log_{10} x$, express y in terms of x.

2 Repeat Question 1, replacing $\log_{10}$ in each part by $\ln$.

3 Population census data for the USA from 1870 to 1910 were as follows.

Year	1870	1880	1890	1900	1910
Population (millions)	38.6	50.2	63.0	76.0	92.0

Investigate how well these figures can be described by an exponential model.

4   For the model in Example 3.9.1, calculate the values given by the equation $E = 70W^{-0.2}$
    for the four animals, and compare these with the numbers given in the table.

5   The table shows the mean relative distance, X, of some of the planets from the Earth and
    the time, T years, taken for one revolution round the sun. By drawing an appropriate graph
    show that there is an approximate law of the form $T = aX^n$, stating the values of a and n.

	Mercury	Venus	Earth	Mars	Saturn
X	0.39	0.72	1.00	1.52	9.54
T	0.24	0.62	1.00	1.88	29.5

6   Jack takes out a fixed rate savings bond. This means he makes one payment and leaves his
    money for a fixed number of years. The value of his bond, B, is given by the formula
    $B = Ax^n$ where A is the original investment and n is the number of complete years since
    he opened the account. The table gives some values of B and n. By plotting a suitable
    graph find the initial value of Jack's investment and the rate of interest he is receiving.

n	2	3	5	8	10
B	982	1056	1220	1516	1752

7   In a spectacular experiment on cell growth the following data were obtained, where N is
    the number of cells at a time t minutes after the start of the growth.

t	1.5	2.7	3.4	8.1	10
N	9	19	32	820	3100

At $t = 10$ a chemical was introduced which killed off the culture.

The relationship between N and t was thought to be modelled by $N = ab^t$, where a and
b are constants.

(a) Use a graph to determine how these figures confirm the supposition that the
     relationship is of this form. Find the values of a and b, each to the nearest integer.

(b) If the growth had not been stopped at $t = 10$ and had continued according to your
     model, how many cells would there have been after 20 minutes?

(c) An alternative expression for the relationship is $N = me^{kt}$. Find the values of
     m and k.

                                                                      (MEI, adapted)

8 It is believed that two quantities, z and d, are connected by a relationship of the form $z = kd^n$, where k and n are constants, provided that d does not exceed some fixed (but unknown) value, D. An experiment produced the following data.

d	750	810	870	930	990	1050	1110	1170
z	2.1	2.6	3.2	4.0	4.8	5.6	5.9	6.1

(a) Plot the values of $\log_{10} z$ against $\log_{10} d$. Use these points to suggest a value for D.

(b) It is known that, for $d < D$, n is a whole number. Use your graph to find the value of n. Show also that $k \approx 5 \times 10^{-9}$.

(c) Use your value of n and the estimate $k \approx 5 \times 10^{-9}$ to find the value of d for which $z = 3.0$.   (MEI, adapted)

Miscellaneous exercise 3

1 Solve each of the following equations to find x in terms of a where $a > 0$ and $a \neq 100$.

(a) $a^x = 10^{2x+1}$ (b) $2\log(2x) = 1 + \log a$ (OCR, adapted)

2 Solve the equation $3^{2x} = 4^{2-x}$, giving your answer to three significant figures.   (OCR)

3 The function f is given by $f : x \mapsto \log(1 + x)$, where $x \in \mathbb{R}$ and $x > -1$. Express the definition of f^{-1} in a similar form.   (OCR, adapted)

4 Find the root of the equation $10^{2-2x} = 2 \times 10^{-x}$ giving your answer exactly in terms of logarithms.   (OCR, adapted)

5 Given the simultaneous equations

$$2^x = 3^y,$$
$$x + y = 1,$$

show that $x = \dfrac{\log 3}{\log 6}$.   (OCR, adapted)

6 Express $\log(2\sqrt{10}) - \frac{1}{3}\log 0.8 - \log\left(\frac{10}{3}\right)$ in the form $c + \log d$ where c and d are rational numbers and the logarithms are to base 10.   (OCR, adapted)

7* Prove that $\log_b a \times \log_c b \times \log_a c = 1$, where a, b and c are positive numbers.

8 Prove that $\log\left(\dfrac{p}{q}\right) + \log\left(\dfrac{q}{r}\right) + \log\left(\dfrac{r}{p}\right) = 0$.

9 If a, b and c are positive numbers in geometric progression, show that $\log a$, $\log b$ and $\log c$ are in arithmetic progression.

10* If $\log_r p = q$ and $\log_q r = p$, prove that $\log_q p = pq$.

11 Express $\log_2(x + 2) - \log_2 x$ as a single logarithm. Hence solve the equation $\log_2(x + 2) - \log_2 x = 3$.

12 The strength of a radioactive source is said to 'decay exponentially'. Explain briefly what is meant by exponential decay, and illustrate your answer by means of a sketch-graph.

After t years the strength S of a particular radioactive source, in appropriate units, is given by $S = 10\,000 \times 3^{-0.0014t}$. State the value of S when $t = 0$, and find the value of t when the source has decayed to one-half of its initial strength, giving your answer correct to 3 significant figures. (OCR, adapted)

13 Differing amounts of fertiliser were applied to a number of fields of wheat of the same size. The weight of wheat at harvest was recorded. It is believed that the relationship between the amount of fertiliser, x kg, and the weight of wheat, y tonnes, is of the form $y = kx^n$, where k and n are constants.

(a) A plot of $\ln y$ against $\ln x$ is drawn for 8 such fields. It is found that the straight line of best fit passes through the points $(4,0)$ and $(0,-1.6)$. Find the values of k and n.

(b) Estimate how much wheat would be obtained from the use of 250 kg of fertiliser. (MEI, adapted)

14 An experiment was conducted to discover how a heavy beam sagged when a load was hung from it. The results are summarised in a table, where w is the load in tonnes and y is the sag in millimetres.

w	1	2	3	4	5
y	18	27	39	56	82

(a) A suggested model for these data is given by $y = a + bw^2$, where a and b are constants. Use the results for $w = 1$ and $w = 5$ to find estimates of a and b, correct to one decimal place. Calculate the sag predicted by this model when the beam supports a load of 3 tonnes.

(b) A second model is given by $y = kc^w$, where k and c are constants. By plotting $\ln y$ against w, estimate the values of k and c.

(c) Compare the fit of the two models to the data. (MEI, adapted)

4 Differentiating exponentials and logarithms

This chapter deals with exponentials and logarithms as functions which can be differentiated and integrated. When you have completed it, you should

- understand how to find the derivative of b^x from the definition
- understand the reason for selecting e as the exponential base
- know the derivative and integral of e^x
- know the derivative of $\ln x$, and how to obtain it
- know the integral of $\dfrac{1}{x}$, and be able to use it for both positive and negative x
- be able to use the extended methods from P1 Chapter 12 to broaden the range of functions that you can differentiate and integrate.

4.1 Differentiating exponential functions

One characteristic of exponential growth is that a quantity increases at a rate proportional to its current value. For continuous exponential growth, this rate of growth is measured by the derivative.

It will be simplest to begin with a particular value $b = 2$, and to consider $f(x) = 2^x$.

To find the derivative of this function you can use the definition given in P1 Section 6.6,

$$f'(x) = \lim_{h \to 0} \frac{f(x+h) - f(x)}{h}.$$

For this function, as illustrated in Fig. 4.1,

$$f(x+h) - f(x) = 2^{x+h} - 2^x = 2^x 2^h - 2^x$$
$$= 2^x \left(2^h - 1\right).$$

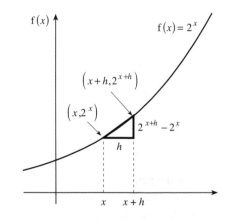

Fig. 4.1

So the definition becomes

$$f'(x) = \lim_{h \to 0} \frac{2^x \left(2^h - 1\right)}{h}.$$

Since 2^x does not involve h, you can write

$$f'(x) = 2^x \lim_{h \to 0} \frac{2^h - 1}{h}.$$

This shows that $f'(x)$ is the product of two factors: 2^x, which is independent of h, and a limit expression which is independent of x.

The limit expression is in fact the gradient of the tangent at the point $(0,1)$. This is because $\dfrac{2^h - 1}{h}$ is the gradient of the chord joining $(0,1)$ to $\left(h, 2^h\right)$, and as h tends to 0 this chord tends to the tangent. So

$$f'(x) = 2^x \times \text{gradient of the tangent at } (0,1) = 2^x \times f'(0);$$

since 2^x is $f(x)$, this can be written as

$$f'(x) = f(x) \times f'(0).$$

This result confirms that the rate of growth of $f(x)$ is proportional to its current value.

The next step is to evaluate the limiting value $f'(0)$. You can do this by calculating $\dfrac{2^h - 1}{h}$ for some small values of h, and setting out the results as in Table 4.2.

h	1	0.1	0.01	0.001	0.000 1
$\dfrac{2^h - 1}{h}$	1	0.717 73	0.695 56	0.693 39	0.693 17

Table 4.2

These are the gradients of chords to the right of $(0,1)$, which you would expect to be greater than the gradient of the tangent. For chords to the left you can take h to be negative, as in Table 4.3.

h	-1	-0.1	-0.01	-0.001	$-0.000\,1$
$\dfrac{2^h - 1}{h}$	0.5	0.669 67	0.690 75	0.692 91	0.693 12

Table 4.3

It follows that, for the function $f(x) = 2^x$, the derived function is

$$f'(x) = \text{constant} \times 2^x, \text{ where the constant is } f'(0) = 0.6931\ldots.$$

The method of finding the derivative for any other exponential function b^x is exactly the same. The only difference is that the numerical value of the constant $f'(0)$ is different for different values of the base b.

> For the general exponential function $f(x) = b^x$, where $b > 0$ and $b \neq 1$, the derived function is $f'(x) = \text{constant} \times b^x$, where the value of the constant, which depends on the base b, is equal to $f'(0)$.

Example 4.1.1

Show that, for any exponential function, the graph of $y = b^x$ bends upwards.

If $y = b^x$, $\dfrac{dy}{dx} = f'(0)b^x$ and $\dfrac{d^2y}{dx^2} = (f'(0))^2 b^x$. Since $b \neq 1$, $f'(0)$ is not zero, so $(f'(0))^2 > 0$. Also, for all x, $b^x > 0$.

Therefore $\dfrac{d^2y}{dx^2} > 0$ for all x, so the graph bends upwards.

If you look back to Figs. 3.1 and 3.2, you can see that $f'(0)$ is positive for $b > 1$ and negative for $0 < b < 1$, but in either case the graph bends upwards throughout its length.

4.2 The number e

If you carry out the limit calculation $\lim\limits_{h \to 0} \dfrac{b^h - 1}{h}$ for values of b other than 2, you get

values for the constant $f'(0)$ like those in Table 4.4 below, reported to 4 decimal places. Since the values of $f'(0)$ depend on b, they have been denoted by $L(b)$.

b	2	3	4	5	6	8	9	10
$L(b)$	0.693 1	1.098 6	1.386 3	1.609 4	1.791 8	2.079 4	2.197 2	2.302 6

Table 4.4

Before reading on, try working out one or two of these for yourself. If you are working in a group, you could share the work and verify the whole table. It is also interesting to find $L(b)$ for a few values of b less than 1, such as 0.1, 0.2, 0.25 and 0.5. Look at the answers and keep a record of anything you notice for future reference.

None of these limits works out to a nice recognisable number; in fact they are all irrational numbers. But Table 4.4 shows that between 2 and 3 there should be a number for which $L(b)$ is 1. This is the number denoted by the letter e, and it turns out to be one of the most important numbers in mathematics.

You can find the value of e more precisely by decimal search. For example, the limit calculation shows that $L(2.71) = 0.9969\ldots$, which is too small, and $L(2.72) = 1.0006\ldots$, which is too large, so $2.71 < e < 2.72$. However, this is a rather tedious process, and there are far more efficient ways of calculating e to many decimal places.

Note that $L(e) = 1$, and that $L(b)$ is the symbol used for the constant $f'(0)$ in the statement

'if $f(x) = b^x$, then $f'(x) = f'(0)b^x$.'

This means that, if $f(x) = e^x$ then $f'(0) = 1$, so

if $f(x) = e^x$, then $f'(x) = e^x$.

It is this property that makes e^x so much more important than all the other exponential functions. It can be described as the 'natural' exponential function, but usually it is called '*the* exponential function' (to distinguish it from b^x for any other value of b, which is simply '*an* exponential function').

The function e^x is sometimes written as $\exp x$, so that the symbol 'exp' strictly stands for the function itself, rather than the output of the function. Thus, in formal function notation,

$\exp : x \mapsto e^x$.

> For the (natural) exponential function e^x, or $\exp x$, $\dfrac{d}{dx}e^x = e^x$.

Many calculators have a special key, often labelled $[e^x]$, for finding values of this function. If you want to know the numerical value of e, you can use this key with an input of 1, so that the output is $e^1 = e$. This gives $e = 2.718\,281\,828\ldots$. (But do not assume that this is a recurring decimal; e is in fact an irrational number, and the single repetition of the digits 1828 is just a coincidence.)

Example 4.2.1

Find the equations of the tangents to the graph $y = e^x$ at the points (a) $(0,1)$, (b) $(1,e)$.

(a) Since $\dfrac{dy}{dx} = e^x$, the gradient at $(0,1)$ is $e^0 = 1$. The equation of the tangent is therefore $y - 1 = 1(x - 0)$, which you can simplify to $y = x + 1$.

(b) The gradient at $(1,e)$ is $e^1 = e$. The equation of the tangent is $y - e = e(x - 1)$, which you can simplify to $y = ex$.

It is interesting that the tangent at $(1,e)$ passes through the origin. You can demonstrate this nicely with a graphic calculator, if you have access to one.

Example 4.2.2

Find (a) $\dfrac{d}{dx}\left(e^{2x}\right)$, (b) $\dfrac{d}{dx}\left(e^{-x^2}\right)$, (c) $\dfrac{d}{dx}\left(e^{2+x}\right)$.

These expressions are all of the form $\dfrac{d}{dx}f(F(x))$, with $f(x) = e^x$. So you can use the chain rule (see P1 Section 12.2) to find the answers

(a) $e^{2x} \times 2 = 2e^{2x}$, (b) $e^{-x^2} \times (-2x) = -2xe^{-x^2}$, (c) $e^{2+x} \times 1 = e^{2+x}$.

For (c) you could write e^{2+x} as $e^2 e^x$. Since e^2 is constant, the derivative is $e^2 e^x$, or e^{2+x}.

From $\dfrac{d}{dx}e^x = e^x$ it follows that $\displaystyle\int e^x \, dx = e^x + k$. This is used in the next example.

Example 4.2.3

Find the area under the graph of $y = e^{2x}$ from $x = 0$ to $x = 1$.

$\displaystyle\int e^{2x} \, dx$ is of the form $\displaystyle\int g(ax + b) \, dx$, with $g(x) = e^x$. The indefinite integral is therefore $\dfrac{1}{a}f(ax + b) + k$, where $f(x)$ is the simplest integral of $g(x)$. (See P1 Section 16.7.) In this case, $f(x) = e^x$, so that $\displaystyle\int e^{2x} \, dx = \tfrac{1}{2}e^{2x} + k$.

The area under the graph is therefore $\displaystyle\int_0^1 e^{2x} \, dx = \left[\tfrac{1}{2}e^{2x}\right]_0^1 = \tfrac{1}{2}e^2 - \tfrac{1}{2}e^0 = \tfrac{1}{2}\left(e^2 - 1\right)$.

Exercise 4A

1 Differentiate each of the following functions with respect to x.

(a) e^{3x}
(b) e^{-x}
(c) $3e^{2x}$
(d) $-4e^{-4x}$

(e) e^{3x+4}
(f) e^{3-2x}
(g) e^{1-x}
(h) $3e \times e^{2+4x}$

(i) e^{x^3}
(j) $e^{-\frac{1}{2}x^2}$
(k) $e^{\frac{1}{x}}$
(l) $e^{\sqrt{x}}$

2 Find, in terms of e, the gradients of the tangents to the following curves for the given values of x.

(a) $y = 3e^x$, where $x = 2$
(b) $y = 2e^{-x}$, where $x = -1$

(c) $y = x - e^{2x}$, where $x = 0$
(d) $y = e^{6-2x}$, where $x = 3$

3 Find the equations of the tangents to the given curves for the given values of x.

(a) $y = e^x$, where $x = -1$
(b) $y = 2x - e^{-x}$, where $x = 0$

(c) $y = x^2 + 2e^{2x}$, where $x = 2$
(d) $y = e^{-2x}$, where $x = \ln 2$

4 Use the chain rule to differentiate

(a) $y = 2e^{x^2+x+1}$,
(b) $y = 3\left(e^{-x} + 1\right)^5$,
(c) $y = e^{\sqrt{1-x^2}}$.

5 Given that $y = \dfrac{5}{1+e^{3x}}$, find the value of $\dfrac{dy}{dx}$ when $x = 0$.

6 Find any stationary points of the graphs of

(a) $y = 2 - e^x$,
(b) $y = 2x - e^x$,
(c) $y = 2x^2 + e^{-x^4}$,

and determine whether they are maxima or minima.

7 Find the following indefinite integrals.

(a) $\displaystyle\int e^{3x}\,dx$
(b) $\displaystyle\int e^{-x}\,dx$
(c) $\displaystyle\int 3e^{2x}\,dx$
(d) $\displaystyle\int -4e^{-4x}\,dx$

(e) $\displaystyle\int e^{3x+4}\,dx$
(f) $\displaystyle\int e^{3-2x}\,dx$
(g) $\displaystyle\int e^{1-x}\,dx$
(h) $\displaystyle\int 3e \times e^{2+4x}\,dx$

8 Find each of the following definite integrals in terms of e, or give its exact value.

(a) $\displaystyle\int_1^2 e^{2x}\,dx$
(b) $\displaystyle\int_{-1}^1 e^{-x}\,dx$
(c) $\displaystyle\int_{-2}^0 2e^{1-2x}\,dx$
(d) $\displaystyle\int_4^5 2e^{2x}\,dx$

(e) $\displaystyle\int_{\ln 3}^{\ln 9} e^x\,dx$
(f) $\displaystyle\int_0^{\ln 2} 2e^{1-2x}\,dx$
(g) $\displaystyle\int_0^1 e^{x\ln 2}\,dx$
(h) $\displaystyle\int_{-3}^9 e^{x\ln 3}\,dx$

9 Find the area bounded by the graph of $y = e^{2x}$, the x- and y-axes and the line $x = 2$.

10 Find $\displaystyle\int_0^N e^{-x}\,dx$. Deduce the value of $\displaystyle\int_0^\infty e^{-x}\,dx$.

11 Sketch the graph of $y = xe^{-x^2}$ for $x > 0$. Find the area contained between this graph and the positive x-axis.

4.3 The natural logarithm

The inverse function of the (natural) exponential function exp is the logarithmic function $\log_e$. You know from the last chapter that this is denoted by ln. It is called the **natural logarithm**.

$$y = e^x = \exp x \quad \Leftrightarrow \quad x = \log_e y = \ln y \text{ for } x \in \mathbb{R},\ y \in \mathbb{R},\ y > 0.$$

$$\ln 1 = 0, \quad \ln e = 1, \quad \ln e^n = n \quad \text{for any } n.$$

The most important property of the natural logarithm is the derivative of $\ln x$. This can be deduced from the result $\dfrac{d}{dx} e^x = e^x$ in Section 4.2, but first we need a result from coordinate geometry.

Mini-theorem   If a line with gradient m (where $m \neq 0$) is reflected in the line $y = x$, the gradient of the reflected line is $\dfrac{1}{m}$.

    Proof   The proof is much like that of the perpendicular line property given in P1 Section 1.9. Fig. 4.5 shows the line of gradient m with a 'gradient triangle' ABC. Its reflection in $y = x$ is the triangle DEF. Completing the rectangle $DEFG$, DGF is a gradient triangle for the reflected line. $GF = DE = AB = 1$ and $DG = EF = BC = m$, so the gradient of the reflected line is $\dfrac{GF}{DG} = \dfrac{1}{m}$.

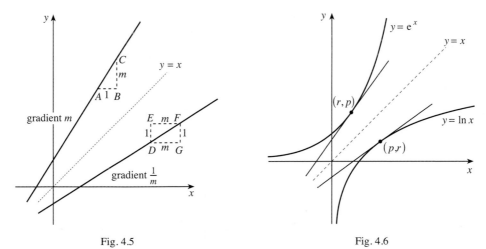

Fig. 4.5   Fig. 4.6

Now consider the graphs of $y = \ln x$ and $y = e^x$ in Fig. 4.6. Since these are graphs of inverse functions, they are reflections of each other in the line $y = x$. The reflection of the tangent at the point (p, r) on $y = \ln x$ is the tangent at the point (r, p) on $y = e^x$, where $p = e^r$.

Since $\dfrac{d}{dx} e^x = e^x$, the gradient of the tangent at (r, p) is $e^r = p$. It follows that the gradient of the tangent to $y = \ln x$ at (p, r) is $\dfrac{1}{p}$.

Since this holds for any point (p,r) on $y = \ln x$, it follows that:

> For $x > 0$, $\dfrac{d}{dx} \ln x = \dfrac{1}{x}$.

Example 4.3.1

Find the minimum value of the function $f(x) = 2x - \ln x$.

The natural domain of $f(x)$ is $x > 0$. Since $f'(x) = 2 - \dfrac{1}{x}$, $f'(x) = 0$ when $x = \frac{1}{2}$.

Also $f''(x) = \dfrac{1}{x^2}$, so $f''(\frac{1}{2}) = 4 > 0$. So the function has a minimum when $x = \frac{1}{2}$.

The minimum value is $f(\frac{1}{2}) = 1 - \ln \frac{1}{2}$. Since $\ln \frac{1}{2} = \ln 2^{-1} = -\ln 2$, it is simpler to write the minimum value as $1 + \ln 2$.

Unless you specifically need a numerical answer, it is better to leave it as $1 + \ln 2$, which is exact, than to use a calculator to convert it into decimal form.

Example 4.3.2

Find   (a) $\dfrac{d}{dx} \ln(3x + 1)$,   (b) $\dfrac{d}{dx} \ln 3x$,   (c) $\dfrac{d}{dx} \ln x^3$,   (d) $\dfrac{d}{dx} \ln\left(x + \dfrac{1}{x} \right)$.

(a) This expression is of the form $f(ax + b)$, with $f(x) = \ln x$, so the derivative is

$$3 \times \frac{1}{3x + 1} = \frac{3}{3x + 1}.$$

(b) You have a choice of method. You can find the derivative as in (a), as

$3 \times \dfrac{1}{3x} = \dfrac{1}{x}$. Or you can note that $\ln 3x = \ln 3 + \ln x$, so that $\dfrac{d}{dx} \ln 3x = \dfrac{d}{dx} \ln x = \dfrac{1}{x}$,

since $\ln 3$ is constant.

(c) Begin by writing $\ln x^3$ as $3 \ln x$. Then $\dfrac{d}{dx} \ln x^3 = \dfrac{d}{dx} (3 \ln x) = 3 \times \dfrac{1}{x} = \dfrac{3}{x}$.

Or use the chain rule, $\dfrac{d}{dx} \ln x^3 = \dfrac{1}{x^3} \times 3x^2 = \dfrac{3}{x}$.

(d) Either write $\ln\left(x + \dfrac{1}{x} \right) = \ln \dfrac{x^2 + 1}{x} = \ln(x^2 + 1) - \ln x$, so

$$\frac{d}{dx} \ln\left(x + \frac{1}{x} \right) = \frac{1}{x^2 + 1} \times 2x - \frac{1}{x} = \frac{2x}{x^2 + 1} - \frac{1}{x};$$

or use the chain rule directly,

$$\frac{d}{dx} \ln\left(x + \frac{1}{x} \right) = \frac{1}{x + \dfrac{1}{x}} \times \left(1 - \frac{1}{x^2} \right) = \frac{x}{x^2 + 1} \times \frac{x^2 - 1}{x^2} = \frac{x^2 - 1}{x(x^2 + 1)}.$$

The two methods give the answer in different forms, but they are equivalent to each other. It doesn't matter in which form you give your answer.

Exercise 4B

1 Differentiate each of the following functions with respect to x.

(a) $\ln 2x$

(b) $\ln(2x-1)$

(c) $\ln(1-2x)$

(d) $\ln x^2$

(e) $\ln(a+bx)$

(f) $\ln\dfrac{1}{x}$

(g) $\ln\dfrac{1}{3x+1}$

(h) $\ln\dfrac{2x+1}{3x-1}$

(i) $3\ln x^{-2}$

(j) $\ln(x(x+1))$

(k) $\ln(x^2(x-1))$

(l) $\ln(x^2+x-2)$

2 Find the equations of the tangents to the following graphs for the given values of x.

(a) $y = \ln x$, where $x = \frac{1}{2}$

(b) $y = \ln 2x$, where $x = \frac{1}{2}$

(c) $y = \ln(-x)$, where $x = -\frac{1}{3}$

(d) $y = \ln 3x$, where $x = e$

3 Find any stationary values of the following curves and determine whether they are maxima or minima. Sketch the curves.

(a) $y = x - \ln x$

(b) $y = \frac{1}{2}x^2 - \ln 2x$

(c) $y = x^2 - \ln x^2$

(d) $y = x^n - \ln x^n$ for $n \geq 1$

4 Use the chain rule to differentiate

(a) $y = \ln(1+x^3)$,

(b) $y = \frac{1}{2}\ln(2+x^4)$,

(c) $y = \ln(x^3+4x)$.

5 Prove that the tangent at $x = e$ to the curve with equation $y = \ln x$ passes through the origin.

6 Find the equation of the normal at $x = 2$ to the curve with equation $y = \ln(2x-3)$.

7 Let $f(x) = \ln(x-2) + \ln(x-6)$. Write down the natural domain of $f(x)$.

Find $f'(x)$ and hence find the intervals for which $f'(x)$ is (a) positive, (b) negative.

Sketch the curve.

8 Repeat Question 7 for the functions

(i) $f(x) = \ln(x-2) + \ln(6-x)$,

(ii) $f(x) = \ln(2-x) + \ln(x-6)$.

4.4 The reciprocal integral

Now that you know that $\dfrac{d}{dx}\ln x = \dfrac{1}{x}$, you also know a new result about integration:

$$\text{For } x > 0, \quad \int \frac{1}{x}\,dx = \ln x + k.$$

This is an important step forward. You may recall that in P1 Section 16.1, when giving the indefinite integral $\int x^n\,dx = \dfrac{1}{n+1}x^{n+1} + k$, an exception had to be made for the case $n = -1$. You can now see why: $\int \dfrac{1}{x}\,dx$ is an entirely different kind of function, the natural logarithm.

Example 4.4.1

Find the area under the graph of $y = \dfrac{1}{x}$ from $x = 2$ to $x = 4$.

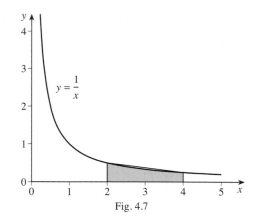

Fig. 4.7

Before working out the exact answer, notice from Fig. 4.7 that the area should be less than the area of the trapezium formed by joining $(2, 0.5)$ and $(4, 0.25)$ with a chord. This area is

$$\tfrac{1}{2} \times 2 \times (0.5 + 0.25) = 0.75.$$

The exact area is given by the integral

$$\int_2^4 \frac{1}{x}\,dx = \left[\ln x\right]_2^4 = \ln 4 - \ln 2$$

$$= \ln \tfrac{4}{2} = \ln 2.$$

The calculator gives this as $0.693\,14\ldots$, which is less than 0.75, as expected.

Example 4.4.2

Find the indefinite integral $\displaystyle\int \frac{1}{3x-1}\,dx$.

This is of the form $\displaystyle\int g(ax + b)\,dx$, so the integral is $\dfrac{1}{a}f(ax + b) + k$, where $f(x)$ is the simplest integral of $g(x)$. Here, $g(x)$ is $\dfrac{1}{x}$, so $f(x) = \ln x$. Therefore

$$\int \frac{1}{3x-1}\,dx = \frac{1}{3}\ln(3x-1) + k.$$

Note that this integral is only valid if $x > \tfrac{1}{3}$, since $\ln(3x - 1)$ only exists if $3x - 1 > 0$.

Exercise 4C

1   Carry out the following indefinite integrations, and state the values of x for which your answer is valid.

(a) $\displaystyle\int \frac{1}{2x}\,dx$ (b) $\displaystyle\int \frac{1}{x-1}\,dx$ (c) $\displaystyle\int \frac{1}{1-x}\,dx$ (d) $\displaystyle\int \frac{1}{4x+3}\,dx$

(e) $\displaystyle\int \frac{4}{1-2x}\,dx$ (f) $\displaystyle\int \frac{4}{1+2x}\,dx$ (g) $\displaystyle\int \frac{4}{-1-2x}\,dx$ (h) $\displaystyle\int \frac{4}{2x-1}\,dx$

2   Calculate the area under the graph of $y = \dfrac{1}{x}$ from

(a) $x = 3$ to $x = 6$,   (b) $x = 4$ to $x = 8$,

(c) $x = \tfrac{1}{2}$ to $x = 1$,   (d) $x = a$ to $x = 2a$, $a > 0$.

3 Calculate the areas under the following graphs.

(a) $y = \dfrac{1}{x+2}$ from $x = -1$ to $x = 0$

(b) $y = \dfrac{1}{2x-1}$ from $x = 2$ to $x = 5$

(c) $y = \dfrac{2}{3x-5}$ from $x = 4$ to $x = 6$

(d) $y = \dfrac{e}{ex-7}$ from $x = 4$ to $x = 5$

(e) $y = \dfrac{1}{-x-1}$ from $x = -3$ to $x = -2$

(f) $y = 2 + \dfrac{1}{x-1}$ from $x = 2$ to $x = 6$

4 Sketch $y = \dfrac{2}{x+1}$, and use your sketch to make a rough estimate of the area under the graph between $x = 3$ and $x = 5$. Compare your answer with the exact answer.

5 The region under the curve with equation $y = \dfrac{1}{\sqrt{x}}$ is rotated through four right angles about the x -axis to form a solid. Find the volume of the solid between $x = 2$ and $x = 5$.

6 The region under the curve with equation $y = \dfrac{1}{\sqrt{2x-1}}$ is rotated through four right angles about the x -axis to form a solid. Find the volume of the solid between $x = 3$ and $x = 8$.

7 Given that $\dfrac{dy}{dx} = \dfrac{3}{2x+1}$ and that the graph of y against x passes through the point $(1,0)$, find y in terms of x .

8 A curve has the property that $\dfrac{dy}{dx} = \dfrac{8}{4x-3}$, and it passes through $(1,2)$. Find its equation.

9 The graph of $y = \dfrac{1}{x^2}$ between $x = 1$ and $x = 2$ is rotated about the y -axis. Find the volume of the solid formed.

4.5 Extending the reciprocal integral

On a first reading of this chapter you may prefer to skip ahead to Miscellaneous exercise 4 and come back to this section later.

You will have noticed that the statements

$$\frac{d}{dx}\ln x = \frac{1}{x} \quad \text{and} \quad \int \frac{1}{x}\,dx = \ln x$$

both contain the condition 'for $x > 0$ '. In the case of the derivative the reason is obvious, since $\ln x$ is only defined for $x > 0$. But no such restriction applies to the function $\dfrac{1}{x}$.

This then raises the question, what is $\displaystyle\int \frac{1}{x}\,dx$ when $x < 0$?

A good guess might be that it is $\ln(-x)$. This has a meaning if x is negative, and you can differentiate it as a special case of $\dfrac{d}{dx}f(ax+b)$ with $a = -1$, $b = 0$ and $f(x) = \ln x$.

This gives $\dfrac{d}{dx}\ln(-x) = -\dfrac{1}{(-x)} = \dfrac{1}{x}$, as required.

So the full statement of the reciprocal integral is:

$$\int \frac{1}{x}\,dx = \begin{cases} \ln x + k & \text{if } x > 0, \\ \ln(-x) + k & \text{if } x < 0. \end{cases}$$

Notice that the possibility $x = 0$ is still excluded. You should expect this, as 0 is not in the domain of the function $\dfrac{1}{x}$. Using the function $|x|$, the result can also be stated in the form:

$$\text{For } x \neq 0, \ \int \frac{1}{x}\,dx = \ln|x| + k.$$

The function $|x|$ is an even function, with a graph symmetrical about the y-axis. It follows that the graph of $\ln|x|$ is symmetrical about the y-axis; it is shown in Fig. 4.8. For positive x it is the same as that of $\ln x$; for negative x, this is reflected in the y-axis.

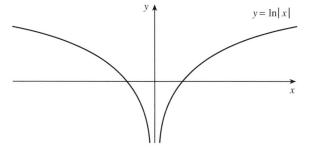

Fig. 4.8

You can see that the gradient of the graph is positive for $x > 0$ and negative for $x < 0$, which is as you would expect since $\dfrac{d}{dx}\ln|x| = \dfrac{1}{x}$.

Example 4.5.1

The graphs of $y = \dfrac{2}{x-2}$ and $y = -x - 1$ intersect where $x = 0$ and $x = 1$. Find the area of the region between them.

You can check from a sketch that the curve lies above the line, so that the area is

$$\int_0^1 \left(\frac{2}{x-2} - (-x-1)\right) dx = \int_0^1 \left(\frac{2}{x-2} + x + 1\right) dx.$$

The trap which you have to avoid is writing the integral of $\dfrac{2}{x-2}$ as $2\ln(x-2)$. Over the interval $0 < x < 1$, $x - 2$ is negative, so $\ln(x-2)$ has no meaning.

There are two ways of avoiding this difficulty. One is to write $\dfrac{2}{x-2}$ as $\dfrac{-2}{2-x}$.

The integral of $\dfrac{1}{2-x}$ is $-\ln(2-x)$, so the integral of $\dfrac{-2}{2-x}$ is $2\ln(2-x)$.

The area is then

$$\left[2\ln(2-x)+\tfrac{1}{2}x^2+x\right]_0^1=\left(2\ln 1+\tfrac{1}{2}+1\right)-(2\ln 2)$$
$$=\tfrac{3}{2}-2\ln 2.$$

The alternative is to use the modulus form of the integral, and to find the area as

$$\left[2\ln\left|x-2\right|+\tfrac{1}{2}x^2+x\right]_0^1=\left(2\ln\left|-1\right|+\tfrac{1}{2}+1\right)-\left(2\ln\left|-2\right|\right)$$
$$=\tfrac{3}{2}-2\ln 2.$$

You might think from this example that the modulus method has the edge. But it has to be used intelligently, as the following 'bogus' example shows.

Example 4.5.2
Find the area under the graph of $y=\dfrac{1}{x}$ from $x=-2$ to $x=+4$.

False solution

$$\int_{-2}^{4}\frac{1}{x}\,dx=\left[\ln\left|x\right|\right]_{-2}^4=\ln\left|4\right|-\ln\left|-2\right|=\ln 4-\ln 2$$
$$=\ln\tfrac{4}{2}=\ln 2.$$

You only have to draw the graph of $y=\dfrac{1}{x}$ to see that there is a problem here. The area does not exist for either of the intervals $-2<x<0$ and $0<x<4$, so it certainly cannot exist for $-2<x<4$. The interval of integration contains $x=0$, for which the rule $\dfrac{d}{dx}\ln\left|x\right|-\dfrac{1}{x}$ breaks down.

4.6* The derivative of b^x

In Sections 4.1 and 4.2 the derivative of b^x was found in the form

$$\frac{d}{dx}b^x=L(b)b^x,$$

where $L(b)$ is a constant whose value depends on b.

It is now possible to find this constant. As exp and ln are inverse functions, the composite function 'exp ln' is an identity function, with domain the positive real numbers. Therefore

$$e^{\ln b}=b.$$

Raising both sides to the power x gives

$$b^x=\left(e^{\ln b}\right)^x=e^{x\ln b},$$

by the power-on-power rule. This is of the form e^{ax}, where a is constant, so

$$\frac{d}{dx}b^x=\frac{d}{dx}\left(e^{x\ln b}\right)=(\ln b)\,e^{x\ln b}=(\ln b)\,b^x.$$

Comparing this with the earlier form of the derivative, you find that

$$L(b) = \ln b.$$

You can check this by using your calculator to compare the values of $L(b)$ in Table 4.4 in Section 4.2 with the corresponding values of $\ln b$. Notice also that Table 4.4 gives a number of examples of the rules for logarithms listed in Section 3.4. For example,

$$L(4) = 2L(2), \qquad L(6) = L(2) + L(3), \qquad L(8) = 3L(2),$$
$$L(9) = 2L(3), \qquad L(10) = L(2) + L(5).$$

The reason for this is now clear.

Exercise 4D

1 Calculate the following.

(a) $\displaystyle\int_{-6}^{-3} \frac{1}{x+2}\,dx$

(b) $\displaystyle\int_{-1}^{0} \frac{1}{2x-1}\,dx$

(c) $\displaystyle\int_{-1}^{0} \frac{2}{3x-5}\,dx$

(d) $\displaystyle\int_{1}^{2} \frac{e}{ex-7}\,dx$

(e) $\displaystyle\int_{2}^{4} \frac{1}{-x-1}\,dx$

(f) $\displaystyle\int_{-1}^{0} \left(2 + \frac{1}{x-1}\right)dx$

2 Calculate the value of $y = \ln|2x-3|$ for $x = -2$, and find $\dfrac{dy}{dx}$ when $x = -2$. Sketch the graph of $y = \ln|2x-3|$.

3* Find the derivatives with respect to x of 2^x, 3^x, 10^x and $\left(\frac{1}{2}\right)^x$.

Miscellaneous exercise 4

1 Differentiate each of the following expressions with respect to x.

(a) $\ln(3x-4)$

(b) $\ln(4-3x)$

(c) $e^x \times e^{2x}$

(d) $e^x \div e^{2x}$

(e) $\ln\dfrac{2-x}{3-x}$

(f) $\ln(3-2x)^3$

2 Use a calculator to find a number a for which $e^x > x^5$ for all $x > a$.

3 Find the coordinates of the points of intersection of $y = \dfrac{4}{x}$ and $2x + y = 9$. Sketch both graphs for values of x such that $x > 0$. Calculate the area between the graphs.

4 A curve is given by the equation $y = \frac{2}{3}e^x + \frac{1}{3}e^{-2x}$.

(a) Evaluate a definite integral to find the area between the curve, the x-axis and the lines $x = 0$ and $x = 1$, showing your working.

(b) Use calculus to determine whether the turning point at the point where $x = 0$ is a maximum or a minimum.

(OCR, adapted)

5 Find $\int \left(2 + e^{-x}\right) dx$.   (OCR)

6 The equation of a curve is $y = 2x^2 - \ln x$, where $x > 0$. Find by differentiation the x-coordinate of the stationary point on the curve, and determine whether this point is a maximum point or a minimum point.   (OCR)

7 Show that $\int_0^1 \left(e^x - e^{-x}\right) dx = \dfrac{(e-1)^2}{e}$.   (OCR)

8 Using differentiation, find the equation of the tangent to the curve $y = 4 + \ln(x+1)$ at the point where $x = 0$.   (OCR)

9 The equation of a curve is $y = \ln(2x)$. Find the equation of the normal at the point $\left(\frac{1}{2}, 0\right)$, giving your answer in the form $y = mx + c$.   (OCR)

10* (a) Express $\dfrac{x-7}{(x-4)(x-1)} + \dfrac{1}{x-4}$ as a single fraction.

   (b) Sketch $y = \dfrac{x-7}{(x-4)(x-1)}$, and calculate the area under the graph between $x = 2$ and $x = 3$.

11 Find the coordinates of the stationary point of the curve $y = \ln\left(x^2 - 6x + 10\right)$ and show that this stationary point is a minimum.

12 (a) Show that e^x is an increasing function of x for all x. Deduce that $e^x \geqslant 1$ for $x \geqslant 0$.

   (b) By finding the area under the graphs of $y = e^x$ and $y = 1$ between 0 and X, where $X \geqslant 0$, deduce that $e^X \geqslant 1 + X$ for $X \geqslant 0$, and that $e^X \geqslant 1 + X + \frac{1}{2}X^2$ for $X \geqslant 0$.

13 (a) Find the stationary value of $y = \ln x - x$, and deduce that $\ln x \leqslant x - 1$ for $x > 0$ with equality only when $x = 1$.

   (b) Find the stationary value of $\ln x + \dfrac{1}{x}$, and deduce that $\dfrac{x-1}{x} \leqslant \ln x$ for $x > 0$ with equality only when $x = 1$.

   (c) By putting $x = \dfrac{z}{y}$ where $0 < y < z$, deduce Napier's inequality, $\dfrac{1}{z} < \dfrac{\ln z - \ln y}{z - y} < \dfrac{1}{y}$.

14 The diagram shows sketches of the graphs of $y = 2 - e^{-x}$ and $y = x$. These graphs intersect at $x = a$ where $a > 0$.

   (a) Write down an equation satisfied by a. (Do not attempt to solve the equation.)

   (b) Write down an integral which is equal to the area of the shaded region.

   (c) Use integration to show that the area is equal to $1 + a - \frac{1}{2}a^2$.   (OCR, adapted)

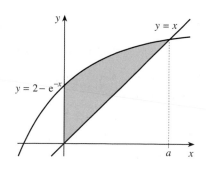

15 Find the coordinates of the three stationary points of the curve $y = e^{x^2(x^2-18)}$.

16 Draw a sketch of the curve $y = e^{-2x} - 3x$. The curve crosses the x-axis at $A(a,0)$ and the y-axis at $B(0,1)$. O is the origin.

    (a) Write down an equation satisfied by a.

    (b) Show that the tangent at A meets the y-axis at the point whose y-coordinate is

        $2ae^{-2a} + 3a$.

    (c) Show that $\dfrac{d^2y}{dx^2} > 0$, and using the results from parts (a) and (b), deduce that

        $6a^2 + 3a < 1$.

    (d) Find, in terms of a, the area of the region bounded by the curve and the line segments OA and OB.

    (e) By comparing this area with the area of the triangle OAB, show that $3a^2 + 4a > 1$.

    Hence show that $\frac{1}{3}\sqrt{7} - \frac{2}{3} < a < \frac{1}{12}\sqrt{33} - \frac{1}{4}$.   (OCR, adapted)

17 Find the exact value of $\displaystyle\int_0^\infty e^{1-2x}\, dx$.

18 The number of bacteria present in a culture at time t hours after the beginning of an experiment is denoted by N. The relation between N and t is modelled by $N = 100e^{\frac{3}{2}t}$.

    (a) After how many hours will the number of bacteria be 9000?

    (b) At what rate per hour will the number of bacteria be increasing when $t = 6$?   (OCR)

19 Show that, if $x > 0$, x^n can be written as $e^{n\ln x}$. Differentiate this last expression by the chain rule, and deduce that $\dfrac{d}{dx}x^n = nx^{n-1}$ if n is any real number.

20* The expression x^n only has a meaning when $x < 0$ if n is a rational number $\dfrac{p}{q}$ and q is an odd integer.

    Make the substitution $u = -x$ so that, when $x < 0$, $u > 0$.

    (a) Show that, if $x < 0$ and q is odd, then $x^{\frac{p}{q}} = -u^{\frac{p}{q}}$ if p is odd, and $x^{\frac{p}{q}} = u^{\frac{p}{q}}$ if p is even.

    (b) Use the chain rule and the result in Question 19 to show that, if p is odd, then

        $\dfrac{d}{dx}\left(x^{\frac{p}{q}}\right) = \dfrac{p}{q}u^{\frac{p-q}{q}} = \dfrac{p}{q}x^{\frac{p}{q}-1}$; and that, if p is even, $\dfrac{d}{dx}\left(x^{\frac{p}{q}}\right) = -\dfrac{p}{q}u^{\frac{p-q}{q}} = \dfrac{p}{q}x^{\frac{p}{q}-1}$.

    (c) Deduce that, for all the values of n for which x^n has a meaning when x is negative,

        $\dfrac{d}{dx}\left(x^n\right) = nx^{n-1}$ for $x < 0$.

5 Trigonometry

This chapter takes further the ideas about trigonometry introduced in unit P1. When you have completed it, you should

- know the definitions, properties and graphs of secant, cosecant and cotangent, including the associated Pythagorean identities
- know the addition and double angle formulae for sine, cosine and tangent, and be able to use these results for calculations, solving equations and proving identities
- know how to express $a\sin\theta + b\cos\theta$ in the forms $R\sin(\theta \pm \alpha)$ and $R\cos(\theta \pm \alpha)$.

5.1 Radians or degrees

All through your work in mathematics, you have probably thought of degrees as the natural unit for angle, but in P1 Chapter 18 a new unit, the radian, was introduced. This unit is important in differentiating and integrating trigonometric functions. For this reason, a new convention about angle will be adopted in this book.

If no units are given for trigonometric functions, you should assume that the units are radians, or that it doesn't matter whether the units are radians or degrees.

For example, if you see the equation $\sin x = 0.5$, then x is in radians. If you are asked for the smallest positive solution of the equation, you should give $x = \frac{1}{6}\pi$. Remember the relation:

$$\pi \text{ rad} = 180°.$$

Identities such as $\cos^2 A + \sin^2 A \equiv 1$ and $\dfrac{\sin\theta}{\cos\theta} \equiv \tan\theta$, or the cosine formula $a^2 = b^2 + c^2 - 2bc\cos A$, are true whatever the units of angle. Formulae such as these, for which it doesn't matter whether the unit is degrees or radians, will be shown without units for angles.

If, however, it is important that degrees are being used, then notation such as $\cos A°$ and $\sin\theta°$ will be used. Thus one solution of the equation $\cos\theta° = -0.5$ is $\theta = 120$.

This may seem complicated, but the context will usually make things clear.

5.2 Secant, cosecant and cotangent

It is occasionally useful to be able to write the functions $\dfrac{1}{\cos x}$, $\dfrac{1}{\sin x}$ and $\dfrac{1}{\tan x}$ in shorter forms. These functions, called respectively the secant, cosecant and cotangent (written and pronounced 'sec', 'cosec' and 'cot') are not defined when the denominators are zero, so their domains contain holes.

The **secant** and **cosecant** are defined by

$$\sec x = \frac{1}{\cos x} \qquad \text{provided that } \cos x \neq 0,$$

$$\operatorname{cosec} x = \frac{1}{\sin x} \qquad \text{provided that } \sin x \neq 0.$$

It is a little more complicated to define the cotangent in this way, since there are values of x for which $\tan x$ is undefined. But you can use the fact that $\tan x = \dfrac{\sin x}{\cos x}$, so $\dfrac{1}{\tan x} = \dfrac{\cos x}{\sin x}$, except where the denominators are zero. This can be used as the definition of $\cot x$.

The **cotangent** is defined by

$$\cot x = \frac{\cos x}{\sin x} \qquad \text{provided that } \sin x \neq 0.$$

Note that $\cot x = \dfrac{1}{\tan x}$ except where $\tan x = 0$ or is undefined.

You won't find sec, cosec or cot keys on your calculator. To find their values you have to use the cos, sin or tan keys, followed by the reciprocal key.

The graphs of $y = \sec x$, $y = \operatorname{cosec} x$ and $y = \cot x$ are shown in Figs. 5.1, 5.2 and 5.3. The functions $\sec x$ and $\operatorname{cosec} x$ have period 2π, and the period of $\cot x$ is π.

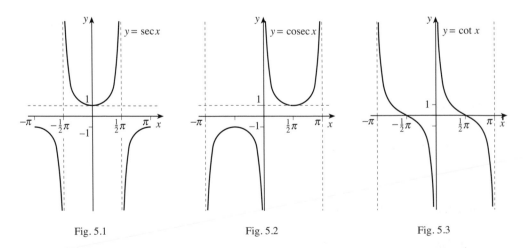

Fig. 5.1   Fig. 5.2   Fig. 5.3

Example 5.2.1

Find the exact values of   (a) $\sec \frac{2}{3}\pi$,   (b) $\operatorname{cosec} \frac{5}{6}\pi$,   (c) $\cot\left(-\frac{2}{3}\pi\right)$.

You need to find the values of $\cos \frac{2}{3}\pi$, $\sin \frac{5}{6}\pi$ and $\tan\left(-\frac{2}{3}\pi\right)$, using the symmetry properties in P1 Section 18.3 together with the exact values in P1 Section 10.3.

(a) $\cos\frac{2}{3}\pi = -\cos\left(\pi - \frac{2}{3}\pi\right) = -\cos\frac{1}{3}\pi = -\frac{1}{2}$, so $\sec\frac{2}{3}\pi = -2$.

(b) $\sin\frac{5}{6}\pi = \sin\left(\pi - \frac{5}{6}\pi\right) = \sin\frac{1}{6}\pi = \frac{1}{2}$, so $\operatorname{cosec}\frac{5}{6}\pi = 2$.

(c) $\tan\left(-\frac{2}{3}\pi\right) = \tan\left(-\frac{2}{3}\pi + \pi\right) = \tan\frac{1}{3}\pi = \sqrt{3}$, so $\cot\left(-\frac{2}{3}\pi\right) = \dfrac{1}{\sqrt{3}} = \frac{1}{3}\sqrt{3}$.

There are new forms of Pythagoras' theorem in trigonometry using these new trigonometric functions. For example, if you divide every term in the identity $\cos^2\theta + \sin^2\theta \equiv 1$ by $\cos^2\theta$, you get

$$\frac{\cos^2\theta}{\cos^2\theta} + \frac{\sin^2\theta}{\cos^2\theta} \equiv \frac{1}{\cos^2\theta}, \quad \text{that is,} \quad 1 + \tan^2\theta \equiv \sec^2\theta.$$

Similarly, if you divide every term of $\cos^2\theta + \sin^2\theta \equiv 1$ by $\sin^2\theta$, you get

$$\frac{\cos^2\theta}{\sin^2\theta} + \frac{\sin^2\theta}{\sin^2\theta} \equiv \frac{1}{\sin^2\theta}, \quad \text{that is,} \quad 1 + \cot^2\theta \equiv \operatorname{cosec}^2\theta.$$

Summarising:

$$1 + \tan^2\theta \equiv \sec^2\theta,$$
$$1 + \cot^2\theta \equiv \operatorname{cosec}^2\theta.$$

Example 5.2.2

Prove the identity $\dfrac{1}{\sec\theta - \tan\theta} \equiv \sec\theta + \tan\theta$ provided that $\sec\theta - \tan\theta \neq 0$.

There are four ways to approach proving identities: you can start with the left side and work towards the right; you can start with the right side and work towards the left; you can subtract one side from the other and try to show that the result is 0; or you can divide one side by the other and try to show that the result is 1. If you use one of the first two methods, you should generally start with the more complicated side.

Use the fourth method, and consider the right side divided by the left side.

You need to show that $(\sec\theta + \tan\theta) \div \dfrac{1}{\sec\theta - \tan\theta}$ is equal to 1.

$$(\sec\theta + \tan\theta) \div \frac{1}{\sec\theta - \tan\theta} \equiv (\sec\theta + \tan\theta)(\sec\theta - \tan\theta)$$
$$\equiv \sec^2\theta - \tan^2\theta \equiv 1,$$

using the first line in the box above.

Therefore $\dfrac{1}{\sec\theta - \tan\theta} \equiv \sec\theta + \tan\theta$.

The condition $\sec\theta - \tan\theta \neq 0$ is necessary, because if $\sec\theta - \tan\theta = 0$ the left side is not defined, and therefore the identity has no meaning.

Exercise 5A

1 Find, giving your answers to 3 decimal places,

    (a) $\cot 304°$,   (b) $\sec(-48)°$,   (c) $\operatorname{cosec} 62°$.

2 Simplify the following.

    (a) $\sec\left(\tfrac{1}{2}\pi - x\right)$ (b) $\dfrac{\cos x}{\sin x}$ (c) $\sec(-x)$

    (d) $1 + \tan^2 x$ (e) $\cot(\pi + x)$ (f) $\operatorname{cosec}(\pi + x)$

3 Find the exact values of

    (a) $\sec\tfrac{1}{4}\pi$,   (b) $\operatorname{cosec}\tfrac{1}{2}\pi$,   (c) $\cot\tfrac{5}{6}\pi$,   (d) $\operatorname{cosec}\left(-\tfrac{3}{4}\pi\right)$,

    (e) $\cot\left(-\tfrac{1}{3}\pi\right)$,   (f) $\sec\tfrac{13}{6}\pi$,   (g) $\cot\left(-\tfrac{11}{2}\pi\right)$,   (h) $\sec\tfrac{7}{6}\pi$.

4 Using a calculator where necessary, find the values of the following, giving any non-exact answers correct to 3 significant figures.

    (a) $\sin\tfrac{2}{5}\pi$ (b) $\sec\tfrac{1}{10}\pi$ (c) $\cot\tfrac{1}{12}\pi$ (d) $\operatorname{cosec}\tfrac{17}{6}\pi$

    (e) $\cos\tfrac{7}{8}\pi$ (f) $\tan\tfrac{5}{12}\pi$ (g) $\sec\left(-\tfrac{11}{12}\pi\right)$ (h) $\cot\left(-\tfrac{1}{6}\pi\right)$

5 Given that $\sin A = \tfrac{3}{5}$, where A is acute, and $\cos B = -\tfrac{1}{2}$, where B is obtuse, find the exact values of

    (a) $\sec A$,   (b) $\cot A$,   (c) $\cot B$,   (d) $\operatorname{cosec} B$.

6 Given that $\operatorname{cosec} C = 7$, $\sin^2 D = \tfrac{1}{2}$ and $\tan^2 E = 4$, find the possible values of $\cot C$, $\sec D$ and $\operatorname{cosec} E$, giving your answers in exact form.

7 Simplify the following.

    (a) $\sqrt{\sec^2 \phi - 1}$ (b) $\dfrac{\tan \phi}{1 + \tan^2 \phi}$ (c) $\dfrac{\tan \phi}{\sec^2 \phi - 1}$

    (d) $\dfrac{1}{\sqrt{1 + \cot^2 \phi}}$ (e) $\dfrac{1}{\sqrt{\operatorname{cosec}^2 \phi - 1}}$ (f) $(\operatorname{cosec} \phi - 1)(\operatorname{cosec} \phi + 1)$

8 (a) Express $3\tan^2 \theta - \sec\theta$ in terms of $\sec\theta$.

    (b) Solve the equation $3\tan^2 \phi - \sec\phi = 1$ for $0 \leqslant \phi \leqslant 2\pi$.

9 Use an algebraic method to find the solution for $0 \leqslant x \leqslant 2\pi$ of the equation $5\cot x + 2\operatorname{cosec}^2 x = 5$.

10 Find, in exact form, all the roots of the equation $2\sin^2 t + \operatorname{cosec}^2 t = 3$ which lie between 0 and 2π.

11 Prove that $\operatorname{cosec} A + \cot A \equiv \dfrac{1}{\operatorname{cosec} A - \cot A}$ provided that $\operatorname{cosec} A \neq \cot A$.

12 Prove that $\dfrac{\sec\theta - 1}{\tan\theta} \equiv \dfrac{\tan\theta}{\sec\theta + 1}$ provided that $\tan\theta \neq 0$.

5.3 The addition formulae for sine and cosine

Suppose that you know the values of $\sin A$, $\cos A$, $\sin B$ and $\cos B$. How could you calculate the values of $\sin(A+B)$, $\sin(A-B)$, $\cos(A+B)$ and $\cos(A-B)$ without using a calculator to find the angles, which, of course, would only give approximations?

One way is to find a general formula which applies to all values of A and B by starting with the formula for $\cos(A-B)$. You may wish to skip the proofs below, and start reading from the next set of results in the box on the next page.

In Fig. 5.4, angles A and B are drawn from the x-axis. The points P and Q then have coordinates $(\cos A, \sin A)$ and $(\cos B, \sin B)$ respectively.

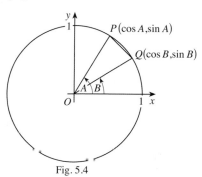

Fig. 5.4

You can write the distance PQ, or rather an expression for PQ^2, in two ways: by using the distance formula in coordinate geometry (see P1 Section 1.1) and by using the cosine formula for the triangle OPQ. These give

$$PQ^2 = (\cos B - \cos A)^2 + (\sin B - \sin A)^2 \text{ and } PQ^2 = 1^2 + 1^2 - 2 \times 1 \times 1 \times \cos(A-B),$$

so $\cos^2 B - 2\cos B \cos A + \cos^2 A + \sin^2 B - 2\sin B \sin A + \sin^2 A = 2 - 2\cos(A-B)$.

Rearrange the left side to get

$$\left(\cos^2 B + \sin^2 B\right) + \left(\cos^2 A + \sin^2 A\right) - 2\cos A \cos B - 2\sin A \sin B = 2 - 2\cos(A-B).$$

But, from Pythagoras' theorem in trigonometry, $\cos^2 B + \sin^2 B = 1$ and $\cos^2 A + \sin^2 A = 1$. So, cancelling and rearranging,

$$\cos(A-B) = \cos A \cos B + \sin A \sin B.$$

Although Fig. 5.4 is drawn with angles A and B acute and $A > B$, the proof in fact holds for angles A and B of any size.

Example 5.3.1

Verify the formula for $\cos(A-B)$ in the cases   (a) $B = A$,   (b) $A = \frac{1}{2}\pi, B = \frac{1}{6}\pi$.

(a) Put $B = A$.

Then $\cos(A-A) = \cos^2 A + \sin^2 A$ and, as $\cos 0 = 1$, you get Pythagoras' theorem, $\cos^2 A + \sin^2 A = 1$.

(b) Put $A = \frac{1}{2}\pi$ and $B = \frac{1}{6}\pi$.

Then $\cos A = 0$, $\sin A = 1$, $\cos B = \frac{1}{2}\sqrt{3}$ and $\sin B = \frac{1}{2}$. The formula then gives $\cos A \cos B + \sin A \sin B = 0 \times \frac{1}{2}\sqrt{3} + 1 \times \frac{1}{2} = \frac{1}{2}$, which is consistent with $\cos(A-B) = \cos\left(\frac{1}{2}\pi - \frac{1}{6}\pi\right) = \cos\frac{1}{3}\pi = \frac{1}{2}$.

If you replace B by $(-B)$ in the formula for $\cos(A - B)$ you get

$$\cos(A - (-B)) = \cos A \cos(-B) + \sin A \sin(-B).$$

Recall that cosine is an even function (P1 Section 18.3), so $\cos(-B) = \cos B$, and that sine is an odd function, so $\sin(-B) = -\sin B$. Writing $\cos(A - (-B))$ as $\cos(A + B)$,

$$\begin{aligned} \cos(A + B) &= \cos A \cos(-B) + \sin A \sin(-B) \\ &= \cos A \cos B - \sin A \sin B. \end{aligned}$$

To find a formula for $\sin(A + B)$, first recall that $\cos\left(\frac{1}{2}\pi - \theta\right) = \sin\theta$ (see P1 Section 10.4 for the equivalent statement in degrees). Using this with $\theta = A + B$,

$$\begin{aligned} \sin(A + B) &= \cos\left(\frac{1}{2}\pi - (A + B)\right) = \cos\left(\left(\frac{1}{2}\pi - A\right) - B\right) \\ &= \cos\left(\frac{1}{2}\pi - A\right)\cos B + \sin\left(\frac{1}{2}\pi - A\right)\sin B \\ &= \sin A \cos B + \cos A \sin B. \end{aligned}$$

You can obtain the formula for $\sin(A - B)$ in a similar way. (This is Exercise 5B Question 6.) The four formulae are true for all angles A and B, so they are identities.

> For all angles A and B,
>
> $$\sin(A + B) \equiv \sin A \cos B + \cos A \sin B,$$
> $$\sin(A - B) \equiv \sin A \cos B - \cos A \sin B,$$
> $$\cos(A + B) \equiv \cos A \cos B - \sin A \sin B,$$
> $$\cos(A - B) \equiv \cos A \cos B + \sin A \sin B.$$

These formulae, called the **addition formulae**, apply whatever units are used for angle. They are important and you should learn them, but you need not learn how to prove them.

Now that you have these formulae, you have a quick method of simplifying expressions such as $\cos\left(\frac{3}{2}\pi - \theta\right)$:

$$\begin{aligned} \cos\left(\frac{3}{2}\pi - \theta\right) &= \cos\left(\frac{3}{2}\pi\right)\cos\theta + \sin\left(\frac{3}{2}\pi\right)\sin\theta \\ &= 0 \times \cos\theta + (-1) \times \sin\theta = -\sin\theta. \end{aligned}$$

Example 5.3.2

Use the formulae for $\cos(A \pm B)$ to find exact values of $\cos 75°$ and $\cos 15°$.

$$\begin{aligned} \cos 75° = \cos(45 + 30)° &= \cos 45° \cos 30° - \sin 45° \sin 30° \\ &= \tfrac{1}{2}\sqrt{2} \times \tfrac{1}{2}\sqrt{3} - \tfrac{1}{2}\sqrt{2} \times \tfrac{1}{2} = \tfrac{1}{4}\left(\sqrt{6} - \sqrt{2}\right). \end{aligned}$$

$$\begin{aligned} \cos 15° = \cos(45 - 30)° &= \cos 45° \cos 30° + \sin 45° \sin 30° \\ &= \tfrac{1}{2}\sqrt{2} \times \tfrac{1}{2}\sqrt{3} + \tfrac{1}{2}\sqrt{2} \times \tfrac{1}{2} = \tfrac{1}{4}\left(\sqrt{6} + \sqrt{2}\right). \end{aligned}$$

Check these results for yourself with a calculator.

Example 5.3.3

You are given that $\sin A = \frac{8}{17}$, that $\sin B = \frac{12}{13}$, and that $0 < B < \frac{1}{2}\pi < A < \pi$. Find the exact value of $\tan(A + B)$.

From the Pythagoras identity, $\cos^2 A + \left(\frac{8}{17}\right)^2 = 1$, $\cos^2 A = 1 - \frac{64}{289} = \frac{225}{289}$, so $\cos A = \pm\frac{15}{17}$. As $\frac{1}{2}\pi < A < \pi$, $\cos A$ is negative, so $\cos A = -\frac{15}{17}$.

Similarly, $\cos^2 B + \left(\frac{12}{13}\right)^2 = 1$, so $\cos B = \pm\frac{5}{13}$. As $0 < B < \frac{1}{2}\pi$, $\cos B$ is positive, so $\cos B = \frac{5}{13}$. Then

$$\sin(A + B) = \frac{8}{17} \times \frac{5}{13} + \left(-\frac{15}{17}\right) \times \frac{12}{13} = \frac{40 - 180}{17 \times 13} = \frac{-140}{17 \times 13}$$

and $\quad \cos(A + B) = \left(-\frac{15}{17}\right) \times \frac{5}{13} - \frac{8}{17} \times \frac{12}{13} = \frac{-75 - 96}{17 \times 13} = \frac{-171}{17 \times 13}$,

so $\quad \tan(A + B) = \dfrac{\sin(A + B)}{\cos(A + B)} = \dfrac{-140 / 17 \times 13}{-171 / 17 \times 13} = \dfrac{140}{171}$.

Example 5.3.4

Prove that $\sin(A + B) + \sin(A - B) \equiv 2\sin A \cos B$.

Starting from the left side, and 'expanding' both terms,

$$\sin(A + B) + \sin(A - B) \equiv (\sin A \cos B + \cos A \sin B) + (\sin A \cos B - \cos A \sin B)$$
$$\equiv \sin A \cos B + \cos A \sin B + \sin A \cos B - \cos A \sin B$$
$$\equiv 2\sin A \cos B.$$

Hence $\sin(A + B) + \sin(A - B) \equiv 2\sin A \cos B$.

Example 5.3.5

Find the value of $\tan x°$, given that $\sin(x + 30)° = 2\cos(x - 30)°$.

Use the addition formulae to write the equation as

$$\sin x° \cos 30° + \cos x° \sin 30° = 2\cos x° \cos 30° + 2\sin x° \sin 30°.$$

Collect the terms involving $\sin x°$ on the left, and those involving $\cos x°$ on the right, substituting the values of $\sin 30°$ and $\cos 30°$:

$$\sin x° \times \frac{1}{2}\sqrt{3} - 2\sin x° \times \frac{1}{2} = 2\cos x° \times \frac{1}{2}\sqrt{3} - \cos x° \times \frac{1}{2},$$

which can be rearranged as

$$\left(\frac{1}{2}\sqrt{3} - 1\right)\sin x° = \left(\sqrt{3} - \frac{1}{2}\right)\cos x°.$$

Hence $\quad \tan x° = \dfrac{\sin x°}{\cos x°} = \dfrac{\sqrt{3} - \frac{1}{2}}{\frac{1}{2}\sqrt{3} - 1} = \dfrac{2\sqrt{3} - 1}{\sqrt{3} - 2}$.

5.4 The addition formulae for tangents

To find a formula for $\tan(A+B)$, use $\tan(A+B) \equiv \dfrac{\sin(A+B)}{\cos(A+B)}$ together with the identities for $\cos(A+B)$ and $\sin(A+B)$. Thus

$$\tan(A+B) \equiv \frac{\sin(A+B)}{\cos(A+B)} \equiv \frac{\sin A \cos B + \cos A \sin B}{\cos A \cos B - \sin A \sin B}.$$

You can get a neater formula by dividing the top and the bottom of the fraction on the right by $\cos A \cos B$. The numerator then becomes

$$\frac{\sin A \cos B + \cos A \sin B}{\cos A \cos B} = \frac{\sin A \cos B}{\cos A \cos B} + \frac{\cos A \sin B}{\cos A \cos B} = \tan A + \tan B,$$

and the denominator becomes

$$\frac{\cos A \cos B - \sin A \sin B}{\cos A \cos B} = \frac{\cos A \cos B}{\cos A \cos B} - \frac{\sin A \sin B}{\cos A \cos B} = 1 - \tan A \tan B.$$

Therefore, putting the fraction together, $\tan(A+B) \equiv \dfrac{\tan A + \tan B}{1 - \tan A \tan B}.$

A similar derivation, or the fact that $\tan(-B) = -\tan B$, yields a formula for $\tan(A-B)$.

$$\tan(A+B) \equiv \frac{\tan A + \tan B}{1 - \tan A \tan B}, \quad \tan(A-B) \equiv \frac{\tan A - \tan B}{1 + \tan A \tan B}.$$

Note that these identities have no meaning if $\tan A$ or $\tan B$ is undefined, of if the denominator is zero.

Example 5.4.1

Given that $\tan(x+y) = 1$ and that $\tan x = \frac{1}{2}$, find $\tan y$.

$$\tan y = \tan((x+y) - x) = \frac{\tan(x+y) - \tan x}{1 + \tan(x+y)\tan x} = \frac{1 - \frac{1}{2}}{1 + 1 \times \frac{1}{2}} = \frac{\frac{1}{2}}{\frac{3}{2}} = \frac{1}{3}.$$

Example 5.4.2

Find the tangent of the angle between the lines $7y = x + 2$ and $x + y = 3$.

The gradients of the lines are $\frac{1}{7}$ and -1, so if they make angles A and B with the x-axis respectively, $\tan A = \frac{1}{7}$ and $\tan B = -1$. Then

$$\tan(A-B) = \frac{\tan A - \tan B}{1 + \tan A \tan B} = \frac{\frac{1}{7} - (-1)}{1 + \frac{1}{7} \times (-1)} = \frac{\frac{8}{7}}{\frac{6}{7}} = \frac{4}{3}.$$

Exercise 5B

1 By writing 75 as $30+45$, find the exact values of $\sin 75°$ and $\tan 75°$.

2 Find the exact values of
   (a) $\cos 105°$,   (b) $\sin 105°$,   (c) $\tan 105°$.

3 Express $\cos\left(x+\frac{1}{3}\pi\right)$ in terms of $\cos x$ and $\sin x$.

4 Use the expansions for $\sin(A+B)$ and $\cos(A+B)$ to simplify $\sin\left(\frac{3}{2}\pi+\phi\right)$ and $\cos\left(\frac{1}{2}\pi+\phi\right)$.

5 Express $\tan\left(\frac{1}{3}\pi+x\right)$ and $\tan\left(\frac{5}{6}\pi-x\right)$ in terms of $\tan x$.

6 Use $\sin(A-B)\equiv\cos\left(\frac{1}{2}\pi-(A-B)\right)\equiv\cos\left(\left(\frac{1}{2}\pi-A\right)+B\right)$ to derive the formula for $\sin(A-B)$.

7 Given that $\cos A=\frac{3}{5}$ and $\cos B=\frac{24}{25}$, where A and B are acute, find the exact values of
   (a) $\tan A$,   (b) $\sin B$,   (c) $\cos(A-B)$,   (d) $\tan(A+B)$.

8 Given that $\sin A=\frac{3}{5}$ and $\cos B=\frac{12}{13}$, where A is obtuse and B is acute, find the exact values of $\cos(A+B)$ and $\cot(A-B)$.

9 Prove that $\cos(A+B)-\cos(A-B)\equiv-2\sin A\sin B$.

5.5 Double angle formulae

If you put $A=B$ in the addition formulae, you obtain identities for the sine, cosine and tangent of $2A$. The first comes from $\sin(A+B)\equiv\sin A\cos B+\cos A\sin B$, which gives

$$\sin(A+A)\equiv\sin A\cos A+\cos A\sin A, \quad\text{or}\quad \sin 2A\equiv 2\sin A\cos A.$$

From $\cos(A+B)\equiv\cos A\cos B-\sin A\sin B$ you get $\cos 2A\equiv\cos^2 A-\sin^2 A$. There are two other useful forms for this which come from replacing $\cos^2 A$ by $1-\sin^2 A$, giving

$$\cos 2A\equiv\cos^2 A-\sin^2 A\equiv\left(1-\sin^2 A\right)-\sin^2 A\equiv 1-2\sin^2 A,$$

or from replacing $\sin^2 A$ by $1-\cos^2 A$, giving

$$\cos 2A\equiv\cos^2 A-\sin^2 A\equiv\cos^2 A-\left(1-\cos^2 A\right)\equiv 2\cos^2 A-1.$$

Finally, the formula $\tan(A+B)\equiv\dfrac{\tan A+\tan B}{1-\tan A\tan B}$ becomes $\tan 2A\equiv\dfrac{2\tan A}{1-\tan^2 A}$.

These formulae are called the **double angle formulae**.

$$\sin 2A\equiv 2\sin A\cos A,$$
$$\cos 2A\equiv\cos^2 A-\sin^2 A\equiv 1-2\sin^2 A\equiv 2\cos^2 A-1,$$
$$\tan 2A\equiv\frac{2\tan A}{1-\tan^2 A}.$$

Example 5.5.1

Given that $\cos A = \frac{1}{3}$, find the exact value of $\cos 2A$.

$$\cos 2A = 2\cos^2 A - 1 = 2 \times \left(\frac{1}{3}\right)^2 - 1 = 2 \times \frac{1}{9} - 1 = -\frac{7}{9}.$$

Example 5.5.2

Given that $\cos A = \frac{1}{3}$, find the possible values of $\cos\frac{1}{2}A$.

Using $\cos 2A \equiv 2\cos^2 A - 1$, with $\frac{1}{2}A$ written in place of A, gives

$\cos A \equiv 2\cos^2\frac{1}{2}A - 1$. In this case, $\frac{1}{3} = 2\cos^2\frac{1}{2}A - 1$, giving $2\cos^2\frac{1}{2}A = \frac{4}{3}$.

This simplifies to $\cos^2\frac{1}{2}A = \frac{2}{3}$, so $\cos\frac{1}{2}A = \pm\sqrt{\frac{2}{3}} = \pm\frac{1}{3}\sqrt{6}$.

Example 5.5.3

Solve the equation $2\sin 2\theta° = \sin\theta°$, giving values of θ such that $0 \le \theta \le 360$ correct to 1 decimal place.

Using the identity $\sin 2\theta° \equiv 2\sin\theta°\cos\theta°$,

$$2 \times 2\sin\theta°\cos\theta° = \sin\theta°, \text{ so } \sin\theta°(4\cos\theta° - 1) = 0.$$

At least one of these factors must be 0. Therefore either

$$\sin\theta° = 0, \quad \text{giving} \quad \theta = 0, 180, 360$$

or $\quad 4\cos\theta° - 1 = 0$, giving $\quad \cos\theta° = 0.25$, so $\theta = 75.52\ldots$ or $284.47\ldots$.

Therefore the required roots are $\theta = 0, 75.5, 180, 284.5, 360$ correct to 1 decimal place.

Example 5.5.4

Prove the identity $\cot A - \tan A \equiv 2\cot 2A$.

Method 1 Put everything in terms of $\tan A$. Starting with the left side,

$$\cot A - \tan A \equiv \frac{1}{\tan A} - \tan A \equiv \frac{1 - \tan^2 A}{\tan A}$$

$$\equiv 2 \times \left(\frac{1 - \tan^2 A}{2\tan A}\right) \equiv 2 \times \frac{1}{\tan 2A} \equiv 2\cot 2A.$$

Method 2 Put everything in terms of $\sin A$ and $\cos A$. Starting with the left side,

$$\cot A - \tan A \equiv \frac{\cos A}{\sin A} - \frac{\sin A}{\cos A} \equiv \frac{\cos^2 A - \sin^2 A}{\sin A \cos A} \equiv \frac{\cos 2A}{\frac{1}{2}\sin 2A} \equiv 2\cot 2A.$$

Example 5.5.5

Prove that $\operatorname{cosec} x + \cot x \equiv \cot \frac{1}{2} x$.

Starting with the left side, and putting everything in terms of sines and cosines,

$$\operatorname{cosec} x + \cot x \equiv \frac{1}{\sin x} + \frac{\cos x}{\sin x} \equiv \frac{1 + \cos x}{\sin x}$$

$$\equiv \frac{1 + \left(2\cos^2 \frac{1}{2} x - 1\right)}{2\sin \frac{1}{2} x \cos \frac{1}{2} x} \equiv \frac{2\cos^2 \frac{1}{2} x}{2\sin \frac{1}{2} x \cos \frac{1}{2} x}$$

$$\equiv \frac{\cos \frac{1}{2} x}{\sin \frac{1}{2} x} \equiv \cot \frac{1}{2} x.$$

Exercise 5C

1 If $\sin A = \frac{2}{3}$ and A is obtuse, find the exact values of $\cos A$, $\sin 2A$ and $\tan 2A$.

2 If $\cos B = \frac{3}{4}$, find the exact values of $\cos 2B$ and $\cos \frac{1}{2} B$.

3 By expressing $\sin 3A$ as $\sin(2A + A)$, find an expression for $\sin 3A$ in terms of $\sin A$.

4 Express $\cos 3A$ in terms of $\cos A$.

5 By writing $\cos x$ in terms of $\frac{1}{2} x$, find an alternative expression for $\dfrac{1 - \cos x}{1 + \cos x}$.

6 Prove that $4\sin\left(x + \frac{1}{6}\pi\right)\sin\left(x - \frac{1}{6}\pi\right) \equiv 3 - 4\cos^2 x$.

7 If $\cos 2A = \frac{7}{18}$, find the possible values of $\cos A$ and $\sin A$.

8 If $\tan 2A = \frac{12}{5}$, find the possible values of $\tan A$.

9 If $\tan 2A = 1$, find the possible values of $\tan A$. Hence state the exact value of $\tan 22\frac{1}{2}^{\circ}$.

10 Solve these equations for values of A between 0 and 2π inclusive.

(a) $\cos 2A + 3 + 4\cos A = 0$ (b) $2\cos 2A + 1 + \sin A = 0$

(c) $\tan 2A + 5\tan A = 0$

5.6 The form $a\sin x + b\cos x$

If you draw the graph of $y = \sin\left(x + \frac{1}{6}\pi\right)$, you will see that it is the graph of $y = \sin x$ moved by $\frac{1}{6}\pi$ in the negative x-direction. Similarly the graph of $y = \sin(x + \alpha)$ has been moved by α in the negative x-direction.

If you compare the graph of $y = 2\sin x$ with that of $y = \sin x$, you will see that the graph of $y = \sin x$ has been stretched in the y-direction by a factor of 2. Similarly the graph of $y = R\sin x$, where $R > 0$, stretches $y = \sin x$ by a factor of R.

Draw the graphs of $y = 3\sin x + 2\cos x$ and $y = \sin x - 4\cos x$, using an interval of values for x of either 2π or $360°$, depending on whether you are working in radians or degrees. What you see may surprise you: it shows that both these graphs are either cosine or sine graphs, first shifted in the x-direction, and then stretched in the y-direction.

These observations suggest that you can write $y = 3\sin x + 2\cos x$ in the form $y = R\sin(x + \alpha)$, where the graph $y = \sin x$ has been shifted by α in the negative x-direction, and then stretched in the y-direction by the factor R, where $R > 0$. But how do you find the values of R and α?

If you equate the two expressions $y = 3\sin x + 2\cos x$ and $y = R\sin(x + \alpha)$, you find that

$$3\sin x + 2\cos x \equiv R\sin x\cos\alpha + R\cos x\sin\alpha.$$

Since these are to be identical, they certainly agree for $x = \frac{1}{2}\pi$ and $x = 0$. This gives

$$3 = R\cos\alpha \quad \text{and} \quad 2 = R\sin\alpha.$$

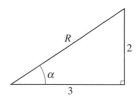

Fig. 5.5

You can find R and α from these equations. Imagine a right-angled triangle, which you might think of as a set-square, with adjacent sides 2 units and 3 units, and hypotenuse R units. Then α is the angle shown in Fig. 5.5.

Therefore $\tan\alpha = \frac{2}{3}$, and $R = \sqrt{2^2 + 3^2} = \sqrt{13}$. It is important to remember that $R > 0$. The equations $2 = R\sin\alpha$ and $3 = R\cos\alpha$ then show that $\cos\alpha$ and $\sin\alpha$ are positive, so that the angle α is acute; in radians $\alpha = 0.58\ldots$. Then

$$3\sin x + 2\cos x \equiv \left(\sqrt{13}\cos 0.58\ldots\right)\sin x + \left(\sqrt{13}\sin 0.58\ldots\right)\cos x$$
$$\equiv \sqrt{13}\sin x\cos 0.58\ldots + \sqrt{13}\cos x\sin 0.58\ldots$$
$$\equiv \sqrt{13}\sin(x + 0.58\ldots).$$

The form $\sqrt{13}\sin(x + 0.58\ldots)$, while it may look less friendly than $3\sin x + 2\cos x$, is in many ways more convenient. Example 5.6.1 shows two applications.

Example 5.6.1
(a) Find the maximum and minimum values of $3\sin x + 2\cos x$, and find in radians to 2 decimal places the smallest positive values of x at which they occur.

(b) Solve the equation $3\sin x + 2\cos x = 1$, for $-\pi \leqslant x \leqslant \pi$, to 2 decimal places.

(a) Since $3\sin x + 2\cos x \equiv \sqrt{13}\sin(x + 0.58\ldots)$, and the maximum and minimum values of the sine function are 1 and -1, the maximum and minimum values of $3\sin x + 2\cos x$ are $\sqrt{13}$ and $-\sqrt{13}$. And since the maximum and minimum values of the sine function occur at $\frac{1}{2}\pi$ and $\frac{3}{2}\pi$, the relevant values of x are given by $x + 0.58\ldots = \frac{1}{2}\pi$ and $x + 0.58\ldots = \frac{3}{2}\pi$. Therefore the maximum $\sqrt{13}$ occurs when $x = \frac{1}{2}\pi - 0.58\ldots = 0.98$, and the minimum $-\sqrt{13}$ occurs when $x = \frac{3}{2}\pi - 0.58\ldots = 4.12$, both correct to 2 decimal places.

(b) $3\sin x + 2\cos x = 1 \iff \sqrt{13}\sin(x + 0.58\ldots) = 1 \iff \sin(x + 0.58\ldots) = \dfrac{1}{\sqrt{13}}$.

Using the methods of P1 Section 18.5, the solutions (between $-\pi$ and π) are

$$x + 0.58\ldots = 0.28\ldots \quad \text{or} \quad x + 0.58\ldots = 2.86\ldots,$$

so $x = -0.31$ or 2.27, correct to 2 decimal places.

In the general case, you can write $a \sin x + b \cos x$ in the form $R \sin(x + \alpha)$ where $R = \sqrt{a^2 + b^2}$ and α is given by the equations $R \cos \alpha = a$ and $R \sin \alpha = b$.

There is nothing special about the form $R \sin(x + \alpha)$. It is often more convenient to use $R \cos(x + \alpha)$, $R \sin(x - \alpha)$ or $R \cos(x - \alpha)$. Thus, for the values of R and α in Example 5.6.1, with $3 = R \cos \alpha$ and $2 = R \sin \alpha$,

$$3 \cos x + 2 \sin x \equiv (R \cos \alpha) \cos x + (R \sin \alpha) \sin x \equiv R \cos(x - \alpha).$$

Always try to choose the form which produces the terms in the right order with the correct sign. For example, write $3 \cos x - 2 \sin x$ in the form $R \cos(x + \alpha)$, and $3 \sin x - 2 \cos x$ in the form $R \sin(x - \alpha)$.

Summarising all this:

If a and b are positive,

$a \sin x \pm b \cos x$ can be written in the form $R \sin(x \pm \alpha)$,

$a \cos x \pm b \sin x$ can be written in the form $R \cos(x \mp \alpha)$,

where $R = \sqrt{a^2 + b^2}$ and $R \cos \alpha = a$, $R \sin \alpha = b$, with $0 < \alpha < \frac{1}{2}\pi$.

You do not need to memorise the results in the box, although it is useful to know that $R = \sqrt{a^2 + b^2}$. Learn how to find α and work it out each time you come to it.

Example 5.6.2

Express $\sin \theta^\circ - 4 \cos \theta^\circ$ in the form $R \sin(\theta - \alpha)^\circ$ giving the values of R and α. Explain why the equation $\sin \theta^\circ - 4 \cos \theta^\circ = 5$ has no solutions.

Identifying $\sin \theta^\circ - 4 \cos \theta^\circ$ with $R \sin(\theta - \alpha)^\circ$ gives

$$\sin \theta^\circ - 4 \cos \theta^\circ \equiv R \sin \theta^\circ \cos \alpha^\circ - R \cos \theta^\circ \sin \alpha^\circ,$$

so $R \cos \alpha^\circ = 1$ and $R \sin \alpha^\circ = 4$.

Therefore $R = \sqrt{1^2 + 4^2} = \sqrt{17}$, with $\cos \alpha^\circ = \dfrac{1}{\sqrt{17}}$ and $\sin \alpha^\circ = \dfrac{4}{\sqrt{17}}$, giving $\tan \alpha^\circ = 4$ and $\alpha = 75.9\ldots$.

Then $\sin \theta^\circ - 4 \cos \theta^\circ \equiv \sqrt{17} \sin(\theta - \alpha)^\circ$, where $\alpha = 75.9\ldots$.

The equation $\sin \theta^\circ - 4 \cos \theta^\circ = 5$ has no solution since if $\sin \theta^\circ - 4 \cos \theta^\circ = 5$, then $\sqrt{17} \sin(\theta - \alpha)^\circ = 5$, so $\sin(\theta - \alpha)^\circ = \dfrac{5}{\sqrt{17}} > 1$.

As there are no values for which the sine function is greater than 1, there is no solution to the equation $\sin(\theta - \alpha)^\circ = \dfrac{5}{\sqrt{17}}$, and therefore no solution to the equation $\sin \theta^\circ - 4 \cos \theta^\circ = 5$.

Exercise 5D

1. Find the value of α between 0 and $\frac{1}{2}\pi$ for which $3\sin x + 2\cos x \equiv \sqrt{13}\sin(x + \alpha)$

2. Find the value of ϕ between 0 and 90 for which $3\cos x° - 4\sin x° \equiv 5\cos(x + \phi)°$.

3. Find the value of R such that, if $\tan\beta = \frac{3}{5}$, then $5\sin\theta + 3\cos\theta \equiv R\sin(\theta + \beta)$.

4. Find the value of R and the value of β between 0 and $\frac{1}{2}\pi$ correct to 3 decimal places such that $6\cos x + \sin x \equiv R\cos(x - \beta)$.

5. Find the value of R and the value of α between 0 and $\frac{1}{2}\pi$ in each of the following cases, where the given expression is written in the given form.

   (a) $\sin x + 2\cos x;$ $\quad R\sin(x + \alpha)$ (b) $\sin x + 2\cos x;$ $\quad R\cos(x - \alpha)$

   (c) $\sin x - 2\cos x;$ $\quad R\sin(x - \alpha)$ (d) $2\cos x - \sin x;$ $\quad R\cos(x + \alpha)$

6. Express $5\cos\theta + 6\sin\theta$ in the form $R\cos(\theta - \beta)$ where $R > 0$ and $0 < \beta < \frac{1}{2}\pi$. State

   (a) the maximum value of $5\cos\theta + 6\sin\theta$ and the least positive value of θ which gives this maximum,

   (b) the minimum value of $5\cos\theta + 6\sin\theta$ and the least positive value of θ which gives this minimum.

7. Express $8\sin x° + 6\cos x°$ in the form $R\sin(x + \phi)°$, where $R > 0$ and $0 < \phi < 90$. Deduce the number of roots for $0 < x < 180$ of the following equations.

   (a) $8\sin x° + 6\cos x° = 5$ (b) $8\sin x° + 6\cos x° = 12$

8. Solve $3\sin x - 2\cos x = 1$ for values of x between 0 and 2π by

   (a) expressing $3\sin x - 2\cos x$ in the form $R\sin(x - \beta)$,

   (b) using a graphical method.

Miscellaneous exercise 5

1. (a) Starting from the identity $\sin^2\phi + \cos^2\phi \equiv 1$, prove that $\sec^2\phi \equiv 1 + \tan^2\phi$.

   (b) Given that $180 < \phi < 270$ and that $\tan\phi° = \frac{7}{24}$, find the exact value of $\sec\phi°$. (OCR)

2. Solve the equation $\tan x° = 3\cot x°$, giving all solutions between 0 and 360. (OCR)

3. (a) State the value of $\sec^2 x - \tan^2 x$.

   (b) The angle A is such that $\sec A + \tan A = 2$. Show that $\sec A - \tan A = \frac{1}{2}$, and hence find the exact value of $\cos A$. (OCR)

4. Let $f(A) = \dfrac{\cos A°}{1 + \sin A°} + \dfrac{1 + \sin A°}{\cos A°}$.

   (a) Prove that $f(A) = 2\sec A°$.

   (b) Solve the equation $f(A) = 4$, giving your answers for A in the interval $0 < A < 360$. (OCR)

5 You are given that $\cos 30° = \dfrac{\sqrt{3}}{2}$ and $\cos 45° = \dfrac{1}{\sqrt{2}}$. Determine the exact value of $\cos 75°$.

(OCR)

6 Prove that $\sin\left(\theta + \tfrac{1}{2}\pi\right) \equiv \cos\theta$.

(OCR)

7 The angle α is obtuse, and $\sin\alpha = \tfrac{3}{5}$.

   (a) Find the value of $\cos\alpha$.

   (b) Find the values of $\sin 2\alpha$ and $\cos 2\alpha$, giving your answers as fractions in their lowest terms.

(OCR, adapted)

8 Given that $\sin\theta° = 4\sin(\theta - 60)°$, show that $2\sqrt{3}\cos\theta° = \sin\theta°$. Hence find the value of θ such that $0 < \theta < 180$.

(OCR)

9 Solve the equation $\sin 2\theta° - \cos^2\theta° = 0$, giving values of θ in the interval $0 < \theta < 360$.

(OCR, adapted)

10 (a) Prove the identity $\cot\tfrac{1}{2}A - \tan\tfrac{1}{2}A \equiv 2\cot A$.

    (b) By choosing a suitable numerical value for A, show that $\tan 15°$ is a root of the quadratic equation $t^2 + 2\sqrt{3}\,t - 1 = 0$.

(OCR)

11 (a) By using the substitution $t = \tan\tfrac{1}{2}x$, prove that $\operatorname{cosec} x - \cot x = \tan\tfrac{1}{2}x$.

    (b) Use this result to show that $\tan 15° = 2 - \sqrt{3}$.

(OCR)

12 Express $\sin\theta° + \sqrt{3}\cos\theta°$ in the form $R\sin(\theta + \alpha)°$, where $R > 0$ and $0 < \alpha < 90$.

Hence find all values of θ, for $0 < \theta < 360$, which satisfy the equation $\sin\theta° + \sqrt{3}\cos\theta° = 1$.

(OCR)

13 The function f is defined for all real x by $f(x) = \cos x° - \sqrt{3}\sin x°$.

    (a) Express $f(x)$ in the form $R\cos(x + \phi)°$, where $R > 0$ and $0 < \phi < 90$.

    (b) Solve the equation $|f(x)| = 1$, giving your answers in the interval $0 \leqslant x \leqslant 360$. (OCR)

14 (a) Express $12\cos x + 9\sin x$ in the form $R\cos(x - \theta)$, where $R > 0$ and $0 < \theta < \tfrac{1}{2}\pi$.

    (b) Use the method of part (a) to find the smallest positive root α of the equation $12\cos x + 9\sin x = 14$, giving your answer correct to three decimal places. (OCR)

15 Express $2\cos x° + \sin x°$ in the form $R\cos(x - \alpha)°$, where $R > 0$ and $0 < \alpha < 90$. Hence

    (a) solve the equation $2\cos x° + \sin x° = 1$, giving all solutions between 0 and 360,

    (b) find the exact range of values of the constant k for which the equation $2\cos x° + \sin x° = k$ has real solutions for x.

(OCR)

16 If $\cos^{-1}(3x + 2) = \tfrac{1}{3}\pi$, find the value of x.

17 (a) Express $5\sin x° + 12\cos x°$ in the form $R\sin(x + \theta)°$, where $R > 0$ and $0 < \theta < 90$.

    (b) Hence, or otherwise, find the maximum and minimum values of $f(x)$ where

    $$f(x) = \dfrac{30}{5\sin x° + 12\cos x° + 17}.$$ State also the values of x, in the range $0 < x < 360$, at which they occur.

(OCR)

18 Express $3\cos x° - 4\sin x°$ in the form $R\cos(x+\alpha)°$, where $R > 0$ and $0 < \alpha < 90$. Hence

   (a) solve the equation $3\cos x° - 4\sin x° = 2$, giving all solutions between 0 and 360,

   (b) find the greatest and least values, as x varies, of the expression $\dfrac{1}{3\cos x° - 4\sin x° + 8}$.

(OCR)

19 (a) Find the value of $\tan^{-1}\sqrt{3} + \tan^{-1}\left(-\dfrac{1}{\sqrt{3}}\right)$.

   (b) If $x = \tan^{-1} A$ and $y = \tan^{-1} B$, find $\tan(A+B)$ in terms of x and y.

20 If $A = \sin^{-1} x$, where $x > 0$,

   (a) show that $\cos A = \sqrt{1 - x^2}$,

   (b) find expressions in terms of x for $\operatorname{cosec} A$ and $\cos 2A$.

21 (a) Find the equation of the straight line joining the points $A(0, 1.5)$ and $B(3, 0)$.

   (b) Express $\sin\theta° + 2\cos\theta°$ in the form $r\sin(\theta+\alpha)°$, where r is a positive number and $\alpha°$ is an acute angle.

   (c) The figure shows a map of a moor-land. The units of the coordinates are kilometres, and the y-axis points due north. A walker leaves her car somewhere on the straight road between A and B. She walks in a straight line for a distance of 2 km to a monument at the origin O. While she is looking at it the fog comes down, so that she cannot see the way back to her car. She needs to work out the bearing on which she should walk.

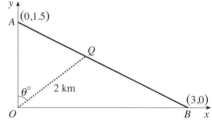

Write down the coordinates of a point Q which is 2 km from O on a bearing of $\theta°$. Show that, for Q to be on the road between A and B, θ must satisfy the equation $2\sin\theta° + 4\cos\theta° = 3$. Calculate the value of θ between 0 and 90 which satisfies this equation.

(OCR)

22 Let a and b be the straight lines with equations $y = m_1 x + c_1$ and $y = m_2 x + c_2$ where $m_1 m_2 \neq 0$. Use appropriate trigonometric formulae to prove that a and b are perpendicular if and only if $m_1 m_2 = -1$.

23 The figure shows the graphs

    {1} $\ y = 5\cos 2x° + 2$

       and

    {2} $\ y = \cos x°$

       for $\ 0 \leqslant x \leqslant 180$.

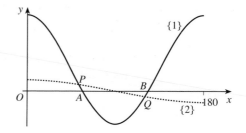

(a) Find the coordinates of the points A and B where the graph {1} meets the x-axis.

(b) By solving a suitable trigonometric equation, find the x-coordinates of the two points P and Q where the graphs {1} and {2} intersect. Hence find the coordinates of the points P and Q.

                                                     (OCR)

6 Differentiating trigonometric functions

This chapter shows how to differentiate the functions $\sin x$ and $\cos x$. When you have completed it, you should

- be familiar with a number of inequalities and limits involving trigonometric functions, and their geometrical interpretations
- know the derivatives and indefinite integrals of $\sin x$ and $\cos x$
- be able to differentiate a variety of trigonometric functions using the chain rule
- be able to integrate a variety of trigonometric functions, using identities where necessary.

6.1 Some inequalities and limits

Fig. 6.1 shows a sector OAB of a circle with radius r units and angle θ radians $\left(<\frac{1}{2}\pi\right)$. The tangent at B meets OA produced at D. Comparing areas,

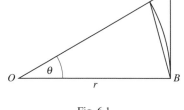

Fig. 6.1

triangle $OAB <$ sector $OAB <$ triangle ODB,

so $\quad \frac{1}{2}r^2\sin\theta < \frac{1}{2}r^2\theta < \frac{1}{2}r\times r\tan\theta$.

Dividing by $\frac{1}{2}r^2$, it follows that, for $0 < \theta < \frac{1}{2}\pi$,

$$\sin\theta < \theta < \tan\theta.$$

If θ is small, Fig. 6.1 suggests that the three numbers $\sin\theta$, θ and $\tan\theta$ will be very close to each other.

Check this by setting $\theta = 0.1$ on your calculator. Remember to put it into radian mode.

So you can also write, if θ is small,

$$\sin\theta \approx \theta \quad \text{and} \quad \tan\theta \approx \theta.$$

One useful form of the inequality can be found by taking the left and right parts separately. First, since $\theta > 0$, you can divide the inequality $\sin\theta < \theta$ by θ to obtain

$$\frac{\sin\theta}{\theta} < 1.$$

Secondly, you can write $\theta < \tan\theta$ as $\theta < \dfrac{\sin\theta}{\cos\theta}$. Multiplying this by $\cos\theta$ and dividing by θ, both of which are positive since $0 < \theta < \frac{1}{2}\pi$, gives

$$\cos\theta < \frac{\sin\theta}{\theta}.$$

Putting these new inequalities together again gives, for $0 < \theta < \frac{1}{2}\pi$,

$$\cos\theta < \frac{\sin\theta}{\theta} < 1.$$

This is illustrated in Fig. 6.2. Notice that the graphs have been extended to the left to cover the interval $-\frac{1}{2}\pi < \theta < \frac{1}{2}\pi$.

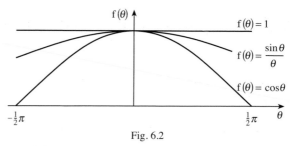

Fig. 6.2

Since $\cos(-\theta) = \cos\theta$

and $\dfrac{\sin(-\theta)}{(-\theta)} = \dfrac{-\sin\theta}{-\theta} = \dfrac{\sin\theta}{\theta}$, both

$\cos\theta$ and $\dfrac{\sin\theta}{\theta}$ are even functions. This shows that the inequality holds also for $-\frac{1}{2}\pi < \theta < 0$.

However, Fig. 6.2 obscures an important point, that $\dfrac{\sin\theta}{\theta}$ is not defined when $\theta = 0$, since the fraction then becomes the meaningless $\dfrac{0}{0}$. But the graph does show that $\dfrac{\sin\theta}{\theta}$ approaches the limit 1 as $\theta \to 0$.

> If $0 < \theta < \frac{1}{2}\pi$, $\sin\theta < \theta < \tan\theta$.
>
> As $\theta \to 0$, $\dfrac{\sin\theta}{\theta} \to 1$.

Example 6.1.1

Show that the graph of $y = \sin x$, shown in Fig. 6.3, has gradient 1 at the origin.

Let the point P on the graph have coordinates $(\theta, \sin\theta)$, so $\dfrac{\sin\theta}{\theta}$ is the gradient of the chord OP. As $\theta \to 0$, this tends to the gradient of the tangent at the origin. But as $\theta \to 0$, $\dfrac{\sin\theta}{\theta} \to 1$, so the sine graph has gradient 1 at the origin.

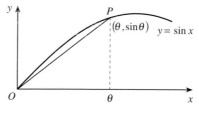

Fig. 6.3

The inequality $\sin\theta < \theta < \tan\theta$ also says something about lengths. In Fig. 6.4, the sector in Fig. 6.1 has been reflected in the radius OB; C and E are the reflections of A and D. Then $AC = 2r\sin\theta$, arc $ABC = r(2\theta) = 2r\theta$, and $DE = 2r\tan\theta$. So the inequality states that

chord $AC <$ arc $ABC <$ tangent DE.

Note also that, in Fig. 6.4,

$$\frac{\text{chord } AC}{\text{arc } ABC} = \frac{2r\sin\theta}{2r\theta} = \frac{\sin\theta}{\theta},$$

so that the ratio of the chord to the arc tends to 1 as θ tends to 0. This result will be needed in the next section.

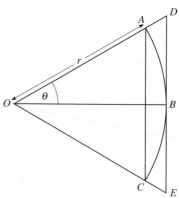

Fig. 6.4

> In a circular sector, as the angle at the centre tends to 0, the ratio of the chord to the arc tends to 1.

6.2* Derivatives of sine and cosine functions

This section shows how the limits established in Section 6.1 can be used to differentiate sines and cosines. You can if you like skip this for a first reading, and pick up the chapter at Section 6.3.

The proof is based on the definitions of $\cos\theta$ and $\sin\theta$ (given in P1 Sections 10.1 and 10.2) as the x- and y-coordinates of a point on a circle of radius 1 unit.

In P1 $\cos\theta$ and $\sin\theta$ were defined with θ in degrees, but the definitions work just as well with θ in radians.

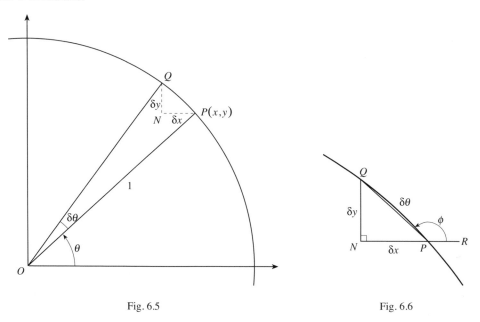

Fig. 6.5   Fig. 6.6

In Fig. 6.5, the point P has coordinates $x = \cos\theta$ and $y = \sin\theta$. If the angle is increased by $\delta\theta$, x increases by δx (which is actually a negative increase if θ is an acute angle, as shown here) and y by δy. The increases in x and y are represented in the figure by the displacements PN and NQ.

Fig. 6.6 is an enlargement of the part of Fig. 6.5 around PNQ. Because the circle has unit radius, the arc PQ has length $\delta\theta$. Extend the line NP to R, parallel to the x-axis, and let ϕ be the angle RPQ. Then

$$\delta x = PQ\cos\phi \quad \text{and} \quad \delta y = PQ\sin\phi.$$

Note that, since ϕ is an obtuse angle, these equations make δx negative and δy positive, as you would expect from the diagrams. If P were located in another quadrant of the circle, the signs would be different, but the equations for δx and δy would still be correct.

The aim is to find $\dfrac{\mathrm{d}x}{\mathrm{d}\theta}$ and $\dfrac{\mathrm{d}y}{\mathrm{d}\theta}$, which are defined as

$$\lim_{\delta\theta \to 0} \frac{\delta x}{\delta\theta} \quad \text{and} \quad \lim_{\delta\theta \to 0} \frac{\delta y}{\delta\theta}.$$

Now $\quad \dfrac{\delta x}{\delta\theta} = \dfrac{PQ}{\delta\theta} \times \cos\phi = \cos\phi \times \dfrac{\text{chord } PQ}{\text{arc } PQ}$,

and $\quad \dfrac{\delta y}{\delta\theta} = \dfrac{PQ}{\delta\theta} \times \sin\phi = \sin\phi \times \dfrac{\text{chord } PQ}{\text{arc } PQ}$.

The proof can be completed by considering the limits of the two parts of these expressions separately. As $\delta\theta \to 0$, the chord PQ becomes the tangent to the circle at P, and Fig. 6.7 shows that the angle ϕ tends to $\theta + \frac{1}{2}\pi$. Also it was shown in the last section that $\dfrac{\text{chord } PQ}{\text{arc } PQ}$ tends to 1.

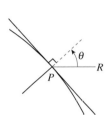

Assuming (as is true) that the limit of the product is equal to the product of the two limits, it follows that

Fig. 6.7

$$\frac{\mathrm{d}x}{\mathrm{d}\theta} = \cos\left(\theta + \tfrac{1}{2}\pi\right) \times 1 \quad \text{and} \quad \frac{\mathrm{d}y}{\mathrm{d}\theta} = \sin\left(\theta + \tfrac{1}{2}\pi\right) \times 1.$$

That is, $\quad \dfrac{\mathrm{d}x}{\mathrm{d}\theta} = -\sin\theta \quad \text{and} \quad \dfrac{\mathrm{d}y}{\mathrm{d}\theta} = \cos\theta.$

The relation $\phi = \theta + \frac{1}{2}\pi$ applies whichever quadrant θ is in, so these results hold for all values of $\theta \in \mathbb{R}$.

6.3 Working with trigonometric derivatives

You have seen the emphasis in trigonometry shift from calculations about triangles to properties of the sine and cosine as functions with domain the real numbers and range the interval $-1 \leqslant y \leqslant 1$. This trend is given a further boost by finding the derivatives, so you can now treat trigonometric functions much like other functions in the mathematical store-cupboard, such as polynomials, power functions, exponential functions and logarithms.

Putting the results of the last section into the usual notation, replacing θ by x:

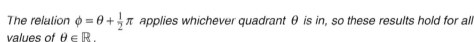

$$\frac{\mathrm{d}}{\mathrm{d}x}\cos x = -\sin x, \quad \frac{\mathrm{d}}{\mathrm{d}x}\sin x = \cos x.$$

Fig. 6.8 shows the graph of $f(x) = \sin x$ for $0 < x < 2\pi$ and, below it, the graphs of $f'(x) = \cos x$ and $f''(x) = -\sin x$.

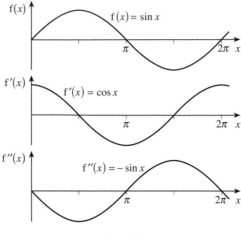

You can see from these that the graph of $f(x)$ is increasing when $f'(x)$ is positive and decreasing when $f'(x)$ is negative. There is a maximum at $\frac{1}{2}\pi$ where $\cos x$ is zero, and $f''\left(\frac{1}{2}\pi\right) = -\sin\left(\frac{1}{2}\pi\right) = -1 < 0$. Also, the graph of $f(x)$ is bending downwards between 0 and π, where $f''(x) = -\sin x$ is negative. What is different in this example from other similar diagrams (like Fig. 15.1 in P1) is that the three graphs are simply translations of each other parallel to the x-axis.

Fig. 6.8

Once you know the derivatives of $\sin x$ and $\cos x$, you can use the chain rule to find the derivatives of many other trigonometric functions.

Example 6.3.1

Differentiate with respect to x (a) $\sin\left(3x - \frac{1}{4}\pi\right)$,   (b) $\cos^4 x$,   (c) $\sec x$.

(a) $\dfrac{\mathrm{d}}{\mathrm{d}x}\sin\left(3x - \frac{1}{4}\pi\right) = \cos\left(3x - \frac{1}{4}\pi\right) \times 3 = 3\cos\left(3x - \frac{1}{4}\pi\right).$

(b)   Remember that $\cos^4 x$ is the conventional way of writing $(\cos x)^4$.

$$\frac{\mathrm{d}}{\mathrm{d}x}\cos^4 x = 4\cos^3 x \times (-\sin x) = -4\cos^3 x \sin x.$$

(c) $\dfrac{\mathrm{d}}{\mathrm{d}x}\sec x = \dfrac{\mathrm{d}}{\mathrm{d}x}\left(\dfrac{1}{\cos x}\right) = -\dfrac{1}{\cos^2 x} \times (-\sin x) = \dfrac{\sin x}{\cos^2 x}.$

The answer to part (c) can be written in several ways. For example,

$$\frac{\sin x}{\cos^2 x} = \frac{\sin x/\cos x}{\cos x} = \frac{\tan x}{\cos x}$$

or $\dfrac{\sin x}{\cos^2 x} = \sin x \times \left(\dfrac{1}{\cos^2 x}\right) = \sin x \sec^2 x$

or $\dfrac{\sin x}{\cos^2 x} = \left(\dfrac{1}{\cos x}\right) \times \left(\dfrac{\sin x}{\cos x}\right) = \sec x \tan x.$

The most usual form is the last one, and it is a result worth remembering.

$$\frac{\mathrm{d}}{\mathrm{d}x}\sec x = \sec x \tan x.$$

Example 6.3.2

Find $\dfrac{d}{dx}\ln\sec x$.

$$\frac{d}{dx}\ln\sec x = \frac{1}{\sec x}\times\sec x\tan x = \tan x.$$

In most practical applications, the independent variable for the sine and cosine functions represents time rather than angle. Just as e^x is the natural function for describing exponential growth and decay, so sine and cosine are the natural functions for describing periodic phenomena. They can be used to model situations as different as the trade cycle, variation in insect populations, seasonal variation in sea temperature, rise and fall of tides (as in P1 Example 10.1.2), motion of a piston in a car cylinder and propagation of radio waves.

Example 6.3.3

The height in metres of the water in a harbour is given approximately by the formula $h = 6 + 3\cos\frac{1}{6}\pi t$ where t is the time measured in hours from noon. Find an expression for the rate at which the water is rising at time t. When is it rising fastest?

Using the chain rule, the rate at which the water is rising is

$$\frac{dh}{dt} = \left(-3\sin\frac{1}{6}\pi t\right)\times\frac{1}{6}\pi = -\frac{1}{2}\pi\sin\frac{1}{6}\pi t\ .$$

The water rises fastest when $\sin\frac{1}{6}\pi t = -1$, that is when $\frac{1}{6}\pi t = \frac{3}{2}\pi,\frac{7}{2}\pi,\frac{11}{2}\pi\ldots$, so $t = 9, 21, 33,\ldots$. The water is rising fastest at 9 p.m. and again at 9 a.m. (This is exactly half-way between low and high tide.)

Example 6.3.4

Find the minima and maxima of $f(x) = 4\cos x + \cos 2x$.

Although the domain is $\mathbb{R}$, you only need to consider the interval $0 \leqslant x < 2\pi$. Since the period of $\cos x$ is 2π, and the period of $\cos 2x$ is π, the graph of $f(x)$ repeats itself after each interval of length 2π.

$f'(x) = -4\sin x - 2\sin 2x = -4\sin x - 4\sin x\cos x = -4\sin x(1+\cos x)$, so $f'(x) = 0$ when $\sin x = 0$ or $\cos x = -1$, that is when $x = 0$ or π.

$f''(x) = -4\cos x - 4\cos 2x$, so $f''(0) = -4-4 = -8$ and $f''(\pi) = 4-4 = 0$. There is therefore a maximum at $x = 0$, but the $f''(x)$ method does not work at $x = \pi$. You must instead consider the sign of $f'(x)$ below and above π.

The factor $1+\cos x$ is always positive except at $x = \pi$, where it is 0; the factor $\sin x$ is positive for $0 < x < \pi$ and negative for $\pi < x < 2\pi$. So $f'(x) = -4\sin x(1+\cos x)$ is negative for $0 < x < \pi$ and positive for $\pi < x < 2\pi$. There is therefore a minimum of $f(x)$ at π.

Over the whole domain there are maxima at 0, $\pm 2\pi$, $\pm 4\pi$, ...; the maximum value is $4+1 = 5$. There are minima at $\pm\pi$, $\pm 3\pi$, $\pm 5\pi$, ..., with minimum value $-4+1 = -3$. If you have access to a graphic calculator, use it to check these results.

Notice that, although it is periodic, the graph is not a simple transformation of a sine graph.

Exercise 6A

1   Use the inequalities $\sin\theta < \theta < \tan\theta$ for a suitable value of θ to show that π lies between 3 and $2\sqrt{3}$.

2   Differentiate the following with respect to x.

   (a) $-\sin x$ (b) $-\cos x$ (c) $\sin 4x$ (d) $2\cos 3x$

   (e) $\sin\frac{1}{2}\pi x$ (f) $\cos 3\pi x$ (g) $\cos(2x-1)$ (h) $5\sin\left(3x+\frac{1}{4}\pi\right)$

   (i) $\cos\left(\frac{1}{2}\pi-5x\right)$ (j) $-\sin\left(\frac{1}{4}\pi-2x\right)$ (k) $-\cos\left(\frac{1}{2}\pi+2x\right)$ (l) $\sin\left(\frac{1}{2}\pi(1+2x)\right)$

3   Differentiate the following with respect to x.

   (a) $\sin^2 x$ (b) $\cos^2 x$ (c) $\cos^3 x$ (d) $5\sin^2\frac{1}{2}x$

   (e) $\cos^4 2x$ (f) $\sin x^2$ (g) $7\cos 2x^3$ (h) $\sin^2\left(\frac{1}{2}x-\frac{1}{3}\pi\right)$

   (i) $\cos^3 2\pi x$ (j) $\sin^3 x^2$ (k) $\sin^2 x^2 + \cos^2 x^2$ (l) $\cos^2\frac{1}{2}x$

4   Show that $\dfrac{d}{dx}\operatorname{cosec} x = -\operatorname{cosec} x\cot x$. Use this result, together with $\dfrac{d}{dx}\sec x = \sec x\tan x$, to differentiate the following with respect to x.

   (a) $\sec 2x$ (b) $\operatorname{cosec} 3x$ (c) $\operatorname{cosec}\left(3x+\frac{1}{5}\pi\right)$ (d) $\sec\left(x-\frac{1}{3}\pi\right)$

   (e) $4\sec^2 x$ (f) $\operatorname{cosec}^3 x$ (g) $\operatorname{cosec}^4 3x$ (h) $\sec^2\left(5x-\frac{1}{4}\pi\right)$

5   Show that $\dfrac{d}{dx}\ln\operatorname{cosec} x = -\cot x$, $\dfrac{d}{dx}\ln\cos x = -\tan x$. Use these and other similar results to differentiate the following with respect to x.

   (a) $\ln\sin 2x$ (b) $\ln\cos 3x$ (c) $\ln\operatorname{cosec}(x-\pi)$

   (d) $\ln\sec 4x$ (e) $\ln\sin^2 x$ (f) $\ln\cos^3 2x$

6   Differentiate the following with respect to x.

   (a) $e^{\sin x}$ (b) $e^{\cos 3x}$ (c) $5e^{\sin^2 x}$

7   Show that the inequality $\sin\theta < \theta$ holds for all values of θ greater than 0. By writing $\cos\theta$ as $\cos 2\left(\frac{1}{2}\theta\right)$, prove that $\cos\theta > 1-\frac{1}{2}\theta^2$ for all values of θ except 0. Sketch graphs illustrating the inequalities $1-\frac{1}{2}\theta^2 < \cos\theta < 1$.

8   (a) Find the equation of the tangent where $x = \frac{1}{3}\pi$ on the curve $y = \sin x$.

   (b) Find the equation of the normal where $x = \frac{1}{4}\pi$ on the curve $y = \cos 3x$.

   (c) Find the equation of the normal where $x = \frac{1}{4}\pi$ on the curve $y = \sec x$.

   (d) Find the equation of the tangent where $x = \frac{1}{4}\pi$ on the curve $y = \ln\sec x$.

   (e) Find the equation of the tangent where $x = \frac{1}{2}\pi$ on the curve $y = 3\sin^2 2x$.

9   Find any stationary points in the interval $0 \leqslant x < 2\pi$ on each of the following curves, and find out whether they are maxima, minima or neither.

   (a) $y = \sin x + \cos x$ (b) $y = x + \sin x$ (c) $y = \sin^2 x + 2\cos x$

   (d) $y = \cos 2x + x$ (e) $y = \sec x + \operatorname{cosec} x$ (f) $y = \cos 2x - 2\sin x$

10 Find $\dfrac{d}{dx}\sin(a+x)$, first by using the chain rule, and secondly by using the addition formula to expand $\sin(a+x)$ before differentiating. Verify that you get the same answer by both methods.

11 Find $\dfrac{d}{dx}\cos\left(\frac{3}{2}\pi-x\right)$. Check your answer by simplifying $\cos\left(\frac{3}{2}\pi-x\right)$ before you differentiate, and $\sin\left(\frac{3}{2}\pi-x\right)$ after you differentiate.

12 Show that $\dfrac{d}{dx}\left(2\cos^2 x\right)$, $\dfrac{d}{dx}\left(-2\sin^2 x\right)$ and $\dfrac{d}{dx}\cos 2x$ are all the same. Explain why.

13 Find $\dfrac{d}{dx}\sin^2\left(x+\frac{1}{4}\pi\right)$, and write your answer in its simplest form.

14 Find whether the tangent to $y=\cos x$ at $x=\frac{5}{6}\pi$ cuts the y-axis above or below the origin.

15 Sketch the graphs with the following equations, and find expressions for $\dfrac{dy}{dx}$.

    (a) $y=\sin\sqrt{x}$ (b) $y=\sqrt{\cos x}$ (c) $y=\sin\dfrac{1}{x}$

16 Show that, if $y=\sin nx$, where n is constant, then $\dfrac{d^2 y}{dx^2}=-n^2 y$. What can you deduce about the shape of the graph of $y=\sin nx$? Give a more general equation which has the same property.

17 Show that $\dfrac{d}{dx}\dfrac{1}{\cos x}$ can be written as $\dfrac{1}{\operatorname{cosec}x-\sin x}$. Hence find an expression for $\dfrac{d^2}{dx^2}\sec x$, and write your answer in as simple a form as possible.

18 By writing $\tan x$ as $\dfrac{\sin x}{\cos x}$, show that $\dfrac{d}{dx}\ln\tan x=2\operatorname{cosec}2x$.

19 The gross national product (GNP) of a country, P billion dollars, is given by the formula $P=1+0.02t+0.05\sin 0.6t$, where t is the time in years after the year 2000. At what rate is the GNP changing

    (a) in the year 2000,   (b) in the year 2005?

20 A tuning fork sounding A above middle C oscillates 440 times a second. The displacement of the tip of the tuning fork is given by $0.02\cos(2\pi\times 440t)$ millimetres, where t is the time in seconds after it is activated. Find

    (a) the greatest speed,   (b) the greatest acceleration of the tip as it oscillates.

    (For the calculation of speed and acceleration, see M1 Chapter 11.)

6.4   Integrating trigonometric functions

The results $\dfrac{d}{dx}\sin x = \cos x$ and $\dfrac{d}{dx}\cos x = -\sin x$ give you two indefinite integrals:

$$\int \cos x \, dx = \sin x + k, \quad \int \sin x \, dx = -\cos x + k,$$

where x is in radians.

Be very careful to get the signs correct when you differentiate or integrate sines and cosines. The minus sign appears when you differentiate $\cos x$, and when you integrate $\sin x$. If you forget which way round the signs go, draw for yourself sketches of the $\sin x$ and $\cos x$ graphs from 0 to $\frac{1}{2}\pi$. You can easily see that it is the $\cos x$ graph which has the negative gradient.

Example 6.4.1

Find the area under the graph of $y = \sin\left(2x + \frac{1}{3}\pi\right)$ from $x = 0$ as far as the first point at which the graph cuts the positive x-axis.

$\sin\left(2x + \frac{1}{3}\pi\right) = 0$ when $2x + \frac{1}{3}\pi = 0, \pi, 2\pi, \dots$. The first positive root is $x = \frac{1}{3}\pi$.

$$\int_0^{\frac{1}{3}\pi} \sin\left(2x + \tfrac{1}{3}\pi\right) dx = \left[\tfrac{1}{2} \times \left(-\cos\left(2x + \tfrac{1}{3}\pi\right)\right)\right]_0^{\frac{1}{3}\pi}$$

$$= \tfrac{1}{2} \times \left(-\cos\pi - \left(-\cos\tfrac{1}{3}\pi\right)\right)$$

$$= \tfrac{1}{2} \times \left(1 + \tfrac{1}{2}\right) = \tfrac{3}{4}.$$

So the area is $\frac{3}{4}$.

You can adapt the addition and double angle formulae found in Chapter 5 to integrate more complicated trigonometric functions. The most useful results are:

$$2\sin A \cos A \equiv \sin 2A.$$

$$2\cos^2 A \equiv 1 + \cos 2A, \quad 2\sin^2 A \equiv 1 - \cos 2A.$$

Other results which are sometimes useful are

$$2\sin A \cos B \equiv \sin(A + B) + \sin(A - B),$$

$$2\cos A \cos B \equiv \cos(A - B) + \cos(A + B),$$

$$2\sin A \sin B \equiv \cos(A - B) - \cos(A + B).$$

It is easy to prove all of these formulae by starting on the right side and using the formulae in the boxes in Sections 5.3 and 5.5.

Example 6.4.2

Let R be the region under the graph of $y = \sin^2 x$ over the interval $0 \leqslant x \leqslant \pi$. Find
(a) the area of R, (b) the volume of revolution formed by rotating R about the x-axis.

(a) The area is given by

$$\int_0^\pi \sin^2 x \, dx = \int_0^\pi \tfrac{1}{2}(1 - \cos 2x) \, dx = \left[\tfrac{1}{2}\left(x - \tfrac{1}{2}\sin 2x\right)\right]_0^\pi$$

$$= \left(\tfrac{1}{2}\pi - 0\right) - (0 - 0) = \tfrac{1}{2}\pi.$$

(b) The volume of revolution is given by $\displaystyle\int_0^\pi \pi\left(\sin^2 x\right)^2 dx.$

Now $\left(\sin^2 x\right)^2 \equiv \left(\tfrac{1}{2}(1 - \cos 2x)\right)^2 \equiv \tfrac{1}{4}\left(1 - 2\cos 2x + \cos^2 2x\right)$

$$= \tfrac{1}{4}\left(1 - 2\cos 2x + \tfrac{1}{2}(1 + \cos 4x)\right) \qquad \text{using } 2\cos^2 A \equiv 1 + \cos 2A$$
$$\text{with } 2x \text{ instead of } A$$
$$\equiv \tfrac{3}{8} - \tfrac{1}{2}\cos 2x + \tfrac{1}{8}\cos 4x.$$

So $\displaystyle\int_0^\pi \pi\left(\sin^2 x\right)^2 dx = \left[\pi\left(\tfrac{3}{8}x - \tfrac{1}{4}\sin 2x + \tfrac{1}{32}\sin 4x\right)\right]_0^\pi = \tfrac{3}{8}\pi^2.$

The area is $\tfrac{1}{2}\pi$ and the volume is $\tfrac{3}{8}\pi^2$.

Example 6.4.3*

Find (a) $\displaystyle\int \sin 2x \cos 3x \, dx,$ (b) $\displaystyle\int \cos^3 x \, dx.$

(a) Writing $A = 2x$ and $B = 3x$ in the formula for $2\sin A\cos B$,

$$2\sin 2x \cos 3x \equiv \sin(2x - 3x) + \sin(2x + 3x) \equiv \sin(-x) + \sin 5x$$
$$\equiv -\sin x + \sin 5x.$$

So $\displaystyle\int \sin 2x \cos 3x \, dx = \tfrac{1}{2}\left(\cos x - \tfrac{1}{5}\cos 5x\right) + k = \tfrac{1}{2}\cos x - \tfrac{1}{10}\cos 5x + k.$

(b) None of the formulae given above can be used directly, but $\cos^3 x$ can be written in other forms which can be integrated.

Method 1 $\cos^3 x \equiv \cos^2 x \cos x \equiv \tfrac{1}{2}(1 + \cos 2x)\cos x$

$$\equiv \tfrac{1}{2}\cos x + \tfrac{1}{2}\cos 2x \cos x$$
$$\equiv \tfrac{1}{2}\cos x + \tfrac{1}{4}(\cos x + \cos 3x) \qquad \text{using } 2\cos A \cos B$$
$$\equiv \cos(A - B) + \cos(A + B)$$
$$\equiv \tfrac{3}{4}\cos x + \tfrac{1}{4}\cos 3x.$$

Therefore $\displaystyle\int \cos^3 x \, dx = \tfrac{3}{4}\sin x + \tfrac{1}{12}\sin 3x + k.$

Method 2 $\cos^3 x \equiv \cos^2 x \cos x \equiv \left(1 - \sin^2 x\right) \cos x$

$$\equiv \cos x - \sin^2 x \cos x.$$

You can integrate the first term directly. To see how to integrate $\sin^2 x \cos x$, look back to Example 6.3.1(b). When $\cos^4 x$ was differentiated using the chain rule, a factor $\dfrac{d}{dx} \cos x = -\sin x$ appeared in the answer. In a similar way,

$$\frac{d}{dx} \sin^3 x = 3\sin^2 x \times \cos x = 3\sin^2 x \cos x.$$

Therefore $\displaystyle\int \cos^3 x \, dx = \int \left(\cos x - \sin^2 x \cos x\right) dx$

$$= \sin x - \tfrac{1}{3} \sin^3 x + k.$$

In (b) it is not obvious that the two methods have given the same answer, but if you work out some values, or use a calculator to draw the graphs, you will find that they are in agreement. The reason for this is that

$$\sin 3x \equiv \sin(2x + x) \equiv \sin 2x \cos x + \cos 2x \sin x$$

$$\equiv (2 \sin x \cos x) \cos x + \left(1 - 2\sin^2 x\right) \sin x$$

$$\equiv 2 \sin x\left(1 - \sin^2 x\right) + \left(1 - 2\sin^2 x\right) \sin x \equiv 3\sin x - 4\sin^3 x.$$

Therefore, $\tfrac{3}{4} \sin x + \tfrac{1}{12} \sin 3x \equiv \tfrac{3}{4} \sin x + \tfrac{1}{4} \sin x - \tfrac{1}{3} \sin^3 x \equiv \sin x - \tfrac{1}{3} \sin^3 x.$

Exercise 6B

1   Integrate the following with respect to x.

   (a) $\cos 2x$

   (b) $\sin 3x$

   (c) $\cos(2x + 1)$

   (d) $\sin(3x - 1)$

   (e) $\sin(1 - x)$

   (f) $\cos\left(4 - \tfrac{1}{2}x\right)$

   (g) $\sin\left(\tfrac{1}{2}x + \tfrac{1}{3}\pi\right)$

   (h) $\cos\left(3x - \tfrac{1}{4}\pi\right)$

   (i) $-\sin\tfrac{1}{2}x$

2   Evaluate the following.

   (a) $\displaystyle\int_0^{\frac{1}{2}\pi} \sin x \, dx$

   (b) $\displaystyle\int_0^{\frac{1}{4}\pi} \cos x \, dx$

   (c) $\displaystyle\int_0^{\frac{1}{4}\pi} \sin 2x \, dx$

   (d) $\displaystyle\int_{\frac{1}{4}\pi}^{\frac{1}{3}\pi} \cos 3x \, dx$

   (e) $\displaystyle\int_{\frac{1}{6}\pi}^{\frac{1}{3}\pi} \sin\left(3x + \tfrac{1}{6}\pi\right) dx$

   (f) $\displaystyle\int_0^{\frac{1}{2}\pi} \sin\left(\tfrac{1}{4}\pi - x\right) dx$

   (g) $\displaystyle\int_0^1 \cos(1 - x) \, dx$

   (h) $\displaystyle\int_0^{\frac{1}{2}} \sin\left(\tfrac{1}{2}x + 1\right) dx$

   (i) $\displaystyle\int_0^{2\pi} \sin\tfrac{1}{2}x \, dx$

3   Integrate the following with respect to x.

   (a) $\tan 2x$

   (b) $\cot 5x$

   (c) $\sec 3x \tan 3x$

   (d) $\operatorname{cosec} 4x \cot 4x$

   (e) $\tan\left(\tfrac{1}{4}\pi - x\right)$

   (f) $\cot\left(\tfrac{1}{3}\pi - 2x\right)$

   (g) $\sec\left(\tfrac{1}{2}x + 1\right)\tan\left(\tfrac{1}{2}x + 1\right)$

   (h) $\operatorname{cosec}(1 - 2x)\cot(1 - 2x)$

   (i) $\dfrac{\sin 2x}{\cos^2 2x}$

4 Evaluate the following.

(a) $\displaystyle\int_0^{\frac{1}{4}\pi} \tan x\,dx$

(b) $\displaystyle\int_0^{\frac{1}{12}\pi} \tan 3x\,dx$

(c) $\displaystyle\int_{\frac{1}{6}\pi}^{\frac{1}{4}\pi} \operatorname{cosec} 3x \cot 3x\,dx$

(d) $\displaystyle\int_{\frac{1}{3}\pi}^{\frac{1}{2}\pi} \cot \tfrac{1}{2}x\,dx$

(e) $\displaystyle\int_{\frac{1}{4}}^{\frac{1}{2}} \operatorname{cosec} 2x \cot 2x\,dx$

(f) $\displaystyle\int_{0.1}^{0.3} \sec \tfrac{1}{4}x \tan \tfrac{1}{4}x\,dx$

5 Integrate the following with respect to x.

(a) $\cos^2 x$

(b) $\cos^2 \tfrac{1}{2}x$

(c) $\sin^2 2x$

(d) $\sin^3 x \cos x$

(e) $\sec^3 x \tan x$ (write as $\sec^2 x(\sec x \tan x)$)

(f) $\operatorname{cosec}^5 2x \cot 2x$

(g) $\sin 3x \cos 4x$

(h) $\sin^3 x$ (write as $\left(1-\cos^2 x\right)\sin x$)

(i) $\sin^5 x$ (write as $\left(1-\cos^2 x\right)^2 \sin x$)

(j) $\sin 2x \sin 6x$

6 (a) Find $\dfrac{d}{dx}\sec^2 x$. Use $\sec^2 x \equiv 1 + \tan^2 x$ to show that $\dfrac{d}{dx}\tan^2 x = 2\tan x \sec^2 x$.

(b) Explain why $\dfrac{d}{dx}\tan^2 x$ is equal to $2\tan x \times \dfrac{d}{dx}\tan x$. Deduce $\dfrac{d}{dx}\tan x$.

(c) Write down $\displaystyle\int \sec^2 x\,dx$, and hence find $\displaystyle\int \tan^2 x\,dx$.

(d) Use a similar method to find $\displaystyle\int \operatorname{cosec}^2 x\,dx$ and $\displaystyle\int \cot^2 x\,dx$.

7 Find the area of the region between the curve $y = \cos x$ and the x-axis from $x = 0$ to $x = \tfrac{1}{2}\pi$.

Find also the volume generated when this area is rotated about the x-axis.

8 Find the area of the region bounded by the curve $y = 1 + \sin x$, the x-axis and the lines $x = 0$ and $x = \pi$.

Find also the volume generated when this area is rotated about the x-axis.

9 The curves $y = \sin x$, $y = \cos x$ and the x-axis enclose a region shown shaded in the sketch.

(a) Find the area of the shaded region.

(b) Find the volume generated when this region is rotated about the x-axis.

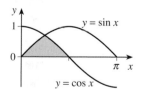

10 In the interval $0 \leqslant x \leqslant \pi$ the curve $y = \sin x + \cos x$ meets the y-axis at P and the x-axis at Q. Find the coordinates of P and Q.

Calculate the area of the region enclosed between the curve and the axes bounded by P and Q.

Calculate also the volume generated when this area is rotated about the x-axis.

Miscellaneous exercise 6

1 (a) Differentiate $\ln \sin 2x$ with respect to x, simplifying your answer.

(b) Find $\displaystyle\int \sin\frac{3}{2}x \cos\frac{1}{2}x \, dx$.

(c) Given that $y = \cos^2 x$, find $\dfrac{dy}{dx}$.

(d) Differentiate $\sin(t^3 + 4)$ with respect to t.

(e) Find $\displaystyle\int \cos^2 3x \, dx$.

(f) Differentiate $\cos\sqrt{x}$ with respect to x.

(g) Find $\displaystyle\int \sin^2 \frac{1}{3}x \, dx$.

2 (a) Express $\sin^2 x$ in terms of $\cos 2x$.

(b) The region R is bounded by the part of the curve $y = \sin x$ between $x = 0$ and $x = \pi$ and the x-axis. Show that the volume of the solid formed when R is rotated completely about the x-axis is $\frac{1}{2}\pi^2$. (OCR)

3 (a) Use the addition formulae to find expressions involving surds for $\sin\frac{1}{12}\pi$ and $\tan\frac{1}{12}\pi$.

(b) Use the fact that $\left(\sqrt{3}+1\right)\left(\sqrt{3}-1\right) = 2$ to show that $\tan\frac{1}{12}\pi = 2 - \sqrt{3}$.

(c) Show that π lies between $3\sqrt{2}\left(\sqrt{3}-1\right)$ and $12\left(2 - \sqrt{3}\right)$. Use a calculator to evaluate these expressions correct to 3 decimal places.

4 Show that, if $0 < x < \frac{1}{2}\pi$, $\tan x = +\sqrt{\sec^2 x - 1}$. Use the chain rule to find $\dfrac{d}{dx}\sqrt{\sec^2 x - 1}$, and hence find $\dfrac{d}{dx}\tan x$ for $0 < x < \frac{1}{2}\pi$ in as simple a form as possible.

Use a similar method to find $\dfrac{d}{dx}\tan x$ for $\frac{1}{2}\pi < x < \pi$.

(You will meet a much simpler way of finding $\dfrac{d}{dx}\tan x$ in Section 7.2.)

5 P, Q and R are the points on the graph of $y = \cos x$ for which $x = 0$, $x = \frac{1}{4}\pi$ and $x = \frac{1}{2}\pi$ respectively. Find the point S where the normal at Q meets the y-axis. Compare the distances SP, SQ and SR. Use your answers to draw a sketch showing how the curve $y = \cos x$ over the interval $-\frac{1}{2}\pi < x < \frac{1}{2}\pi$ is related to the circle with centre S and radius SQ.

6 By writing $\cos\theta$ as $\cos 2\left(\frac{1}{2}\theta\right)$, and using the approximation $\sin\theta \approx \theta$ when θ is small, show that $\cos\theta \approx 1 - \frac{1}{2}\theta^2$ when θ is small.

Since sine is an odd function, it is suggested that a better approximation for sine might have the form $\sin\theta \approx \theta - k\theta^3$ when θ is small. By writing $\sin\theta$ as $\sin 2\left(\frac{1}{2}\theta\right)$, using the approximation $\cos\theta \approx 1 - \frac{1}{2}\theta^2$ and equating the coefficients of θ^3, find an appropriate numerical value for k.

Investigate whether this approximation is in fact better, by evaluating θ and $\theta - k\theta^3$ numerically when $\theta = \frac{1}{6}\pi$.

7 The motion of an electric train on the straight stretch of track between two stations is given
by $x = 11\left(t - \dfrac{45}{\pi}\sin\left(\dfrac{\pi}{45}t\right)\right)$, where x metres is the distance covered t seconds after
leaving the first station. The train stops at these two stations and nowhere between them.

(a) Find the velocity, v m s^{-1} in terms of t. Hence find the time taken for the journey
between the two stations.

(b) Calculate the distance between the two stations. Hence find the average velocity of the
train.

(c) Find the acceleration of the train 30 seconds after leaving the first station.   (OCR)

(For the calculation of velocity and acceleration, see M1 Chapter 11.)

8 A mobile consists of a bird with flapping
wings suspended from the ceiling by two
elastic strings. A small weight A hangs
below it. A is pulled down and then
released. After t seconds, the distance,
y cm, of A below its equilibrium position
is modelled by the periodic function
$y = 5\cos 2t + 10\sin t$.

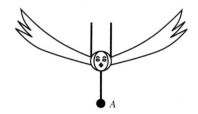

(a) Verify that the (t, y) graph has a stationary point where $t = \tfrac{1}{6}\pi$.

(b) Show that all the stationary points of the graph correspond to solutions of the equation
$\cos t(2\sin t - 1) = 0$. Find the other two solutions in the interval $0 \le t \le \pi$.

(c) State one limitation of the model. Explain why $y = e^{-kt}(5\cos 2t + 10\sin t)$, where k is
a small constant, might give a better model.   (OCR)

9 (a) By first expressing $\cos 4x$ in terms of $\cos 2x$, show that
$$\cos 4x = 8\cos^4 x - 8\cos^2 x + 1,$$
and hence show that
$$8\cos^4 x = \cos 4x + 4\cos 2x + 3.$$

(b) The region R, shown shaded in the diagram,
is bounded by the part of the curve $y = \cos^2 x$
between $x = 0$ and $x = \tfrac{1}{2}\pi$ and by the x- and
y-axes. Show that the volume of the solid
formed when R is rotated completely about the
x-axis is $\tfrac{3}{16}\pi^2$.   (OCR)

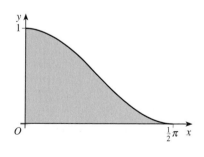

10 In this question $f(x) = \sin\tfrac{1}{2}x + \cos\tfrac{1}{3}x$.

(a) Find $f'(x)$.

(b) Find the values of $f(0)$ and $f'(0)$.

(c) State the periods of $\sin\tfrac{1}{2}x$ and $\cos\tfrac{1}{3}x$.

(d) Write down another value of x (not 0) for which $f(x) = f(0)$ and $f'(x) = f'(0)$.   (OCR)

11   The diagram shows a sketch, not to scale, of part of
the graph of $y = f(x)$, where $f(x) = \sin x + \sin 2x$
and where x is measured in radians.

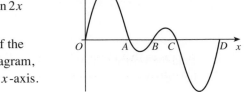

(a) Find, in terms of π, the x-coordinates of the
points A, B, C and D, shown in the diagram,
where the graph of f meets the positive x-axis.

(b) Show that $f(\pi - \theta)$ may be expressed as
$\sin\theta - \sin 2\theta$, and show also that $f(\pi - \theta) + f(\pi + \theta) = 0$ for all values of θ.

(c) Differentiate $f(x)$, and hence show that the greatest value of $f(x)$, for $0 \leqslant x \leqslant 2\pi$,
occurs when

$$\cos x = \frac{-1 + \sqrt{33}}{8}.$$ (OCR)

Revision exercise 1

1 You are given that the equation $f(x) = 0$ has a solution at $x = 3$. Using this information, write down as many solutions as you can to each of the following equations.

(a) $2f(x+5) = 0$

(b) $f(3x) = 0$

(c) $|f(x)| = 0$

(d) $f(|x|) = 0$ (OCR)

2 Use the addition formulae to find an expression for $\cos^2(A+B) + \sin^2(A+B)$. Verify that your expression reduces to 1.

Use a similar method to find an expression for $\cos^2(A+B) - \sin^2(A+B)$. Verify that this reduces to $\cos(2A+2B)$.

3 Differentiate each of the following functions with respect to x.

(a) e^{3x-1}

(b) $\ln(x^2 - 1)$

4 Solve the following equations and inequalities.

(a) $|x-9| = 16$

(b) $|x^2 - 9| = 16$

(c) $|x-9| \leqslant 16$

(d) $|x^2 - 9| \leqslant 16$.

5 The region R is bounded by the x-axis, the y-axis, part of the curve with equation $y = e^{2x}$ and part of the straight line with equation $x = 3$.

Calculate, giving your answers in exact form,

(a) the area of R,

(b) the volume of the solid of revolution generated when R is rotated through four right angles about the x-axis.

6 Express $6\cos x - \sin x$ in the form $R\cos(x+\alpha)$, where $R > 0$ and $0 < \alpha < \frac{1}{2}\pi$.

Hence, or otherwise, solve the equation $6\cos x - \sin x = 5$ for x in the interval $-\frac{1}{2}\pi < x < \frac{1}{2}\pi$, giving your answer in radians, correct to 3 decimal places.   (OCR)

7 The angle made by a wasp's wings with the horizontal is given by the equation $\theta = 0.4\sin 600t$ radians, where t is the time in seconds. How many times a second do its wings oscillate? Find an expression for $\dfrac{d\theta}{dt}$, the angular velocity, in radians per second.

What is the value of θ when the angular velocity has

(a) its greatest magnitude,

(b) its smallest magnitude?

8 Find the following integrals.

(a) $\displaystyle\int \sin\left(2x + \tfrac{1}{6}\pi\right) dx$

(b) $\displaystyle\int \sin^2 3x \, dx$

(c) $\displaystyle\int \sin^2 2x \cos 2x \, dx$

9 A polynomial $P(x)$ is the product of $\left(x^3 + ax^2 - x - 2\right)$ and $x - b$.

(a) The coefficients of x^3 and x in $P(x)$ are zero. Find the values of a and b.

(b) Hence factorise $P(x)$ completely and find the roots of the equation $P(x) = 0$.   (OCR)

10 Find the factors of the polynomial $x^3 - x^2 - 14x + 24$.

11 The number of bacteria in a culture increases exponentially with time. When observation started there were 1000 bacteria, and five hours later there were 10 000 bacteria. Find, correct to 3 significant figures,

   (a) when there were 5000 bacteria,

   (b) when the number of bacteria would exceed one million,

   (c) how may bacteria there would be 12 hours after the first observation.

12 (a) Let $y = e^{3x^2 - 6x}$. Find $\dfrac{dy}{dx}$.

   (b) Find the coordinates of the stationary point on the curve $y = e^{3x^2 - 6x}$, and decide whether it is a maximum or a minimum.

   (c) Find the equation of the normal to the curve $y = e^{3x^2 - 6x}$ at the point where $x = 2$.

13 (a) Find the gradient m of the line segment joining the points A and B with x-coordinates 0 and 1 respectively on the graph of $y = e^x$.

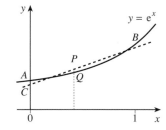

   With this value of m, the line $y = mx + c$ is drawn to meet the y-axis at C and intersect the graph of $y = e^x$ twice between A and B, as shown in the diagram. A line parallel to the y-axis between the points of intersection of the straight line and the curve meets the straight line at P and the curve at Q.

   (b) Find the value of c which makes the maximum value of $|PQ|$ equal to the value of the distance $|AC|$.

14 In the figure A, B, C are the points on the graph of $y = \sin x$ for which $x = \alpha - \tfrac{1}{3}\pi$, α, $\alpha + \tfrac{1}{3}\pi$ respectively. D is the point $(\alpha, 0)$.

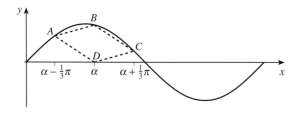

   (a) Sketch separate diagrams showing $ABCD$ in the special cases where $\alpha = \tfrac{1}{3}\pi$, $\alpha = \tfrac{1}{2}\pi$, $\alpha = \tfrac{2}{3}\pi$.

   (b) Use addition formulae to simplify $\sin\!\left(\alpha - \tfrac{1}{3}\pi\right) + \sin\!\left(\alpha + \tfrac{1}{3}\pi\right)$.

   (c) Write down the coordinates of the mid-point of AC.

   (d) Show that $ABCD$ is a parallelogram.   (OCR)

15 Solve the equation $3\cos 2x + 4\sin 2x = 2$, for values of x between 0 and 2π, giving your answers correct to 2 decimal places.

7 Differentiating products

You now know how to differentiate powers of x, and exponential, logarithmic and trigonometric functions. This chapter shows how to differentiate some more complicated functions made up from these four types. When you have completed it, you should

- know and be able to apply the product and quotient rules for differentiation.

7.1 The sum and product rules

If $f(x) = x^2 + \sin x$, then $f'(x) = 2x + \cos x$. You know this because it was proved (in P1 Section 6.6) that $\dfrac{d}{dx} x^2 = 2x$ and (in Chapter 6) that $\dfrac{d}{dx} \sin x = \cos x$. But the statement also depends on another property of differentiation:

Sum rule   If u and v are functions of x, and

$$\text{if } y = u + v, \text{ then } \frac{dy}{dx} = \frac{du}{dx} + \frac{dv}{dx}.$$

This rule was justified in P1 Section 6.4 by means of an example. For a general proof, it is convenient to use 'delta notation'.

Take a particular value of x, and increase x by δx. There will then be corresponding increases in u, v and y of δu, δv and δy:

$$y = u + v \quad \text{and} \quad y + \delta y = (u + \delta u) + (v + \delta v).$$

Subtracting the first equation from the second gives

$$\delta y = \delta u + \delta v; \text{ and, dividing by } \delta x, \; \frac{\delta y}{\delta x} = \frac{\delta u}{\delta x} + \frac{\delta v}{\delta x}.$$

To find $\dfrac{dy}{dx}$, you must take the limit as $\delta x \to 0$:

$$\frac{dy}{dx} = \lim_{\delta x \to 0} \frac{\delta y}{\delta x} = \lim_{\delta x \to 0} \left(\frac{\delta u}{\delta x} + \frac{\delta v}{\delta x} \right) = \lim_{\delta x \to 0} \frac{\delta u}{\delta x} + \lim_{\delta x \to 0} \frac{\delta v}{\delta x} = \frac{du}{dx} + \frac{dv}{dx}, \text{ as required.}$$

You might (rightly) object that the crucial assumption of the proof, that the limit of the sum of two terms is the sum of the limits, has never been justified. This can only be an assumption at this stage, because you don't yet have a mathematical definition of what is meant by a limit. But it *can* be justified, and for the time being you may quote the result with confidence. You may also assume the corresponding result for the limit of the product of two terms: this has already been assumed in earlier chapters, when proving the chain rule and the derivatives of $\sin x$ and $\cos x$.

The rule for differentiating the product of two functions is rather more complicated than the rule for sums.

Example 7.1.1

Show that, if $y = uv$, then in general $\dfrac{dy}{dx}$ does *not* equal $\dfrac{du}{dx} \times \dfrac{dv}{dx}$.

The words 'in general' are put in because there might be special functions for which equality does hold. For example, if u and v are both constant functions, then y is also constant: $\dfrac{du}{dx}$, $\dfrac{dv}{dx}$ and $\dfrac{dy}{dx}$ are all 0, so $\dfrac{dy}{dx}$ does equal $\dfrac{du}{dx} \times \dfrac{dv}{dx}$.

To show that this is not always true, it is sufficient to find a counterexample. For example, if $u = x^2$ and $v = x^3$, then $y = x^5$. In this case $\dfrac{dy}{dx} = 5x^4$, but

$\dfrac{du}{dx} \times \dfrac{dv}{dx} = 2x \times 3x^2 = 6x^3$. These two expressions are not the same.

To find the correct rule, use the same notation as for the sum rule, but with $y = uv$. Then

$$y = uv \quad \text{and} \quad y + \delta y = (u + \delta u)(v + \delta v) = uv + (\delta u)v + u(\delta v) + (\delta u)(\delta v).$$

Subtracting the first equation from the second gives

$$\delta y = (\delta u)v + u(\delta v) + (\delta u)(\delta v); \quad \text{then} \quad \frac{\delta y}{\delta x} = \left(\frac{\delta u}{\delta x}\right)v + u\left(\frac{\delta v}{\delta x}\right) + \left(\frac{\delta u}{\delta x}\right)(\delta v).$$

Taking limits as $\delta x \to 0$, and making the assumptions about limits of sums and products,

$$\frac{dy}{dx} = \lim_{\delta x \to 0}\left(\left(\frac{\delta u}{\delta x}\right)v\right) + \lim_{\delta x \to 0}\left(u\left(\frac{\delta v}{\delta x}\right)\right) + \lim_{\delta x \to 0}\left(\frac{\delta u}{\delta x}\right)\lim_{\delta x \to 0}(\delta v).$$

The last term on the right is 0 because, as $\delta x \to 0$, $\delta v \to 0$.

Therefore:

> **Product rule**   If u and v are functions of x, and
>
> if $y = uv$, then $\dfrac{dy}{dx} = \dfrac{du}{dx}v + u\dfrac{dv}{dx}$.
>
> In function notation, if $y = f(x)g(x)$, then $\dfrac{dy}{dx} = f'(x)g(x) + f(x)g'(x)$.

Example 7.1.2

Verify the product rule when $u = x^2$ and $v = x^3$.

Using the equation in the box, the right side is $2x \times x^3 + x^2 \times 3x^2 = 2x^4 + 3x^4 = 5x^4$, which is the derivative of $y = x^2 \times x^3 = x^5$.

Example 7.1.3

Find the derivatives with respect to x of (a) $x^3 \sin x$, (b) xe^{3x}, (c) $\sin^5 x \cos^3 x$.

(a) $\dfrac{d}{dx}\left(x^3 \sin x\right) = 3x^2 \times \sin x + x^3 \times \cos x = 3x^2 \sin x + x^3 \cos x$.

(b) $\dfrac{d}{dx}\left(xe^{3x}\right) = 1 \times e^{3x} + x \times \left(e^{3x} \times 3\right) = e^{3x} + 3xe^{3x} = (1+3x)e^{3x}$.

(c) $\dfrac{d}{dx}\left(\sin^5 x \cos^3 x\right) = \left(5\sin^4 x \times \cos x\right)\cos^3 x + \sin^5 x\left(3\cos^2 x \times (-\sin x)\right)$

$$= 5\sin^4 x \cos^4 x - 3\sin^6 x \cos^2 x$$

$$= \sin^4 x \cos^2 x\left(5\cos^2 x - 3\sin^2 x\right).$$

Notice that in part (c) the chain rule is used to find $\dfrac{du}{dx}$ and $\dfrac{dv}{dx}$.

Example 7.1.4

Find the points on the graph of $y = x \sin x$ at which the tangent passes through the origin.

The product rule gives $\dfrac{dy}{dx} = \sin x + x \cos x$, so the tangent at a point P
$(p, p\sin p)$ has gradient $\sin p + p\cos p$. This has to equal the gradient of OP,
which is $\dfrac{p\sin p}{p} = \sin p$. Therefore

$$\sin p + p\cos p = \sin p,$$

giving $p\cos p = 0$.

This equation is satisfied by $p = 0$ and
all odd multiples of $\frac{1}{2}\pi$. These points
have coordinates $(0,0)$, $\left(\pm\frac{1}{2}\pi, \frac{1}{2}\pi\right)$,
$\left(\pm\frac{3}{2}\pi, -\frac{3}{2}\pi\right)$, $\left(\pm\frac{5}{2}\pi, \frac{5}{2}\pi\right), \dots$.

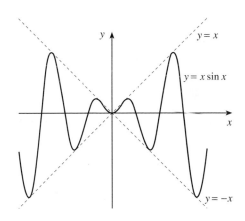

This is illustrated in Fig. 7.1. The graph
oscillates between the lines $y = x$ and $y = -x$,
touching $y = x$ when $\sin|x| = 1$,
and $y = -x$ when $\sin|x| = -1$.

Fig. 7.1

Exercise 7A

1 Differentiate the following functions with respect to x by using the product rule. Verify
your answers by multiplying out the products and then differentiating.

(a) $(x+1)(x-1)$ 　　　　(b) $x^2(x+2)$ 　　　　(c) $\left(x^3+4\right)\left(x^2+3\right)$

(d) $\left(3x^2+5x+2\right)(7x+5)$ 　　(e) $\left(x^2-2x+4\right)(x+2)$ 　　(f) $x^m x^n$

2 Differentiate the following with respect to x.

   (a) xe^x (b) $x^2 \ln x$ (c) $x^3(\sin x + 1)$

   (d) $\sin x \cos x$ (e) $x \cos x$ (f) $e^{-x} \sin x$

3 Find $\dfrac{dy}{dx}$ when

   (a) $y = (x^2 + 3)e^x$,   (b) $y = x^2(\sin x + \cos x)$,   (c) $y = x \sin^2 x$.

4 Find $f'(x)$ when

   (a) $f(x) = x^2(2 + e^x)$,   (b) $f(x) = x^3 e^{2x}$,   (c) $f(x) = (4 + 3x^2)\ln x$.

5 Find the value of the gradient of the following curves when $x = 2$. Give your answers in exact form.

   (a) $y = xe^{-2x}$ (b) $y = e^x \sin x$ (c) $y = x \ln 3x$

6 Find the equations of tangents to the following curves at the given points.

   (a) $y = x \sin x$ when $x = \pi$ (b) $y = x^3 \ln x$ when $x = 1$

   (c) $y = x\sqrt{3x + 1}$ when $x = 5$ (d) $y = x^3 e^{-2x}$ when $x = 0$

7 Find the coordinates of the turning points of the curve $y = x^2 e^{-x}$.

8 Differentiate the following with respect to x.

   (a) $x^2 \sin^3 2x$ (b) $e^x \sqrt{5x^2 + 2}$ (c) $\sin^4 2x \cos^3 5x$

   (d) $(4x + 1)^3 \ln 3x$ (e) $\sqrt{x} \ln 2x$ (f) $e^{ax} \cos(bx + \tfrac{1}{2}\pi)$

9 When $f(x) = x \sin(\tfrac{1}{2}x)$, find the exact value of $f'(4)$.

10 Find the equation of the normal to the curve $y = x \ln(2x - 1)$ at the point on the curve with x-coordinate 1.

11 Find the coordinates of the stationary point on the curve $y = (x^2 - 4)\sqrt{4x - 1}$, $x \geqslant \tfrac{1}{4}$.

12 The volume, V, of a solid is given by $V = x^2\sqrt{8 - x}$. Use calculus to find the maximum value of V and the value of x at which it occurs.

13 Find the x-coordinates of the stationary points on the curve $y = x^n e^{-x}$, where n is a positive integer. Determine the nature of these stationary points, distinguishing between the cases when n is odd and when n is even.

14 Use the product rule to establish the rule, $\dfrac{d}{dx}uvw = \dfrac{du}{dx}vw + u\dfrac{dv}{dx}w + uv\dfrac{dw}{dx}$, for differentiating a 'triple' product uvw. Use the new rule to find

   (a) $\dfrac{d}{dx}xe^x \sin x$,   (b) $\dfrac{d}{dx}x^2 e^{-3x}\cos 4x$.

7.2 Differentiating quotients

Functions of the form $\dfrac{u}{v}$ can often be written in a different form so that they can be differentiated by the product rule.

Example 7.2.1

Differentiate with respect to x (a) $f(x) = \dfrac{\sin x}{e^x}$,   (b) $g(x) = \dfrac{e^x}{\sin x}$.

(a) Since $\dfrac{1}{e^x} = e^{-x}$, you can write $f(x)$ as $e^{-x} \sin x$. Therefore

$$f'(x) = \frac{d}{dx}\left(e^{-x} \sin x\right) = \left(-e^{-x}\right)\sin x + e^{-x} \cos x = e^{-x}(\cos x - \sin x).$$

(b) **Method 1**   You can write $g(x)$ as $e^x \times \dfrac{1}{\sin x}$, so

$$g'(x) = \frac{d}{dx}e^x \times \frac{1}{\sin x} + e^x \times \frac{d}{dx}\left(\frac{1}{\sin x}\right) = e^x \times \frac{1}{\sin x} + e^x \times \left(\frac{-1}{\sin^2 x} \times \cos x\right),$$

using the product rule and then the chain rule.

You can simplify this to $g'(x) = e^x\left(\dfrac{\sin x - \cos x}{\sin^2 x}\right)$.

Method 2   Since $g(x) = \dfrac{1}{f(x)}$, the chain rule gives

$$g'(x) = \left(\frac{-1}{f(x)^2}\right) \times f'(x) = -\frac{f'(x)}{f(x)^2}.$$

Therefore, using the result of part (a),

$$g'(x) = -\frac{e^{-x}(\cos x - \sin x)}{\left(\dfrac{\sin x}{e^x}\right)^2} = e^{2x}e^{-x}\left(\frac{\sin x - \cos x}{\sin^2 x}\right)$$

$$= e^x\left(\frac{\sin x - \cos x}{\sin^2 x}\right).$$

However, it is often useful to have a separate formula for differentiating $\dfrac{u}{v}$. This can be found by applying the product rule to $u \times \dfrac{1}{v}$, using the chain rule to differentiate $\dfrac{1}{v}$. This gives

$$\frac{d}{dx}\left(\frac{u}{v}\right) = \frac{d}{dx}\left(u \times \frac{1}{v}\right) = \frac{du}{dx} \times \frac{1}{v} + u \times \left(-\frac{1}{v^2}\right)\frac{dv}{dx}$$

$$= \frac{du}{dx}\frac{1}{v} - \frac{u}{v^2}\frac{dv}{dx}.$$

This can be conveniently written as:

> **Quotient rule**   If u and v are functions of x, and
>
> $$\text{if } y = \frac{u}{v}, \text{ then } \frac{dy}{dx} = \frac{\dfrac{du}{dx}v - u\dfrac{dv}{dx}}{v^2}.$$
>
> In function notation, if $y = \dfrac{f(x)}{g(x)}$, then $\dfrac{dy}{dx} = \dfrac{f'(x)g(x) - f(x)g'(x)}{g(x)^2}$.

An important application of this rule is to differentiate $\tan x$. Since $\tan x = \dfrac{\sin x}{\cos x}$, the quotient rule with $u = \sin x$, $v = \cos x$ gives

$$\frac{d}{dx}\tan x = \frac{\cos x \times \cos x - \sin x \times (-\sin x)}{(\cos x)^2} = \frac{\cos^2 x + \sin^2 x}{\cos^2 x} = \frac{1}{\cos^2 x}.$$

Since $\dfrac{1}{\cos x}$ is $\sec x$, you can write this as:

> $$\frac{d}{dx}\tan x = \sec^2 x.$$

You will often need this result, so you should remember it. You also need to recognise its integral form:

> $$\int \sec^2 x\, dx = \tan x + k.$$

Example 7.2.2

Find $\displaystyle\int_0^{\frac{1}{4}\pi} \sec^2\left(2x - \tfrac{1}{3}\pi\right) dx$.

This integral is of the form $\displaystyle\int g(ax + b)\, dx$, where $g(x) = \sec^2 x$.

So, using the result in P1 Section 16.7, the integral is

$$\left[\tfrac{1}{2}\tan\left(2x - \tfrac{1}{3}\pi\right)\right]_0^{\frac{1}{4}\pi},$$

with value $\tfrac{1}{2}\tan\tfrac{1}{6}\pi - \tfrac{1}{2}\tan\left(-\tfrac{1}{3}\pi\right) = \tfrac{1}{2}\left(\tfrac{1}{3}\sqrt{3} + \sqrt{3}\right) = \tfrac{2}{3}\sqrt{3}.$

Example 7.2.3

Find the minimum and maximum values of $f(x) = \dfrac{x-1}{x^2+3}$.

The denominator is never zero, so $f(x)$ is defined for all real numbers.

The quotient rule with $u = x - 1$ and $v = x^2 + 3$ gives

$$f'(x) = \frac{1 \times \left(x^2 + 3\right) - (x-1) \times 2x}{\left(x^2 + 3\right)^2} = \frac{-x^2 + 2x + 3}{\left(x^2 + 3\right)^2}$$

$$= \frac{-\left(x^2 - 2x - 3\right)}{\left(x^2 + 3\right)^2} = \frac{-(x+1)(x-3)}{\left(x^2 + 3\right)^2}.$$

So $f'(x) = 0$ when $x = -1$ and $x = 3$.

You could use the quotient rule again to find $f''(x)$, but in this example it is much easier to note that $f'(x)$ is positive when $-1 < x < 3$ and negative when $x < -1$ and when $x > 3$. So there is a minimum at $x = -1$ and a maximum at $x = 3$.

The minimum value is $f(-1) = \frac{-2}{4} = -\frac{1}{2}$, and the maximum value is $\frac{2}{12} = \frac{1}{6}$.

Exercise 7B

1 Differentiate with respect to x

   (a) $\dfrac{x}{1+5x}$,

   (b) $\dfrac{x^2}{3x-2}$,

   (c) $\dfrac{x^2}{1+2x^2}$,

   (d) $\dfrac{e^{3x}}{4x-3}$,

   (e) $\dfrac{x}{1+x^3}$,

   (f) $\dfrac{e^x}{x^2+1}$.

2 By writing $\cot x = \dfrac{\cos x}{\sin x}$, differentiate $\cot x$ with respect to x.

3 Differentiate with respect to x

   (a) $\dfrac{\sin x}{x}$,

   (b) $\dfrac{x}{\sin x}$,

   (c) $\left(\dfrac{x}{\sin x}\right)^2$.

4 Differentiate with respect to x

   (a) $\dfrac{x}{\sqrt{x+1}}$,

   (b) $\dfrac{\sqrt{x-5}}{x}$,

   (c) $\dfrac{\sqrt{3x+2}}{2x}$.

5 Find $\dfrac{dy}{dx}$ when

   (a) $y = \dfrac{\cos x}{\sqrt{x}}$,

   (b) $y = \dfrac{e^x + 5x}{e^x - 2}$,

   (c) $y = \dfrac{\sqrt{1-x}}{\sqrt{1+x}}$.

6 Find $\dfrac{dy}{dx}$ when

(a) $y = \dfrac{\ln x}{x}$,

(b) $y = \dfrac{\ln\left(x^2 + 4\right)}{x}$,

(c) $y = \dfrac{\ln(3x + 2)}{2x - 1}$.

7 Find the equation of the tangent at the point with coordinates $(1,1)$ to the curve with equation $y = \dfrac{x^2 + 3}{x + 3}$. (OCR)

8 (a) If $f(x) = \dfrac{e^x}{2x + 1}$, find $f'(x)$.

(b) Find the coordinates of the turning point of the curve $y = f(x)$.

9 Find the equation of the normal to the curve $y = \dfrac{2x - 1}{x(x - 3)}$ at the point on the curve where $x = 2$.

10 Find the turning points of the curve $y = \dfrac{x^2 + 4}{2x - x^2}$.

11 (a) If $f(x) = \dfrac{x^2 - 3x}{x + 1}$, find $f'(x)$.

(b) Find the values of x for which $f(x)$ is decreasing.

12 Calculate

(a) $\displaystyle\int_{-\frac{1}{4}\pi}^{\frac{1}{4}\pi} \sec^2 x \, dx$,

(b) $\displaystyle\int_{0}^{\frac{1}{6}\pi} \sec^2\left(3x - \tfrac{1}{4}\pi\right) dx$.

Miscellaneous exercise 7

1 (a) Differentiate $x^3 \sin x$.

(b) Differentiate $\dfrac{x}{\sqrt{x + 3}}$ simplifying your answer as far as possible. (OCR)

2 Given that $y = xe^{-3x}$, find $\dfrac{dy}{dx}$.

Hence find the coordinates of the stationary point on the curve $y = xe^{-3x}$. (OCR)

3 A function f is defined by $f(x) = e^x \cos x$ $\qquad (0 \leqslant x \leqslant 2\pi)$.

(a) Find $f'(x)$.

(b) State the values of x between 0 and 2π for which $f'(x) < 0$.

(c) What does the fact that $f'(x) < 0$ in this interval tell you about the shape of the graph of $y = f(x)$? (OCR)

4 Find the gradient of the curve $y = \dfrac{\sin x}{x^2}$ at the point where $x = \pi$, leaving your answer in terms of π. (OCR)

5 Use appropriate rules of differentiation to find $\dfrac{dy}{dx}$ in each of the following cases.

   (a) $y = \sin 2x \cos 4x$ (b) $y = \dfrac{3x^2}{\ln x}$ for $x > 1$ (c) $y = \left(1 - \dfrac{x}{5}\right)^{10}$ (OCR)

6 Use differentiation to find the coordinates of the turning point on the curve whose equation
   is $y = \dfrac{4x + 2}{\sqrt{x}}$.   (OCR)

7 A curve C has equation $y = \dfrac{\sin x}{x}$, where $x > 0$.

   Find $\dfrac{dy}{dx}$, and hence show that the x-coordinate of any stationary point of C satisfies the
   equation $x = \tan x$.   (OCR)

8 A curve has equation $y = \dfrac{x}{\sqrt{2x^2 + 1}}$.

   (a)   Show that $\dfrac{dy}{dx} = \left(2x^2 + 1\right)^{-\frac{3}{2}}$.

   (b)   Hence show that the curve has no turning points.   (OCR)

9 The region R, shown shaded in the diagram, is
   bounded by the x- and y-axes, the line $x = \frac{1}{3}\pi$
   and the curve $y = \sec\frac{1}{2}x$. Show that the volume
   of the solid formed when R is rotated
   completely about the x-axis is $\dfrac{2\pi}{\sqrt{3}}$.

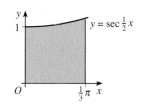

10 A length of channel of given depth d is
   to be made from a rectangular sheet of
   metal of width $2a$. The metal is to be
   bent in such a way that the cross-section
   $ABCD$ is as shown in the figure, with

   $AB + BC + CD = 2a$ and with AB and CD each inclined to the line BC at an angle θ.
   Show that $BC = 2(a - d\operatorname{cosec}\theta)$ and that the area of the cross-section $ABCD$ is
      $2ad + d^2(\cot\theta - 2\operatorname{cosec}\theta)$.
   Show that the maximum value of $2ad + d^2(\cot\theta - 2\operatorname{cosec}\theta)$, as θ varies, is $d(2a - d\sqrt{3})$.
   By considering the length of BC, show that the cross-sectional area can only be made
   equal to this maximum value if $2d \leqslant a\sqrt{3}$.   (OCR)

11 (a)   Find the value of x for which $x^2 e^{-ax}$ has its maximum value, where a is a positive
        constant. Denoting this by c, and the maximum value by M, deduce that

$$xe^{-ax} < \frac{M}{x} \quad \text{if } x > c.$$

      Hence show that $xe^{-ax} \to 0$ as $x \to \infty$.

   (b)   Use a similar method to show that $x^2 e^{-ax} \to 0$ as $x \to \infty$.

8 Solving equations numerically

This chapter is about numerical methods for solving equations when no exact method is available. When you have completed it, you should

- be able to use the sign-change rule to find approximate solutions by decimal search
- know how to use a chord approximation to improve the efficiency of decimal search
- be able to use an iterative method to produce a sequence which converges to a root
- understand that the choice of iterative method affects whether a sequence converges or not, and know what determines its behaviour
- appreciate that it is possible to modify an iterative method to speed up convergence
- appreciate that decisions about choice of method may depend on what sort of calculator or computer software you are using.

How you use this chapter will depend on what calculating aids you have available. It has been written to emphasise the underlying mathematical principles, so that you can follow the procedures with a simple calculator. But if you have a programmable or graphic calculator, if you like to write your own computer programs, or if you have access to a spreadsheet program, you will be able to carry out some of the calculations far more quickly.

8.1 Some basic principles

In mathematical problems the final step is often to solve an equation. If the equation is linear or quadratic, or if it can be reduced to one of these forms, then you have a method for solving it. But for many equations no simple method exists, and then you have to resort to some kind of solution by successive approximation, either numerical or algebraic.

Any equation in x can be rearranged so that it takes the form $f(x) = 0$. A value of x for which $f(x)$ takes the value 0 is called a **root** of the equation. The **solution** of the equation is the set of all the roots.

A useful way of representing the solution of $f(x) = 0$ is to draw the graph of $y = f(x)$. The roots are the x-coordinates of the points of the graph that lie on the x-axis.

This observation leads at once to a very useful rule for locating roots.

> **Sign-change rule**
> If the function $f(x)$ is continuous in an interval $p \leqslant x \leqslant q$ of its domain, and if $f(p)$ and $f(q)$ have opposite signs, then $f(x) = 0$ has at least one root between p and q.

This rule is illustrated in Fig. 8.1. The condition that $f(x)$ is continuous means that the graph cannot jump across the x-axis without meeting it.

The words 'at least one' are important. Fig. 8.2 shows that there may be more than one root between p and q.

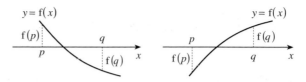

Fig. 8.1

When you have an equation $f(x) = 0$ to solve, it usually helps to begin by finding the shape of the graph of $y = f(x)$. Suppose that you want to solve the cubic equation

$$x^3 - 3x - 5 = 0.$$

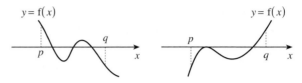

Fig. 8.2

Writing $f(x) = x^3 - 3x - 5$, you can find

$$f'(x) = 3x^2 - 3 = 3(x+1)(x-1) \text{ and } f''(x) = 6x.$$

It follows that $f(x)$ has a maximum where $x = -1$ and a minimum where $x = 1$. The coordinates of the maximum and minimum points are $(-1, -3)$ and $(1, -7)$. From this you can sketch the graph, as in Fig. 8.3.

The graph shows that the equation has only one root, and that it is greater than 1. Also, $f(x)$ is negative for values of x below the root, and positive above the root. This information suggests where to start looking for the root.

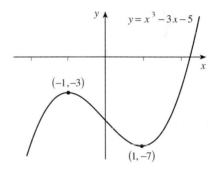

Fig. 8.3

8.2 Decimal search

This section describes how you can use the sign-change rule to find a sequence of approximations to the root, improving the accuracy by 1 decimal place at a time.

Continuing with the equation $x^3 - 3x - 5 = 0$, the graph in Fig. 8.3 suggests calculating $f(2) = -3$ and $f(3) = 13$. It follows that the root is between 2 and 3.

You could now start calculating $f(2.1), f(2.2), \ldots$ until you reach a value of x for which $f(x)$ is positive. But pause to ask if this strategy is sensible. Fig. 8.4 is a sketch of the graph of $f(x)$ between $x = 2$ and $x = 3$. Since $AP = 3$ is about $\frac{1}{4}$ of $BQ = 13$, you might guess that X is about $\frac{1}{5}$ of the distance from P to Q. So it might be best to begin by calculating $f(2.2) = -0.952$. Since this is negative, go on to calculate $f(2.3) = 0.267$.

There is no need to go further. Since $f(2.2)$ is negative and $f(2.3)$ is positive, the root is between 2.2 and 2.3.

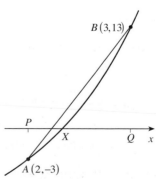

Fig. 8.4

If you have access to a graphic calculator, the equivalent procedure would be to zoom in on the interval $2 \leqslant x \leqslant 3$. You will then see that the graph cuts the x-axis between 2.2 and 2.3.

You now repeat this process to get the second decimal place. Since $|f(2.2)| = 0.952$ is about 4 times $|f(2.3)| = 0.267$, the root is probably about $\frac{4}{5}$ of the distance from 2.2 to 2.3, so begin by calculating $f(2.28) = 0.012...$. Since this is positive, 2.28 is too large, so try calculating $f(2.27) = -0.112...$. This is negative, so the root is between 2.27 and 2.28.

With a graphic calculator you would zoom in on the interval $2.2 \leqslant x \leqslant 2.3$, and see that the graph cuts the x-axis between 2.27 and 2.28.

To make sure you know what to do, continue the calculation for yourself to the third decimal place. You will find that $f(2.279)$ is negative, and you know that $f(2.280)$ is positive, so the root is between 2.279 and 2.280.

This process is called **decimal search**.

What you are finding by this method are the terms in two sequences: one sequence of numbers above the root

$$a_0 = 3, \quad a_1 = 2.3, \quad a_2 = 2.28, \quad a_3 = 2.280, \quad ...,$$

and one sequence of numbers below the root

$$b_0 = 2, \quad b_1 = 2.2, \quad b_2 = 2.27, \quad b_3 = 2.279, \quad ... .$$

Both sequences converge to the root as a limit; the difference between a_r and b_r is 10^{-r}, which tends to 0 as r increases.

Notice that, if you want to find the root correct to 3 decimal places, you need to know whether it is closer to 2.279 or to 2.280, so you need to find $f(2.2795)$, which is $0.006\,05...$. Since this is positive, and $f(2.279)$ is negative, the root lies between 2.279 and 2.2795. That is, its value is 2.279 correct to 3 decimal places.

Example 8.2.1
Solve the equation $x e^x = 1$.

Begin by investigating the equation graphically, using the idea that a root of an equation $f(x) = g(x)$ is the x-coordinate of a point of intersection of the graphs of $y = f(x)$ and $y = g(x)$. Fig. 8.5 shows four different ways of doing this, based on the equation as stated and the three rearrangements

$$e^x = \frac{1}{x}, \quad x = e^{-x}, \quad x = -\ln x .$$

All the graphs show that there is just one root, but the third graph is probably the most informative. The tangent to $y = e^{-x}$ at $(0,1)$ has gradient -1, and so meets $y = x$ at $\left(\frac{1}{2}, \frac{1}{2}\right)$. This shows that the root is slightly greater than 0.5.

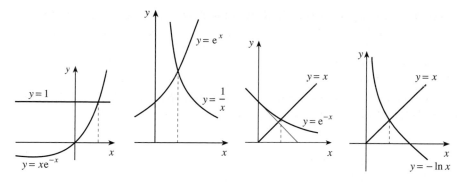

Fig. 8.5

To use the sign-change method you need to write the equation as $f(x) = 0$. There are again several possibilities:

$$x e^x - 1 = 0, \quad e^x - \frac{1}{x} = 0, \quad x - e^{-x} = 0, \quad x + \ln x = 0.$$

If you have a calculator with graphics or programming facilities it makes little difference which equation you use. But with a basic calculator it pays to use the form whose calculation involves the fewest key steps. This is probably the last, so take $f(x) = x + \ln x$.

The calculation then proceeds as follows.

$$f(0.5) = -0.193..., \quad f(0.6) = 0.089....$$

The root is between 0.5 and 0.6. Since $|f(0.5)|$ is about 2 times $|f(0.6)|$, the root is probably about $\frac{2}{3}$ of the distance from 0.5 to 0.6, which is about 0.57. So calculate

$$f(0.57) = 0.0078... \qquad (0.57 \text{ is too large}),$$
$$f(0.56) = -0.0198... \;.$$

The root is between 0.56 and 0.57, and $|f(0.56)|$ is between 2 and 3 times $|f(0.57)|$. This suggests that the root is between $\frac{2}{3}$ and $\frac{3}{4}$ of the distance from 0.56 to 0.57, which is about 0.567. So calculate

$$f(0.567) = -0.000\ 39... \qquad (0.567 \text{ is too small}),$$
$$f(0.568) = 0.002\ 36... \;.$$

The root therefore lies between 0.567 and 0.568.

At each stage the first step is to decide the next x-value at which to begin looking for the root; this corresponds in Fig. 8.4 to estimating where the chord AB cuts the x-axis. There is no point in doing this calculation very accurately; its only purpose is to decide where to begin the next step of the search, and for that you only need to work to 1 significant figure.

Exercise 8A

1 Show that the equation $2x^3 - 3x^2 - 2x + 5 = 0$ has a root between -1.5 and -1.

2 The equation $e^{-x} - x + 2 = 0$ has one root, α. Find an integer N such that $N < \alpha < N+1$.

3 Given $f(x) = 3x + 13 - e^x$, evaluate $f(3)$ and $f(4)$, correct to 3 significant figures. Explain the significance of the answers in relation to the equation $3x + 7 = e^x$.

4 For each of parts (a) to (f),
   (i) use the sign-change rule to determine the integer N such that the equation
       $f(x) = 0$ has a root in the interval $N < x < N+1$;
   (ii) use decimal search to find each root correct to 2 decimal places.

   (a) $f(x) = x^5 - 5x + 6$ (b) $f(x) = x + \sqrt{x^3 + 1} - 7$ (c) $f(x) = e^x - \dfrac{5}{x}$

   (d) $f(x) = 1000 - e^x \ln x$ (e) $f(x) = \ln(x^2 + 1) - 12 - x$ (f) $f(x) = x^5 + x^3 - 1999$

5 The function $f(x)$ is such that $f(a)f(b) < 0$ for real constants a and b with $a < b$, yet $f(x) = 0$ for no value of x such that $a < x < b$. Explain the feature of the function which allows this situation to arise, and illustrate your answer with a suitable example.

8.3 Finding roots by iteration

Example 8.3.1
Find the terms of the sequence defined by the inductive definition $x_0 = 0$, $x_{r+1} = e^{-x_r}$.

In the calculation all the available figures have been retained in the calculator, but the answers are tabulated correct to 5 decimal places.

r	x_r	r	x_r	r	x_r
0	0	9	0.571 14	18	0.567 12
1	1	10	0.564 88	19	0.567 16
2	0.367 88	11	0.568 43	20	0.567 14
3	0.692 20	12	0.566 41	21	0.567 15
4	0.500 47	13	0.567 56	22	0.567 14
5	0.606 24	14	0.566 91	23	0.567 14
6	0.545 40	15	0.567 28	24	0.567 14
7	0.579 61	16	0.567 07	...	...
8	0.560 12	17	0.567 19	...	...

Table 8.6

From $r = 22$ onwards it seems that the values correct to 5 decimal places are all $0.567\,14$. This is especially convincing in this example, since you can see that the terms are alternately below and above $0.567\,14$; so once you have two successive terms with this value, the same value will continue indefinitely. (You met a similar sequence in P1 Section 14.3. The sum sequence for a geometric series with common ratio -0.2 had terms alternately above and below $0.8333\ldots$.)

Notice also that the limit towards which these terms are converging appears to be the same (to the accuracy available) as the root of the equation $x = e^{-x}$ found in Example 8.2.1. The sequence illustrates a process called **iteration**, which can often be used to solve equations of the form $x = F(x)$.

> If the sequence given by the inductive definition $x_{r+1} = F(x_r)$, with some initial value x_0, converges to a limit l, then l is a root of the equation $x = F(x)$.

It is quite easy to see why. Since the sequence is given to be convergent, the left side x_{r+1} tends to l as $r \to \infty$, and the right side $F(x_r)$ tends to $F(l)$. (To be sure of this, the function $F(x)$ must be continuous.)

So $l = F(l)$; that is, l is a root of $x = F(x)$.

For another illustration take the equation $x^3 - 3x - 5 = 0$, for which the root was found earlier by the sign-change method. This can be rearranged as

$$x^3 = 3x + 5, \quad \text{or} \quad x = \sqrt[3]{3x + 5}.$$

This is of the form $x = F(x)$, so you can try to find the root by iteration, using a sequence defined by

$$x_{r+1} = \sqrt[3]{3x_r + 5}.$$

Fig. 8.3 suggests that the root is close to 2, so take $x_0 = 2$. Successive terms, correct to 5 decimal places, are then as in Table 8.7.

r	x_r	r	x_r	r	x_r
0	2	4	2.278 62	8	2.279 02
1	2.223 98	5	2.278 94	9	2.279 02
2	2.268 37	6	2.279 00	…	…
3	2.276 97	7	2.279 02	…	…

Table 8.7

This suggests that the limit is $2.279\,02$, but this time you cannot be quite sure. Since the terms get steadily larger, rather than being alternately too large and too small, it is just possible that if you go on longer there might be another change in the final digit. So for a final check go back to the sign-change method. Writing $f(x) = x^3 - 3x - 5$, calculate $f(2.279\,015) = -0.000\,047\ldots$ and $f(2.279\,025) = 0.000\,078\ldots$. This shows that the root is indeed $2.279\,02$ correct to 5 decimal places.

At each step of this iteration you have to use the key sequence $\left[\times, 3, +, 5, =, \sqrt[3]{}\right]$ to get from one term to the next. If you have a calculator with an [ANS] key, or if you set the process up as a small computer program or a spreadsheet, you can get the answer much more quickly.

8.4 Iterations which go wrong

There is more than one way of rearranging an equation $f(x) = 0$ as $x = F(x)$. For example, $x^3 - 3x - 5 = 0$ could be written as

$$3x = x^3 - 5, \quad \text{or} \quad x = \tfrac{1}{3}(x^3 - 5).$$

But if you perform the iteration

$$x_{r+1} = \tfrac{1}{3}(x_r^3 - 5), \text{ with } x_0 = 2,$$

the first few terms are

$$2, 1, -1.333\,33, -2.456\,79, -6.609\,58, -97.916\,54, \dots .$$

Clearly this is never going to converge to a limit.

Such a sequence is called **divergent**.

The same can happen with the equation $xe^x = 1$. If, instead of constructing an iteration from $x = e^{-x}$, you write it as

$$\ln x = -x, \quad \text{or} \quad x = -\ln x,$$

then the corresponding iteration is

$$x_{r+1} = -\ln x_r .$$

You can't start with $x_0 = 0$ this time, so take x_0 to be 0.5. Then you get

$$0.5, \ 0.693\,15, \ 0.366\,51, \ 1.003\,72, \ -0.003\,71, \ \text{ERROR!}$$

The terms are alternately above and below the root, as they were in Example 8.3.1, but they get further away from it each time until you eventually get a term which is outside the domain of $\ln x$.

So if you have an equation $f(x) = 0$, and rearrange it as $x = F(x)$, then the sequence $x_{r+1} = F(x_r)$ may or may not converge to a limit. If it does, then the limit is a root of the equation. If not, you should try rearranging the equation another way.

Exercise 8B

1   For each of parts (a) to (c), find three possible rearrangements of the equation $f(x) = 0$ into the form $x = F(x)$.

(a) $f(x) = x^5 - 5x + 6$ (b) $f(x) = e^x - \dfrac{5}{x}$ (c) $f(x) = x^5 + x^3 - 1999$

2 Each of parts (a) to (c) defines a sequence by an iteration of the form $x_{r+1} = F(x_r)$.

    (i) Rearrange the equation $x = F(x)$ into the form $f(x) = 0$, where f is a polynomial function.

    (ii) Use the iteration, with the given initial approximation x_0, to find the terms of the sequence $x_0, x_1, \ldots$ as far as x_5.

    (iii) Describe the behaviour of the sequence.

    (iv) If the sequence converges, investigate whether x_5 is an approximate root of $f(x) = 0$.

    (a) $x_0 = 0, x_{r+1} = \sqrt[11]{x_r^7 - 6}$
                        (b) $x_0 = 3, x_{r+1} = \left(\dfrac{17 - x_r^2}{x_r}\right)^2$

    (c) $x_0 = 7, x_{r+1} = \sqrt[3]{500 + \dfrac{10}{x_r}}$

3 Show that the equation $x^5 + x - 19 = 0$ can be arranged into the form $x = \sqrt[3]{\dfrac{19 - x}{x^2}}$ and that the equation has a root α between $x = 1$ and $x = 2$.

Use an iteration based on this arrangement, with initial approximation $x_0 = 2$, to find the values of $x_1, x_2, \ldots, x_6$. Investigate whether this sequence is converging to α.

4 (a) Show that the equation $x^2 + 2x - e^x = 0$ has a root in the interval $2 < x < 3$.

    (b) Use an iterative method based on the rearrangement $x = \sqrt{e^x - 2x}$, with initial approximation $x_0 = 2$, to find the value of x_{10} to 4 decimal places. Describe what is happening to the terms of this sequence of approximations.

5 Show that the equation $e^x = x^3 - 2$ can be arranged into the form $x = \ln(x^3 - 2)$. Show also that it has a root between 2 and 3.

Use the iteration $x_{r+1} = \ln(x_r^3 - 2)$, commencing with $x_0 = 2$ as an initial approximation to the root, to show that this arrangement is not a suitable one for finding this root.

Find an alternative arrangement of $e^x = x^3 - 2$ which can be used to find this root, and use it to calculate the root correct to 2 decimal places.

6 (a) Determine the value of the positive integer N such that the equation $12 - x - \ln x = 0$ has a root α such that $N < \alpha < N + 1$.

    (b) Define the sequence $x_0, x_1, \ldots$ of approximations to α iteratively by $x_0 = N + \frac{1}{2}$, $x_{r+1} = 12 - \ln x_r$.

        Find the number of steps required before two consecutive terms of this sequence are the same when rounded to 4 significant figures. Show that this common value is equal to α to this degree of accuracy.

7 Sketch the graphs of $y = x$ and $y = \cos x$, and state the number of roots of the equation $x = \cos x$.

Use a suitable iteration and starting point to find the positive root of the equation $x = \cos x$, giving your answer correct to 3 decimal places.

Show that the iteration $x = \cos^{-1} x$ starting from $x = 0$ does not converge.

8.5* Choosing convergent iterations

The rest of this chapter is about how to rearrange an equation to ensure that the iterative sequence converges. You may if you like omit it and go on to Miscellaneous exercise 8.

The solution of $x = F(x)$ can be represented graphically by the intersection of the graph of $y = F(x)$ with the line $y = x$. Fig. 8.8 shows this for the equations in the last two sections, each with two alternative forms.

(a) $xe^x = 1$ (b) $x^3 - 3x - 5 = 0$

(i) $x = e^{-x}$ (ii) $x = -\ln x$ (i) $x = \sqrt[3]{3x+5}$ (ii) $x = \dfrac{x^3 - 5}{3}$

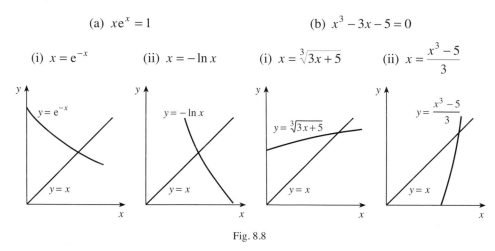

Fig. 8.8

In both cases the sequence converged in version (i), but not in version (ii).

Inspection of the graphs suggests that it is the gradient of the graph of $y = F(x)$ at or near the root which governs the nature of the iteration.

In (a)(i) the gradient is negative, but numerically small (about -0.5). The sequence converges, though quite slowly; it takes 22 steps to reach the root correct to 5 decimal places. The terms alternate above and below the root.

In (a)(ii) the gradient is negative, but numerically larger (about -2). The sequence does not converge, but the terms again alternate above and below the root.

In (b)(i) the gradient is positive, and numerically small (about 0.2). The sequence converges quite fast, taking only 7 steps to reach the root correct to 5 decimal places. The terms get steadily larger, approaching the root from below.

In (b)(ii) the gradient is positive and numerically large (about 5). The sequence does not converge, and the terms get steadily smaller.

This discussion points to the following conclusions, which are generally true. You can test them for yourself using the sequences which you produced in Exercise 8B.

- If the equation $x = F(x)$ has a root, then a sequence defined by $x_{r+1} = F(x_r)$ with a starting value close to the root will converge to the root if the gradient of the graph of $y = F(x)$ at and around the root is not too large (roughly between -1 and 1).
- The smaller the modulus of the gradient, the fewer steps will be needed to reach the root to a given accuracy.
- If the gradient is negative, the terms will be alternately above and below the root; if it is positive, the terms will approach the root steadily from one side.

There is one further point to notice about these examples. The pairs of functions used for $F(x)$ are in fact inverses.

(a) $x \rightarrow [\ +/-\ ] \rightarrow [\ \exp\ ] \rightarrow$ has output e^{-x},

$\quad \leftarrow [\ +/-\ ] \leftarrow [\ \ln\ ] \leftarrow x$ (read from right to left) has output $-\ln x$.

(b) $x \rightarrow [\ \times 3\ ] \rightarrow [\ +5\ ] \rightarrow [\ \sqrt[3]{\ }\ ] \rightarrow$ has output $\sqrt[3]{3x+5}$,

$\quad \leftarrow [\ \div 3\ ] \leftarrow [\ -5\ ] \leftarrow [\ (\ )^3\ ] \leftarrow x$ has output $\frac{1}{3}(x^3 - 5)$.

Their graphs are therefore reflections of each other in the line $y = x$. This is why, if the gradient of one graph is numerically small, the gradient of the other is large. This leads to a useful rule for deciding how to rearrange an equation.

If the function F is one–one, and if $x = F(x)$ has a root, then usually one of the sequences $x_{r+1} = F(x_r)$ and $x_{r+1} = F^{-1}(x_r)$ converges to the root, but the other does not.

Example 8.5.1
Show that the equation $x^3 - 3x - 1 = 0$ has three roots, and find them correct to 4 decimal places.

The graph of $y = x^3 - 3x - 1$ is shown in Fig. 8.9. It is in fact the graph in Fig. 8.3 translated by $+4$ in the y-direction. You can see that it cuts the x-axis in three places: between -2 and -1, -1 and 0, and 1 and 2.

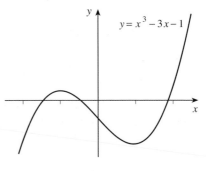

Fig. 8.9

You could use iterations based on a rearrangement $x = F(x)$, where $F(x)$ is either $\frac{1}{3}(x^3 - 1)$ or $\sqrt[3]{3x+1}$. These are illustrated in Fig. 8.10; again, they are inverse functions. To get a small gradient at the intersection, you should use $F(x) = \sqrt[3]{3x+1}$ for the first and last of the roots, and $F(x) = \frac{1}{3}(x^3 - 1)$ for the middle root.

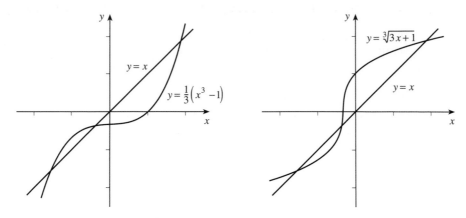

Fig. 8.10

You can check for yourself that:

$x_{r+1} = \sqrt[3]{3x_r + 1}$ with $x_0 = -2$ reaches the root -1.5321 in 11 steps,

$x_{r+1} = \frac{1}{3}(x_r^3 - 1)$ with $x_0 = 0$ reaches the root -0.3473 in 4 steps,

$x_{r+1} = \sqrt[3]{3x_r + 1}$ with $x_0 = 2$ reaches the root 1.8794 in 6 steps.

The next example shows a trick which you can use to reduce the number of steps needed to reach the root, or even to produce a convergent iteration from one which does not converge.

Example 8.5.2

For each of the equations   (a) $x = e^{-x}$,   (b) $x = -\ln x$,   add an extra term kx to both sides, and choose k so that, near the root, the function on the right has a small gradient. Use this to produce a sequence which converges rapidly to the root.

(a) Write $kx + x = kx + e^{-x}$. You saw earlier that near the root the graph of e^{-x} has a gradient of about -0.5, so $kx + e^{-x}$ has a gradient of about $k - 0.5$. To make this small, choose $k = 0.5$. Then the equation becomes

$$1.5x = 0.5x + e^{-x}, \quad \text{or} \quad x = \frac{1}{3}\left(x + 2e^{-x}\right).$$

The iteration $x_{r+1} = \frac{1}{3}\left(x_r + 2e^{-x_r}\right)$ with $x_0 = 1$ reaches the root $0.567\,14$, correct to 5 decimal places, in 4 steps. (In Example 8.3.1 it took 22 steps of the iteration $x_{r+1} = e^{-x_r}$ to achieve the same accuracy.)

(b) Write $kx + x = kx - \ln x$. Near the root the graph of $-\ln x$ has a gradient of about -2, so $kx - \ln x$ has a gradient of about $k - 2$. Choose $k = 2$, so that the equation becomes

$$3x = 2x - \ln x, \quad \text{or} \quad x = \frac{1}{3}(2x - \ln x).$$

The iteration $x_{r+1} = \frac{1}{3}(2x_r - \ln x_r)$ with $x_0 = 1$ reaches the root in 6 steps. (You saw in Section 8.4 that the iteration $x_{r+1} = -\ln x_r$ does not even converge.)

1 In parts (a) to (c), sketch the graph of $y = F(x)$, and hence decide whether the iteration $x_{r+1} = F(x_r)$, with initial approximation x_0, is suitable for finding the root of the equation $x = F(x)$ near to $x = x_0$.

Where the process leads to a convergent sequence of approximations to the required root, find this root. Where the process is unsuitable, find $F^{-1}(x)$ and use it to find the root.

In parts (a) and (b) give your answers correct to 3 decimal places; in parts (c) to (e) give your answers correct to 4 decimal places.

(a) $F(x) = \dfrac{3}{x} - 1, \quad x_0 = 1$

(b) $F(x) = 5 - e^{3x}, \quad x_0 = 0$

(c) $F(x) = \frac{1}{2}\tan x, \quad x_0 = 1$

(d) $F(x) = 30 - \frac{1}{10}x^6, \quad x_0 = -2$

(e) $F(x) = 2\sin x, \quad x_0 = \frac{1}{2}\pi$

2 In each of parts (a) to (d) find a constant k for which
$$k x_{r+1} + x_{r+1} = k x_r + F(x_r)$$
is a better form than $x_{r+1} = F(x_r)$ to use to find the root of the equation $x = F(x)$ near x_0. In each case, find this root correct to 4 significant figures.

(a) $F(x) = 2 - 5\ln x, \quad x_0 = 1$

(b) $F(x) = x^2 + 6\ln x - 50, \quad x_0 = 6$

(c) $F(x) = \frac{1}{48}(x^4 - x^7 - 192), \quad x_0 = -2$

(d) $F(x) = x\ln x - e^{-x} - 20, \quad x_0 = 12$

1 Given that $f(x) = 2^x + 3^x$, evaluate $f(1)$ and $f(2)$. Using these values,

(a) state what this tells you about the root of the equation $f(x) = 10$,

(b) suggest a suitable initial approximation to this root.

2 Find the positive integer N such that $40e^{-x} = x^2$ has a root between N and $N+1$.

3 Show that there exists a root $x = \alpha$ of the equation $x^3 - 6x + 3 = 0$ such that $2 < \alpha < 3$. Use decimal search to find this root correct to 2 decimal places.

4 Show that the equation $2x - \ln(x^2 + 2) = 0$ has a root in the interval $0.3 < x < 0.4$. Use decimal search to find an interval of width 0.001 in which this root lies.

5 The equation $e^x = 50\sqrt{2x - 1}$ has two positive real roots. Use decimal search to find the larger root correct to 1 decimal place.

6 (a) On the same diagram, sketch the graphs of $y = 2^{-x}$ and $y = x^2$.

(b) One of the points of intersection of these graphs has a positive x-coordinate. Find this x-coordinate correct to 2 decimal places and give a brief indication of your method.

(OCR)

7 (a) On a single diagram, sketch the graphs of $y = \tan x°$ and $y = 4\cos x° - 3$ for
         $0 \leqslant x \leqslant 180$. Deduce the number of roots of the equation $f(x) = 0$ which exist for
         $0 \leqslant x \leqslant 180$, where $f(x) = 3 + \tan x° - 4\cos x°$.

   (b) By evaluating $f(x)$ for suitably chosen values of x, show that a root of the equation
         $f(x) = 0$ occurs at $x = 28$ (correct to the nearest integer).

8 Show that there is a root α of the equation $2\sin x° - \cos x° + 1 = 0$ such that
   $230 < \alpha < 240$. Use a decimal search method to determine this root to the nearest 0.1.

9 The points A and B have coordinates $(0, -2)$ and $(-30, 0)$ respectively.

   (a) Find an equation of the line which passes through A and B.

   The function f is defined by $f(x) = 1 + \tan x°$, $-90 < x < 90$.

   (b) Explain why there is just one point where the line in (a) meets the graph of $y = f(x)$.

   (c) Use an appropriate method to find the value of the integer N such that the value of the
         x-coordinate of the point where the graph of $y = f(x)$ meets the line of part (a)
         satisfies $N < x < N + 1$.   (OCR, adapted)

10 Find, correct to 2 decimal places, the x-coordinate of the turning point on the curve with
    equation $y = 5\cos x + x^2$, $x > 0$.

11 The region, R, of the plane enclosed by the axes, the curve $y = e^x + 4$ and the line $x = 2$
    has area A. Find, correct to 4 significant figures, the value of m, $0 < m < 2$, such that the
    portion of R between the y-axis and the line $x = m$ has area $\frac{1}{2}A$.

12 Show that the equation $3.5x = 1.6^x$ has a real solution between 6 and 7. By rearranging
    the equation into the form $x = a + b\ln x$, determine this root correct to 2 decimal places.

13 (a) Given that $f(x) = e^{2x} - 6x$, evaluate $f(0)$ and $f(1)$, giving each answer correct to 3
         decimal places. Explain how the equation $f(x) = 0$ could still have a root in the
         interval $0 < x < 1$ even though $f(0)f(1) > 0$.

    (b) Rewrite the equation $f(x) = 0$ in the form $x = F(x)$, for some suitable function F.
         Taking $x_0 = 0.5$ as an initial approximation, use an iterative method to determine one
         of the roots of this equation correct to 3 decimal places. How could you demonstrate
         that this root has the required degree of accuracy?

    (c) Deduce the value, to 2 decimal places, of one of the roots of the equation $e^x - 3x = 0$.

14 (a) Show that the equation $x^3 - 3x^2 - 1 = 0$ has a root α between $x = 3$ and $x = 4$.

    (b) The iterative formula $x_{r+1} = 3 + \dfrac{1}{x_r^2}$ is used to calculate a sequence of approximations
         to this root. Taking $x_0 = 3$ as an initial approximation to α, determine the values of
         x_1, x_2, x_3 and x_4 correct to 5 decimal places. State the value of α to 3 decimal places
         and justify this degree of accuracy.

15* (a) Show that the equation $x + \ln x - 4 = 0$ has a root α in the interval $2 < x < 3$.

    (b) Find which of the two iterative forms $x_{r+1} = e^{4 - x_r}$ and $x_{r+1} = 4 - \ln x_r$
         is more likely to give a convergent sequence of approximations to α, giving a reason
         for your answer. Use your chosen form to determine α correct to 2 decimal places.

16* (a) Find the positive integer N such that the equation $(t-1)\ln 4 = \ln(9t)$ has a solution $t = r$ in the interval $N < t < N+1$.

(b) Write down two possible rearrangements of this equation in the form $t = F(t)$ and $t = F^{-1}(t)$. Show which of these two arrangements is more suitable for using iteratively to determine an approximation to r to 3 decimal places, and find such an approximation.

17* (a) The equation $x = F(x)$ has a single root α. Find by trial the integer N such that $N < \alpha < N+1$.

(b) By adding a term kx to both sides of $x = F(x)$, where k is a suitably chosen integer, determine α correct to 4 decimal places.

18* (a) Find the coordinates of the points of intersection of the graphs with equations $y = x$ and $y = g(x)$, where $g(x) = \dfrac{5}{x}$.

(b) Show that the iterative process defined by $x_0 = 2$, $x_{r+1} = g(x_r)$ cannot be used to find good approximations to the positive root of the equation $x = \dfrac{5}{x}$.

(c) Describe why the use of the inverse function $g^{-1}(x)$ is also inappropriate in this case.

(d) Use a graph to explain why the iterative process defined by

$$x_0 = 2, \quad x_{r+1} = \frac{1}{2}\left(x_r + \frac{5}{x_r}\right)$$

leads to a convergent sequence of approximations to this root. Find this root correct to 6 decimal places.

19* (a) Use a graph to show that the equation $f(x) = 0$, where $f(x) = x - 10 - 30\cos x°$, has only one root. Denote this root by α.

(b) Find two numbers, a_0 and b_0, such that $b_0 - a_0 = 10$ and $a_0 < \alpha < b_0$.

(c) Evaluate $f(m)$, where $m = \frac{1}{2}(a_0 + b_0)$. Determine whether $a_0 < \alpha < m$ or $m < \alpha < b_0$. Hence write down two numbers, a_1 and b_1, such that $b_1 - a_1 = 5$ and $a_1 < \alpha < b_1$.

(d) Use a method similar to part (c) to find two numbers, a_2 and b_2, such that $b_2 - a_2 = 2.5$ and $a_2 < \alpha < b_2$.

(e) Continuing this way, find two sequences, a_r and b_r, such that $b_r - a_r = 10 \times 2^{-r}$ and $a_r < \alpha < b_r$. Go on until you find two numbers of the sequence which enable you to write down the value of α correct to 1 decimal place. (This is called the 'bisection method'.)

20 Given the one–one function $F(x)$, explain why roots of the equation $F(x) = F^{-1}(x)$ are also roots of the equation $x = F(x)$.

Use this to solve the following equations, giving your answers correct to 5 decimal places.

(a) $x^3 - 1 = \sqrt[3]{1+x}$, \qquad (b) $\frac{1}{10}e^x = \ln(10x)$.

9 The trapezium rule

This chapter is about approximating to integrals. When you have completed it, you should

- be able to use the trapezium rule to estimate the value of a definite integral
- be able to use a sketch, in some cases, to determine whether the trapezium rule approximation is an overestimate or an underestimate.

9.1 The need for approximation

There are times when it is not possible to evaluate a definite integral directly, using the standard method,

$$\int_a^b f(x)\,dx = \left[I(x)\right]_a^b = I(b) - I(a),$$

where $I(x)$ is the simplest function for which $\dfrac{d}{dx}I(x) = f(x)$.

Two examples which you cannot integrate with your knowledge so far are $\displaystyle\int_0^1 \frac{1}{1+x^2}\,dx$ and $\displaystyle\int_0^1 \sqrt{1+x^3}\,dx$. You need to use another method for approximating to the integrals.

9.2 The trapezium rule: simple form

Suppose that you wish to find an estimate for the integral $\displaystyle\int_a^b f(x)\,dx$.

You know, from P1 Section 16.3, that the value of the definite integral $\displaystyle\int_a^b f(x)\,dx$ represents the shaded area in Fig. 9.1. The principle behind the trapezium rule is to approximate to this area by using the shaded trapezium in Fig. 9.2.

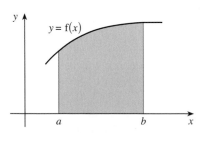

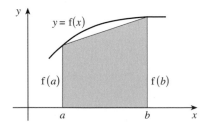

Fig. 9.1   Fig. 9.2

The area of the shaded trapezium is given by

$$\text{area of trapezium} = \tfrac{1}{2} \times (\text{sum of parallel sides}) \times (\text{distance between them}),$$

so $\text{area of trapezium} = \tfrac{1}{2} \times (f(a) + f(b)) \times (b - a).$

So $\displaystyle\int_a^b f(x)\,dx \approx \frac{1}{2}(b-a)(f(a)+f(b))$.

This is the simplest form of the **trapezium rule**.

Example 9.2.1

Use the simplest form of the trapezium rule to find estimates for $\displaystyle\int_0^1 \frac{1}{1+x^2}\,dx$ and $\displaystyle\int_0^1 \sqrt{1+x^3}\,dx$.

$$\int_0^1 \frac{1}{1+x^2}\,dx \approx \frac{1}{2}(1-0)\left(\frac{1}{1+0^2}+\frac{1}{1+1^2}\right)=\frac{1}{2}\times 1\times\left(1+\frac{1}{2}\right)=0.75.$$

$$\int_0^1 \sqrt{1+x^3}\,dx \approx \frac{1}{2}(1-0)\left(\sqrt{1+1^3}+\sqrt{1+0^3}\right)=\frac{1}{2}\times 1\times\left(\sqrt{2}+1\right)\approx 1.21.$$

9.3 The trapezium rule: general form

If you said that the simple form of the trapezium rule is not very accurate, especially over a large interval on the x-axis, you would be correct.

You can improve the accuracy by dividing the large interval from a to b into several smaller ones, and then using the trapezium rule on each interval. The amount of work sounds horrendous but, with good notation and organisation, it is not too bad.

Divide the interval from a to b into n equal intervals, each of width h, so that $nh = b - a$.

Call the x-coordinate of the left side of the first interval x_0, so $x_0 = a$, and then successively let $x_1 = x_0 + h$, $x_2 = x_0 + 2h$ and so on until $x_{n-1} = x_0 + (n-1)h$ and $x_n = x_0 + nh = b$.

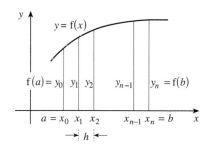

Fig. 9.3

To shorten the amount of writing, use the shorthand $y_0 = f(x_0)$, $y_1 = f(x_1)$ and so on, as in Fig. 9.3.

Then, using the simple form of the trapezium rule on each interval of width h in turn, you find that

$$\int_a^b f(x)\,dx \approx \frac{1}{2}h(y_0+y_1)+\frac{1}{2}h(y_1+y_2)+\frac{1}{2}h(y_2+y_3)+\ldots+\frac{1}{2}h(y_{n-1}+y_n)$$

$$=\frac{1}{2}h(y_0+y_1+y_1+y_2+y_2+y_3+\ldots+y_{n-2}+y_{n-1}+y_{n-1}+y_n)$$

$$=\frac{1}{2}h\{(y_0+y_n)+2(y_1+y_2+\ldots+y_{n-1})\}.$$

The trapezium rule with n intervals is sometimes called the trapezium rule with $n+1$ ordinates. (The term 'ordinate' means y-coordinate.)

The **trapezium rule** with n intervals states that

$$\int_a^b y\,dx \approx \tfrac{1}{2}h\{(y_0 + y_n) + 2(y_1 + y_2 + \ldots + y_{n-1})\}, \quad \text{where} \quad h = \frac{b-a}{n}.$$

Example 9.3.1
Use the trapezium rule with 5 intervals to estimate $\int_0^1 \dfrac{1}{1+x^2}\,dx$, giving your answer correct to 3 decimal places.

The values of y_n in Table 9.4 are given correct to 5 decimal places.

n	x_n	y_n	Sums	Weight	Total
0	0	1			
5	1	0.5	1.5	$\times 1 =$	1.5
1	0.2	0.961 54			
2	0.4	0.862 07			
3	0.6	0.735 29			
4	0.8	0.609 76	3.168 66	$\times 2 =$	6.337 32
					7.837 32

Table 9.4

The factor $\tfrac{1}{2}h$ is $\tfrac{1}{2} \times 0.2 = 0.1$. Therefore the approximation to the integral is $0.1 \times 7.837\,32 = 0.783\,732$. Thus the 5-interval approximation correct to 3 decimal places is 0.784.

The accurate value of $\int_0^1 \dfrac{1}{1+x^2}\,dx$ is $\tfrac{1}{4}\pi$, which correct to 3 decimal places is 0.785, so you can see that the 5-interval version of the trapezium rule is a considerable improvement on the 1-interval version in Example 9.2.1.

How you organise the table to give the value of $\{(y_0 + y_n) + 2(y_1 + y_2 + \ldots + y_{n-1})\}$ is up to you, and may well depend on the kind of software or calculator that you have. It is important, however, that you make clear how you reach your answer.

9.4 Accuracy of the trapezium rule

It is not easy with the mathematics that you know at present to give a quantitative approach to the possible error involved with the trapezium rule.

However, in simple situations you can see whether the trapezium rule answer is too large or too small. If a graph is bending downwards over the whole interval from a to b, as in Fig. 9.5, then you can be certain that the trapezium rule will give you an underestimate of the true area. If on the other hand, a graph is bending upwards over the whole interval from a to b, as in Fig. 9.6, then you can be certain that the trapezium rule will give you an overestimate of the true area.

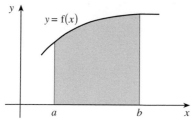

Fig. 9.5

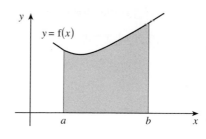

Fig. 9.6

However, if the graph sometimes bends upwards and sometimes downwards over the interval from a to b, you cannot be sure whether your approximation to the integral is an overestimate or an underestimate.

Exercise 9

1 Use the simplest case of the trapezium rule (that is, 1 interval) to estimate the values of

(a) $\displaystyle\int_3^4 \sqrt{1+x}\,dx$,

(b) $\displaystyle\int_2^4 \frac{1}{x}\,dx$.

2 Use the trapezium rule with 3 intervals to estimate the value of $\displaystyle\int_0^3 \sqrt{1+x^2}\,dx$.

3 Use the trapezium rule with 3 ordinates (that is, 2 intervals) to estimate the value of $\displaystyle\int_1^3 \sqrt{1+\sqrt{x}}\,dx$.

4 Find approximations to the value of $\displaystyle\int_1^5 \frac{1}{x^2}\,dx$ by

(a) using the trapezium rule with 2 intervals,

(b) using the trapezium rule with 4 intervals.

(c) Evaluate the integral exactly and compare your answer with those found in parts (a) and (b).

5 The diagram shows the graph of $y = \dfrac{4}{\sqrt{x}}$.

Use the trapezium rule with 6 intervals to find an approximation to the area of the shaded region, and explain why the trapezium rule overestimates the true value.

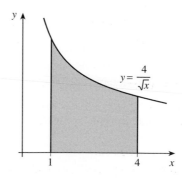

6   Use the trapezium rule with 5 intervals to estimate the value of $\int_0^1 \left(\frac{1}{10}x^2 + 1\right) dx$.

Draw the graph of $y = \frac{1}{10}x^2 + 1$ and explain why the trapezium rule gives an overestimate of the true value of the integral.

7   Draw the graph of $y = x^3 + 8$ and use it to explain why use of the trapezium rule with 4 intervals will give the exact value of $\int_{-2}^2 (x^3 + 8) dx$.

8   Find an approximation to $\int_1^2 \sqrt{x^2 + 4x}\, dx$ by using the trapezium rule with 4 intervals.

9   Find an approximation to $\int_0^4 \frac{x^2}{2^x} dx$ by using the trapezium rule with 8 intervals.

10 The diagram shows part of a circle with its centre at the origin. The curve has equation $y = \sqrt{25 - x^2}$.

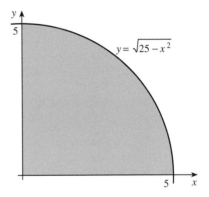

(a) Use the trapezium rule with 10 intervals to find an approximation to the area of the shaded region.

(b) Does the trapezium rule overestimate or underestimate the true area?

(c) Find the exact area of the shaded region.

(d) By comparing your answers to parts (a) and (c), obtain an estimate for π to 2 decimal places.

Miscellaneous exercise 9

1   Use the trapezium rule, with ordinates at $x = 1$, $x = 2$ and $x = 3$, to estimate the value of $\int_1^3 \sqrt{40 - x^3}\, dx$. (OCR)

2   The diagram shows the region R bounded by the curve $y = \sqrt{1 + x^3}$, the axes and the line $x = 2$. Use the trapezium rule with 4 intervals to obtain an approximation for the area of R, showing your working and giving your answer to a suitable degree of accuracy.

Explain, with the aid of a sketch, whether the approximation is an overestimate or an underestimate.   (OCR)

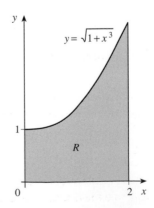

3 Use the trapezium rule with subdivisions at $x = 3$ and $x = 5$ to obtain an approximation to $\displaystyle\int_1^7 \frac{x^3}{1+x^4}\,dx$, giving your answer correct to 3 places of decimals. (OCR)

4 Use the trapezium rule with 5 intervals to estimate the value of $\displaystyle\int_0^{0.5} \sqrt{1+x^2}\,dx$, showing your working. Give your answer correct to 2 decimal places. (OCR)

5 The diagram shows the region R bounded by the axes, the curve $y = \left(x^2 + 1\right)^{-\frac{3}{2}}$ and the line $x = 1$. Use the trapezium rule, with ordinates at $x = 0$, $x = \frac{1}{2}$ and $x = 1$, to estimate the value of

$$\int_0^1 \left(x^2 + 1\right)^{-\frac{3}{2}}\,dx,$$

giving your answer correct to 2 significant figures. (OCR)

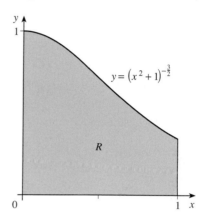

6 The diagram shows a sketch of $y = \sqrt{2 + x^3}$ for values of x between -0.5 and 0.5.

(a) Use the trapezium rule, with ordinates at $x = -0.5$, $x = 0$ and $x = 0.5$ to find an approximate value for $\displaystyle\int_{-0.5}^{0.5} \sqrt{2 + x^3}\,dx$.

(b) Explain briefly, with reference to the diagram, why the trapezium rule can be expected to give a good approximation to the value of the integral in this case. (OCR)

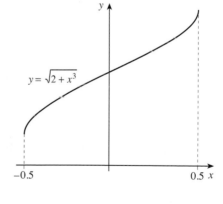

7 A certain function f is continuous and is such that

$$f(2.0) = 15, \quad f(2.5) = 22, \quad f(3.0) = 31, \quad f(3.5) = 28, \quad f(4.0) = 27.$$

Use the trapezium rule to find an approximation to $\displaystyle\int_2^4 f(x)\,dx$.

8 The speeds of an athlete on a training run were recorded at 30-second intervals:

Time after start (s)	0	30	60	90	120	150	180	210	240
Speed $(\mathrm{m\,s^{-1}})$	3.0	4.6	4.8	5.1	5.4	5.2	4.9	4.6	3.8

The area under a speed–time graph represents the distance travelled. Use the trapezium rule to estimate the distance covered by the athlete, correct to the nearest 10 metres.

9 At a time t minutes after the start of a journey, the speed of a car travelling along a main road is v km h^{-1}. The table gives values of v every minute on the 10-minute journey.

t	0	1	2	3	4	5	6	7	8	9	10
v	0	31	46	42	54	57	73	70	68	48	0

Use the trapezium rule to estimate of the length of the 10-minute journey in kilometres.

10 A river is 18 metres wide in a certain region and its depth, d metres, at a point x metres from one side is given by the formula $d = \frac{1}{18}\sqrt{x(18-x)(18+x)}$.

(a) Produce a table showing the depths (correct to 3 decimal places where necessary) at $x = 0, 3, 6, 9, 12, 15$ and 18.

(b) Use the trapezium rule to estimate the cross-sectional area of the river in this region.

(c) Given that, in this region, the river is flowing at a uniform speed of 100 metres per minute, estimate the number of cubic metres of water passing per minute.   (OCR)

11 The diagram shows the curve $y = 4^{-x}$.
Taking subdivisions at $x = 0.25, 0.5, 0.75$,
find an approximation to the shaded area.

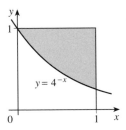

12 The left diagram shows the part of the curve $y = 2.5 - 2^{1-x^2}$ for which $-0.5 \leqslant x \leqslant 0.5$. The shaded region forms the cross-section of a straight concrete drainage channel, as shown in the right diagram. The units involved are metres.

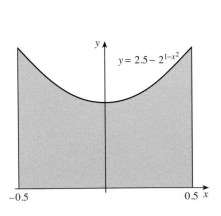
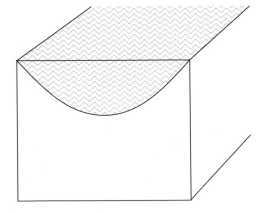

(a) Use the trapezium rule with 4 intervals to estimate the area of the shaded region.

(b) Estimate the volume of concrete in a 20-metre length of channel.

(c) Estimate the volume of water in the 20-metre length of channel when it is full.

(d) Of the estimates in parts (b) and (c), which is an overestimate and which is an underestimate?

13 The integral $\int_{36}^{64} \sqrt{x}\, dx$ is denoted by I.

   (a) Find the exact value of I.

   (b) Use the trapezium rule with 2 intervals to find an estimate for I, giving your answer in terms of $\sqrt{2}$.

   Use your two answers to deduce that $\sqrt{2} \approx \frac{149}{105}$.

14 It is given that $x - 2$ is a factor of $f(x)$, where $f(x) = 2x^3 - 7x^2 + x + a$. Find the value of a and factorise $f(x)$ completely.

   Sketch the graph of $y = f(x)$. (You do not need to find the coordinates of the stationary points.)

   Use the trapezium rule, with ordinates at $x = -1$, $x = 0$, $x = 1$ and $x = 2$ to find an approximation to $\int_{-1}^{2} f(x)\, dx$.

   Find the exact value of the integral and show that the trapezium rule gives a value that is in error by about 11%. $\hspace{2cm}$ (OCR)

15 The trapezium rule, with 2 intervals of equal width, is to be used to find an approximate value for $\int_{1}^{2} \frac{1}{x^2}\, dx$. Explain, with the aid of a sketch, why the approximation will be greater than the exact value of the integral.

   Calculate the approximate value and the exact value, giving each answer correct to 3 decimal places.

   Another approximation to $\int_{1}^{2} \frac{1}{x^2}\, dx$ is to be calculated by using two trapezia of unequal width; the ordinates are at $x = 1$, $x = h$ and $x = 2$. Find, in terms of h, the total area, T, of these two trapezia.

   Find the value of h for which T is a minimum.

16 (a) Calculate the exact value of the integral $\int_{0}^{1} x^2\, dx$.

   (b) Find the trapezium rule approximations to this integral using $1, 2, 4$ and 8 intervals. Call these A_1, A_2, A_4 and A_8.

   (c) For each of your answers in part (b), calculate the error E_i, where

   $$E_i = \int_{0}^{1} x^2\, dx - A_i, \text{ for } i = 1, 2, 4 \text{ and } 8.$$

   (d) Look at your results for part (c), and guess the relationship between the error E_n and the number n of intervals taken.

   (e) How many intervals would you need to approximate to the integral to within 10^{-6}?

10 Parametric equations

This chapter is about a method of describing curves using parameters. When you have completed it, you should

- know how to describe a curve using a parameter
- be able, in simple cases, to convert from a parametric equation of a curve to the cartesian equation of the curve
- be able to use parametric methods to establish properties of curves.

10.1 Introduction

Imagine a person P going round on a turntable, centre the origin O and radius 1 unit, at a constant speed (see Fig. 10.1). Suppose that P starts at the x-axis and moves anticlockwise in such a way that the angle at the centre t seconds after starting is t radians.

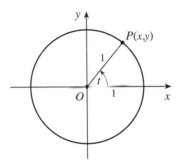

Fig. 10.1

Where is P after t seconds? You can see from Fig. 10.1 that the coordinates of P are given by

$$x = \cos t, \quad y = \sin t.$$

These equations allow you to find the position of P at any time, and they describe the path of P completely.

Fig. 10.2 shows the values of t at various points on the first revolution of the turntable. Notice that for each value of t there is a unique point on the curve.

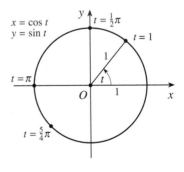

Fig. 10.2

During the first revolution, each point on the curve has a t-value corresponding to the time at which the person is at that point. However, for each additional revolution there will be another t-value associated with each point.

Example 10.1.1
Find the t-value of the starting point, the first time that P returns to it.

The starting point is $(1,0)$. Since $x = \cos t, y = \sin t$, you find that $1 = \cos t$ and $0 = \sin t$. These equations are simultaneously satisfied by $t = 0, \pm 2\pi, \pm 4\pi, \ldots$. The smallest positive solution is $t = 2\pi$.

The equations $x = \cos t$, $y = \sin t$ are an example of **parametric equations**, and the variable t is an example of a **parameter**. In this case the variable t represents time, but in other cases it may not, as you will see in Example 10.1.2.

If you have a graphic calculator, you may be able to use it to draw curves from parametric equations. Put the calculator into parametric mode. You then have to enter the parametric equations into the calculator, and you may have to give an interval of values of t. For example, if you gave an interval of 0 to π for t in Example 10.1.1, you would get only the upper semicircle of the path. If you use the trace key, the calculator will also give you the t-value for any point.

Recall that the curve looks like a circle only if you use the same scale on both axes.

You could also plot the curve using a spreadsheet with graph-plotting facilities.

Here are other examples of curves with parametric equations.

Example 10.1.2
A curve has parametric equations $x = t^2$, $y = 2t$. Sketch the curve for values of t from -3 to 3.

Draw up a table of values, Table 10.3.

t	-3	-2	-1	0	1	2	3
x	9	4	1	0	1	4	9
y	-6	-4	-2	0	2	4	6

Table 10.3

The points $(9,-6)$, $(4,-4)$, $(1,-2)$, $(0,0)$, $(1,2)$, $(4,4)$ and $(9,6)$ lie on the curve shown in Fig. 10.4. The points which are plotted are labelled with the t-values of the parameter.

The idea that a point is defined by the value of its parameter is an important one. Thus, for the curve $x = t^2$, $y = 2t$ you can talk about the point $t = -2$, which means the point $(4,-4)$.

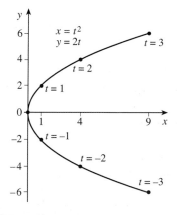

Fig. 10.4

The curve looks like a parabola on its side, and you will, in the next section, be able to prove that it is a parabola.

Example 10.1.3

A curve has parametric equations $x = \sin t$, $y = \sin 2t$, for values of t from 0 to 2π. Plot the curve, and indicate the points corresponding to values of t which are multiples of $\frac{1}{6}\pi$.

Draw up a table of values, Table 10.5.

t	0	$\frac{1}{6}\pi$	$\frac{1}{3}\pi$	$\frac{1}{2}\pi$	$\frac{2}{3}\pi$	$\frac{5}{6}\pi$
x	0	0.5	0.866	1	0.866	0.5
y	0	0.866	0.866	0	−0.866	−0.866

t	π	$\frac{7}{6}\pi$	$\frac{4}{3}\pi$	$\frac{3}{2}\pi$	$\frac{5}{3}\pi$	$\frac{11}{6}\pi$	2π
x	0	−0.5	−0.866	−1	−0.866	−0.5	0
y	0	0.866	0.866	0	−0.866	−0.866	0

Table 10.5

Fig. 10.6 illustrates this curve, with the points from the table labelled with their t-values, except for the origin, which is the point for which $t = 0$, $t = \pi$ and $t = 2\pi$.

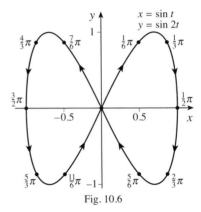

Fig. 10.6

Notice the arrows, which show the way that the values of t are increasing. These are not essential, but you should put them in if you want to show the direction in which the parameter is increasing.

If you go on to use values of t outside the interval from 0 to 2π, the curve will repeat itself.

Finally, you should notice that parametric equations enable you to produce curves whose equations can't be written in the form $y = f(x)$. The circle in Fig. 10.2, and the curves in Fig. 10.4 and Fig. 10.6, cannot be described by such an equation, because none of them have just one value of y for each value of x.

It is time to give a definition of a parameter.

> If $x = f(t)$ and $y = g(t)$, where f and g are functions of a variable t defined for some domain of values of t, then the equations $x = f(t)$ and $y = g(t)$ are called **parametric equations**, and the variable t is a **parameter**.

10.2* From parametric to cartesian equations

A curve that is described parametrically can sometimes also be described by a cartesian equation, by eliminating the parameter between the two parametric equations.

For example, in Example 10.1.2 the curve is given parametrically by $x = t^2$, $y = 2t$. In this case, you can write $t = \frac{1}{2}y$, so that $x = \left(\frac{1}{2}y\right)^2 = \frac{1}{4}y^2$, which you can rewrite as $y^2 = 4x$. The parameter t has been eliminated between the two equations $x = t^2$, $y = 2t$. You can see from Fig. 10.4 that $x = \frac{1}{4}y^2$ is simply $y = \frac{1}{4}x^2$ 'on its side'.

In general:

> If $x = f(t)$ and $y = g(t)$ are parametric equations of a curve C, and you eliminate the parameter between the two equations, each point of the curve C lies on the curve represented by the resulting cartesian equation.

Example 10.2.1

A curve is given parametrically by the equations $x = 2t + 1$, $y = 3t - 2$. Show that the 'curve' is a straight line and find its gradient.

From the first equation $t = \frac{1}{2}(x - 1)$, so $y = 3\left(\frac{1}{2}(x - 1)\right) - 2$; that is, $2y = 3x - 7$.

This is the equation of a straight line. Its gradient is $\frac{3}{2}$.

Example 10.2.2

Let E be the curve given parametrically by $x = a\cos t$, $y = b\sin t$, where a and b are constants and t is a parameter which takes values from 0 to 2π. Find the cartesian equation of E.

Since $x = a\cos t$ and $y = b\sin t$, $\cos t = \dfrac{x}{a}$ and $\sin t = \dfrac{y}{b}$. Then, using $\cos^2 t + \sin^2 t = 1$,

$\left(\dfrac{x}{a}\right)^2 + \left(\dfrac{y}{b}\right)^2 = 1$. This is the equation of the ellipse shown in Fig. 10.7. If a and b are equal, it is a circle of radius a.

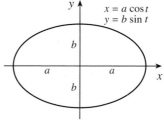

Fig. 10.7

<hr>

Exercise 10A

1 Find the coordinates of the point on the curve $x = 5t^2$, $y = 10t$

    (a) when $t = 6$,   (b) when $t = -1$.

2 Find the coordinates of the point on the curve $x = 1 - \dfrac{1}{t}$, $y = 1 + \dfrac{1}{t}$

    (a) when $t = 3$,   (b) when $t = -1$.

3 The parametric equations of a curve are $x = 2\cos t$, $y = 2\sin t$, for $0 \leqslant t < 2\pi$. What is the value of t at the point $(0,2)$?

4 A curve is given by $x = 5\cos t$, $y = 2\sin t$ for $0 \leqslant t < 2\pi$. Find the value of t at the point $\left(-2\tfrac{1}{2}, \sqrt{3}\right)$.

5 Sketch the curve given by $x = t^2$, $y = \dfrac{1}{t}$ for $t > 0$.

6 Sketch the curve given by $x = 3\cos t$, $y = 2\sin t$ for $0 \leqslant t < 2\pi$.

7 Sketch the graph of $x = 3t^2$, $y = 6t$ for $-4 \leqslant t \leqslant 4$.

8 Sketch the locus given by $x = \cos^2 t$, $y = \sin^2 t$ for $0 \leqslant t < 2\pi$.

9* Find cartesian equations for curves with these parametric equations.

    (a) $x = t^2$, $y = \dfrac{1}{t}$ (b) $x = 3t^2$, $y = 6t$ (c) $x = 2\cos t$, $y = 2\sin t$

10* Find cartesian equations for curves with these parametric equations.

    (a) $x = \cos^2 t$, $y = \sin^2 t$ (b) $x = \cos^3 t$, $y = \sin^3 t$

    (c) $x = 1 - \dfrac{1}{t}$, $y = 1 + \dfrac{1}{t}$ (d) $x = 3t^2$, $y = 2t^3$

11* Show that parametric equations for a circle with centre (p,q) and radius r are
$x = p + r\cos t$, $y = q + r\sin t$. Eliminate the parameter t to obtain the cartesian equation of
the circle in the form $(x - p)^2 + (y - q)^2 = r^2$.

10.3 Differentiation and parametric form

Suppose that a curve is defined parametrically. How can you find the gradient at a point on the curve without first finding the cartesian equation of the curve?

The key observation is that for a point P with parameter t on the curve, the coordinates (x,y) of P are both functions of t, so as t changes, x and y also change.

> If a curve is given parametrically by equations for x and y in terms of a parameter t, then
> $$\frac{dy}{dx} = \frac{dy}{dt} \bigg/ \frac{dx}{dt}.$$

You can now use the result in the box, but if you don't need the proof, skip to Example 10.3.1.

To establish the result, suppose that the value of t is increased by δt; then x increases by δx and y by δy.

Then, provided that $\delta x \neq 0$, $\dfrac{\delta y}{\delta x} = \dfrac{\delta y}{\delta t} \Big/ \dfrac{\delta x}{\delta t}$.

As $\delta t \to 0$, both $\delta x \to 0$ and $\delta y \to 0$, so $\displaystyle\lim_{\delta x \to 0} \dfrac{\delta y}{\delta x} = \lim_{\delta t \to 0} \dfrac{\delta y}{\delta x}$.

Therefore, assuming that $\displaystyle\lim_{\delta t \to 0}\left(\dfrac{\delta y}{\delta t}\Big/\dfrac{\delta x}{\delta t}\right) = \left(\lim_{\delta t \to 0}\dfrac{\delta y}{\delta t}\right)\Big/\left(\lim_{\delta t \to 0}\dfrac{\delta x}{\delta t}\right)$,

$$\frac{dy}{dx} = \lim_{\delta x \to 0}\frac{\delta y}{\delta x} = \lim_{\delta t \to 0}\frac{\delta y}{\delta x} = \lim_{\delta t \to 0}\left(\frac{\delta y}{\delta t}\Big/\frac{\delta x}{\delta t}\right) = \left(\lim_{\delta t \to 0}\frac{\delta y}{\delta t}\right)\Big/\left(\lim_{\delta t \to 0}\frac{\delta x}{\delta t}\right) = \frac{dy}{dt}\Big/\frac{dx}{dt}.$$

Therefore $\dfrac{dy}{dx} = \dfrac{dy}{dt}\Big/\dfrac{dx}{dt}$.

Notice that, just as the chain rule for differentiation is easy to remember because of 'cancelling', so is this rule. However, you should remember that this is no more than a helpful feature of the notation, and cancellation has no meaning in this context.

Example 10.3.1

Use parametric differentiation to find the gradient at $t = 3$ on the parabola $x = t^2$, $y = 2t$.

$\dfrac{dy}{dt} = 2$ and $\dfrac{dx}{dt} = 2t$, so $\dfrac{dy}{dx} = \dfrac{dy}{dt}\Big/\dfrac{dx}{dt} = \dfrac{2}{2t} = \dfrac{1}{t}$. When $t = 3$, the gradient is $\tfrac{1}{3}$.

Example 10.3.2

Find the equation of the normal at $(-8,4)$ to the curve which is given parametrically by $x = t^3$, $y = t^2$. Sketch the curve, showing the normal.

For the point $(-8,4)$, $t^3 = -8$ and $t^2 = 4$. These are both satisfied by $t = -2$.

As $\dfrac{dy}{dt} = 2t$ and $\dfrac{dx}{dt} = 3t^2$, $\dfrac{dy}{dx} = \dfrac{dy}{dt}\Big/\dfrac{dx}{dt} = \dfrac{2t}{3t^2} = \dfrac{2}{3t}$. When $t = -2$ the gradient is

$\dfrac{2}{3 \times (-2)} = -\dfrac{1}{3}$, so the gradient of the normal is $-\dfrac{1}{-1/3} = 3$.

Therefore the equation of the normal is $y - 4 = 3\big(x - (-8)\big)$ or $y = 3x + 28$.

Fig. 10.8 shows a sketch of the curve and the normal; remember that the normal will look perpendicular to the curve only if the scales on both axes are the same.

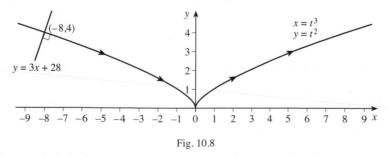

Fig. 10.8

At the origin, where $t = 0$, the curve has a **cusp**. As t increases, the point moves along the curve from left to right, but at the cusp it comes to a stop and starts to move back upwards.

The gradient is not defined when $t = 0$, because the tangent at the origin is the y-axis.

Exercise 10B

1 Find $\dfrac{dy}{dx}$ in terms of t for the following curves.

(a) $x = t^3, y = 2t$

(b) $x = \sin t, y = \cos t$

(c) $x = 2\cos t, y = 3\sin t$

(d) $x = t^3 + t, y = t^2 - t$

2 Find the gradients of the tangents to the following curves, at the specified values of t.

(a) $x = 3t^2, y = 6t$ when $t = 0.5$

(b) $x = t^3, y = t^2$ when $t = 2$

(c) $x = 1 - \dfrac{1}{t}, y = 1 + \dfrac{1}{t}$ when $t = 2$

(d) $x = t^2, y = \dfrac{1}{t}$ when $t = 3$

3 Find the gradients of the normals to the following curves, at the specified values of t.

(a) $x = 5t^2, y = 10t$ when $t = 3$

(b) $x = \cos^2 t, y = \sin^2 t$ when $t = \frac{1}{3}\pi$

(c) $x = \cos^3 t, y = \sin^3 t$ when $t = \frac{1}{6}\pi$

(d) $x = t^2 + 2, y = t - 2$ when $t = 4$

4 Show that the equation of the tangent to the curve $x = 3\cos t, y = 2\sin t$ when $t = \frac{3}{4}\pi$ is $3y = 2x + 6\sqrt{2}$.

5 (a) Find the gradient of the curve $x = t^3, y = t^2 - t$ at the point $(1,0)$.

(b) Hence find the equation of the tangent to the curve at this point.

6 A curve has parametric equations $x = t - \cos t, y = \sin t$. Find the equation of the tangent to the curve when $t = \pi$.

7 Find the equations of the tangents to these curves at the specified values.

(a) $x = t^2, y = 2t$ when $t = 3$

(b) $x = 5\cos t, y = 3\sin t$ when $t = \frac{11}{6}\pi$

8 Find the equations of the normals to these curves at the specified values.

(a) $x = 5t^2, y = 10t$ when $t = 3$

(b) $x = \cos t, y = \sin t$ when $t = \frac{2}{3}\pi$

9 (a) Find the equation of the normal to the hyperbola $x = 4t, y = \dfrac{4}{t}$ at the point $(8,2)$.

(b) Find the coordinates of the point where this normal crosses the curve again.

10 (a) Find the equation of the normal to the parabola $x = 3t^2, y = 6t$ at the point where $t = -2$.

(b) Find the coordinates of the point where this normal crosses the curve again.

10.4* Proving properties of curves

Parameters are a powerful tool for proving properties about curves. Here are two examples which show a general method.

Example 10.4.1

A parabola is given by $x = at^2$, $y = 2at$. The tangent at a point P on the parabola meets the x-axis at T. Prove that PT is bisected by the tangent at the vertex of the parabola.

You may wonder why the parabola in Fig. 10.9 is on its side. This is just a convention. Mathematicians usually express the parabola parametrically as $x = at^2$, $y = 2at$ rather than $x = 2at$, $y = at^2$. In this case, the vertex is still the point where the axis of symmetry meets the parabola, which is the origin, and the tangent at the vertex is the y-axis.

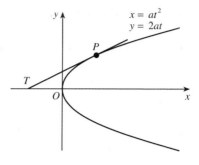

Fig. 10.9

Let P be the point on the parabola, shown in Fig. 10.9, with coordinates $\left(at^2, 2at\right)$. Since

$$\frac{dy}{dx} = \frac{dy}{dt} \bigg/ \frac{dx}{dt} = \frac{2a}{2at} = \frac{1}{t}$$

the gradient at P is $\dfrac{1}{t}$. The equation of the tangent at P is therefore

$$y - 2at = \frac{1}{t}\left(x - at^2\right), \text{ which can be simplified to } ty = x + at^2.$$

This tangent meets the x-axis at the point where $y = 0$, so $x = -at^2$ and T is the point with coordinates $\left(-at^2, 0\right)$.

The mid-point of PT is $\left(\frac{1}{2}\left(at^2 + \left(-at^2\right)\right), \frac{1}{2}(2at + 0)\right)$, which is $(0, at)$. Since the tangent at the vertex has equation $x = 0$, the point $(0, at)$ lies on it. Therefore PT is bisected by the tangent at the vertex.

Example 10.4.2

A curve is given parametrically by $x = a\cos^3 t$, $y = a\sin^3 t$, where a is a positive constant, for $0 \leqslant t < 2\pi$. The tangent at any point P meets the x-axis at A and the y-axis at B. Prove that the length of AB is constant.

Let P be the point on the curve, shown in Fig. 10.10, with parameter t. P has coordinates $\left(a\cos^3 t, a\sin^3 t\right)$.

To find the gradient at P, calculate

$$\frac{dy}{dx} = \frac{dy}{dt} \bigg/ \frac{dx}{dt} = \frac{3a\sin^2 t\cos t}{-3a\cos^2 t\sin t} = -\frac{\sin t}{\cos t}.$$

The gradient at P is $-\dfrac{\sin t}{\cos t}$.

The equation of the tangent at P is

$$y - a\sin^3 t = -\frac{\sin t}{\cos t}\left(x - a\cos^3 t\right).$$

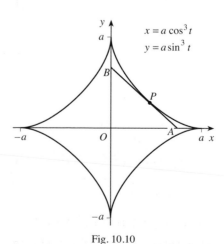

Fig. 10.10

This can be simplified to

$$y\cos t + x\sin t = a\sin^3 t\cos t + a\sin t\cos^3 t$$

$$= a\sin t\cos t\left(\sin^2 t + \cos^2 t\right)$$

$$= a\sin t\cos t.$$

The points A and B have coordinates $(a\cos t, 0)$ and $(0, a\sin t)$. The length AB is

$$\sqrt{(0 - a\cos t)^2 + (a\sin t - 0)^2} = \sqrt{a^2\cos^2 t + a^2\sin^2 t} = a.$$

The length AB is therefore constant.

The curve $x = a\cos^3 t$, $y = a\sin^3 t$ is called an *astroid*. If you think of the tangent as a ladder of length a sliding down the 'wall and floor' made by the y-axis and the x-axis, then the ladder always touches the astroid.

Exercise 10C*

1 Let P be a point on the curve $x = t^2$, $y = \dfrac{1}{t}$. If the tangent to the curve at P meets the x- and y-axes at A and B respectively, prove that $PA = 2BP$.

2 A parabola is given parametrically by $x = at^2$, $y = 2at$. If P is any point on the parabola, let F be the foot of the perpendicular from P onto the axis of symmetry. Let G be the point where the normal from P crosses the axis of symmetry.

   Prove that $FG = 2a$.

3 P is a point on the parabola given parametrically by $x = at^2$, $y = 2at$, where a is a constant. Let S be the point $(a, 0)$, Q be the point $(-a, 2at)$ and T be the point where the tangent at P to the parabola crosses the axis of symmetry of the parabola.

   (a) Show that $SP = PQ = QT = ST = at^2 + a$.

   (b) Prove that angle QPT is equal to angle SPT.

   (c) If PM is parallel to the axis of the parabola, with M to the right of P, and PN is the normal to the parabola at P, show that angle MPN is equal to angle NPS.

4 P, Q, R and S are four points on the hyperbola $x = ct$, $y = \dfrac{c}{t}$ with parameters p, q, r and s respectively. Prove that, if the chord PQ is perpendicular to the chord RS, then $pqrs = -1$.

5 Let P be a point on the ellipse with parametric equations $x = 5\cos t$, $y = 3\sin t$ for $0 \leqslant t < 2\pi$, and let F and G be the points $(-4, 0)$ and $(4, 0)$ respectively. Prove that

   (a) $FP = 5 + 4\cos t$,   (b) $FP + PG = 10$.

   Let the normal at P make angles θ and ϕ with FP and GP respectively. Prove that

   (c) $\tan\theta = \dfrac{4}{3}\sin t$,   (d) $\theta = \phi$.

6 Let H be the curve with parametric equations $x = t$, $y = \dfrac{1}{t}$, and let P be a point on H. Let the tangent at P meet the x-axis at T, and let O be the origin. Prove that $OP = PT$.

7 For the curve H in Question 6, let S be the point $\left(\sqrt{2}, \sqrt{2}\right)$. Let N be the point on the tangent to H at P such that SN is perpendicular to PN.

(a) Show that the coordinates of N satisfy the equations $t^2 y + x = 2t$ and $y - t^2 x = \sqrt{2}\left(1 - t^2\right)$.

(b) If you square and add the equations in part (a), show that you obtain $x^2 + y^2 = 2$. Interpret this result geometrically.

8 Let P and Q be the points with parameters t and $t + \pi$ on the curve, called a *cardioid*, with parametric equations $x = 2\cos t - \cos 2t$, $y = 2\sin t - \sin 2t$. Let A be the point $(1,0)$. Prove that

(a) the gradient of AP is $\tan t$, (b) PAQ is a straight line,

(c) the length of the line segment PQ is constant.

<hr>

Miscellaneous exercise 10

1 The parametric equations of a curve are $x = \cos t$, $y = 2\sin t$ where the parameter t takes all values such that $0 \leqslant t \leqslant \pi$.

(a) Find the value of t at the point A where the line $y = 2x$ intersects the curve.

(b) Show that the tangent to the curve at A has gradient -2 and find the equation of this tangent in the form $ax + by = c$, where a and b are integers. (OCR)

2 The parametric equations of a curve are $x = 2\cos t$, $y = 5 + 3\cos 2t$, where $0 < t < \pi$.

Express $\dfrac{dy}{dx}$ in terms of t, and hence show that the gradient at any point of the curve is less than 6. (OCR)

3 A curve is defined by the parametric equations: $x = t - \dfrac{1}{t}$, $y = t + \dfrac{1}{t}$, $t \neq 0$.

(a) Use parametric differentiation to determine $\dfrac{dy}{dx}$ as a function of the parameter t.

(b) Show that the equation of the normal to the curve at the point where $t = 2$ may be written as $3y + 5x = 15$.

(c) Determine the cartesian equation of the curve. (OCR)

4 A curve is defined parametrically by $x = t^3 + t$, $y = t^2 + 1$.

(a) Find $\dfrac{dy}{dx}$ in terms of t.

(b) Find the equation of the normal to this curve at the point where $t = 1$. (OCR)

5 A curve is defined by the parametric equations $x = \sin t, y = \sqrt{3}\cos t$.

    (a) Determine $\dfrac{dy}{dx}$ in terms of t for points on the curve where t is not an odd multiple of $\frac{1}{2}\pi$.

    (b) Find an equation for the tangent to the curve at the point where $t = \frac{1}{6}\pi$.

    (c) Show that all points on the curve satisfy the equation $x^2 + \frac{1}{3}y^2 = 1$.   (OCR)

6 The parametric equations of a curve are $x = t + e^{-t}$, $y = 1 - e^{-t}$, where t takes all real values. Express $\dfrac{dy}{dx}$ in terms of t, and hence find the value of t for which the gradient of the curve is 1, giving your answer in logarithmic form.   (OCR)

7 A curve is defined by the parametric equations $x = 3\sin t, y = 2\cos t$.

    (a) Show that the cartesian equation of the curve is $4x^2 + 9y^2 = 36$.

    (b) Determine an equation of the normal to the curve at the point with parameter $t = \alpha$ where $\sin\alpha = 0.6$ and $\cos\alpha = 0.8$.

    (c) Find the cartesian coordinates of the point where the normal in part (b) meets the curve again.   (OCR)

8 A curve is defined parametrically for $0 \leqslant t \leqslant \pi$ by $x = 2(1 + \cos t), y = 4\sin^2 t$.

    (a) Determine the equation of the tangent to the curve at the point where $t = \frac{1}{3}\pi$.

    (b) Obtain the cartesian equation of the curve in simplified form.   (OCR)

9 The curves in parts (a) to (h) are examples of *Lissajous figures*. By first finding the coordinates of the points where either x or y takes the values -1, 0 or 1, sketch the curves completely. Indicate on your sketches, with arrows, the direction on each curve in which t is increasing. Check your sketches with a graphic calculator, if you have one.

    (a) $x = \cos t, y = \cos 2t$ (b) $x = \sin t, y = \cos 2t$

    (c) $x = \sin t, y = \sin 3t$ (d) $x = \sin t, y = \cos 3t$

    (e) $x = \cos 2t, y = \sin 3t$ (f) $x = \cos 2t, y = \cos 3t$

    (g) $x = \sin 2t, y = \sin 3t$ (h) $x = \sin 2t, y = \cos 3t$

10* A curve is defined parametrically by $x = t^2$, $y = t^2$ where t is real.

    (a) Describe the curve.

    (b) Eliminate the parameter to find the cartesian equation of the curve. Describe the curve resulting from the cartesian equation.

    (c) Reconcile what you find with the result in the box in Section 10.2.

11 Curves defined implicitly

This chapter shows how to find gradients of curves which are described by implicit equations. When you have completed it, you should

- recognise the form of the equation of a circle
- understand the nature of implicit equations, and be able to differentiate them.

11.1 The equation of a circle

You have met the cartesian equations of many curves in this course, but it may seem surprising that these have not included the simplest curve of all, a circle. It will be useful to know this as an example of the type of equation discussed in this chapter.

Remember that the equation of a curve is a rule satisfied by the coordinates (x, y) of any point which lies on it, and not by any points which do not lie on it. For the circle, the rule expresses the fact that it consists of all points which are a fixed distance (the radius) from a fixed point (the centre).

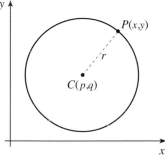

Let P with coordinates (x, y) be a point on the circumference of a circle with centre $C(p, q)$ and radius r, where, of course, $r > 0$ (see Fig. 11.1). Then, for all possible positions of P on the circle, the distance $CP = r$.

But, from the distance formula in P1 Section 1.1, the distance CP is $\sqrt{(x - p)^2 + (y - q)^2}$, so the equation of the circle is

Fig. 11.1

$$\sqrt{(x - p)^2 + (y - q)^2} = r, \quad \text{or} \quad (x - p)^2 + (y - q)^2 = r^2.$$

> The equation of a circle with centre (p, q) and radius r is
> $$(x - p)^2 + (y - q)^2 = r^2.$$
> When the centre is $(0, 0)$, the equation is $x^2 + y^2 = r^2$.

Example 11.1.1
Find the equation of the circle with centre $(1, 2)$ and radius 3.

Using the formula, the equation is $(x - 1)^2 + (y - 2)^2 = 9$.

You can also multiply out the brackets to get

$$x^2 - 2x + 1 + y^2 - 4y + 4 = 9, \quad \text{which is} \quad x^2 + y^2 - 2x - 4y - 4 = 0.$$

Either of the forms $(x - 1)^2 + (y - 2)^2 = 9$ and $x^2 + y^2 - 2x - 4y - 4 = 0$ is usually acceptable.

Example 11.1.1 shows that the circle has an equation of the form $x^2 + y^2 + ax + by + c = 0$, where a, b and c are constants. The next example reverses the argument, and shows how you can find the centre and radius when you know the values of a, b and c.

Example 11.1.2

Find the centre and radius of the circle $x^2 + y^2 - 2x + 4y - 7 = 0$.

Writing the equation as $(x^2 - 2x) + (y^2 + 4y) = 7$, completing the squares inside the brackets and compensating the right side gives

$$(x^2 - 2x + 1) + (y^2 + 4y + 4) = 7 + 1 + 4,$$

that is,

$$(x - 1)^2 + (y + 2)^2 = 12.$$

This equation expresses the property that the square of the distance of (x, y) from $(1, -2)$ is equal to 12. It is therefore the equation of a circle with centre $(1, -2)$ and radius $\sqrt{12}$.

11.2 Equations of curves

In this course you have used coordinates and graphs in two ways: for understanding functions, and for obtaining geometrical results.

The graph of a function provides a visual representation of an equation $y = f(x)$. The variables x and y play different roles: for each x there is a unique y, but the reverse need not be true. The graph shows properties of the function such as whether it is increasing or decreasing, and where it has its maximum value. It is usually unnecessary to have equal scales in the x- and y-directions. Indeed, in many applications, the two variables may represent quite different kinds of quantity, measured in different units.

When you use coordinates in geometry, the x- and y-coordinates have equal status. You must use the same scales in both directions, otherwise circles will not look circular and perpendicular lines will not appear perpendicular. Equations are often written not as $y = f(x)$, but in forms such as $ax + by + c = 0$ or $x^2 + y^2 + ax + by + c = 0$, which emphasise that x and y are equal partners. These are **implicit equations** which define the relation between x and y.

Sometimes you can put such equations into the $y = f(x)$ form: for example, you can write $3x - 2y + 6 = 0$ as $y = \frac{3}{2}x + 3$.

However, the circle $(x - 1)^2 + (y - 2)^2 = 9$ has two values of y for each x between -2 and 4, given by $y = 2 \pm \sqrt{9 - (x - 1)^2}$. So the equation of the circle cannot be written as an equation of the form $y = f(x)$.

Similarly, the curve in Fig. 11.2, whose equation is

$$x^3 + y^3 + x^2 - y = 0,$$

cannot be put into either of the forms $y = f(x)$ or $x = f(y)$.

If you take a particular value for x, it gives a cubic equation for y, and if you take a particular value for y, it gives a cubic equation for x. For some values of y there are three values of x, and for some values of x there are three values of y, so the equation cannot be expressed in function form, as $y = f(x)$ or as $x = f(y)$.

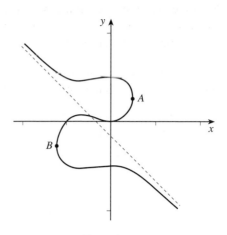

Fig. 11.2

You can easily find a few features of the curve from the equation:

- The equation is satisfied by $x = 0$, $y = 0$, so the curve contains the origin.
- The curve cuts the x-axis, $y = 0$, where $x^3 + x^2 = 0$ so $x = 0$ or -1.
- The curve cuts the y-axis, $x = 0$, where $y^3 - y = 0$ so $y = -1$, 0 or 1.

11.3 Finding gradients from implicit equations

When differentiation was first introduced in P1, the method used was to take two points P and Q close together on the graph of $y = f(x)$, and to find the gradient of the chord joining them. Denoting their coordinates by (x, y) and $(x + \delta x, y + \delta y)$, you can write $y = f(x)$ and $y + \delta y = f(x + \delta x)$, so the gradient is

$$\frac{\delta y}{\delta x} = \frac{f(x + \delta x) - f(x)}{\delta x}.$$

Then, letting Q move round the curve towards P, you get the limiting value

$$\frac{dy}{dx} = \lim_{\delta x \to 0} \frac{\delta y}{\delta x} = \lim_{\delta x \to 0} \frac{f(x + \delta x) - f(x)}{\delta x}.$$

If you want to find $\dfrac{dy}{dx}$ for a curve like the one in Fig. 11.2, the same principles apply, but the algebra is different because you don't have an equation in the form $y = f(x)$. The coordinates therefore have to be substituted into the implicit equation, giving (for this example) the two equations

$$x^3 + y^3 + x^2 - y = 0, \qquad\qquad \text{Equation P}$$

and $\quad (x + \delta x)^3 + (y + \delta y)^3 + (x + \delta x)^2 - (y + \delta y) = 0. \qquad \text{Equation Q}$

Using the binomial theorem, the terms of Equation Q can be expanded to give

$$\left(x^3 + 3x^2(\delta x) + 3x(\delta x)^2 + (\delta x)^3\right) + \left(y^3 + 3y^2(\delta y) + 3y(\delta y)^2 + (\delta y)^3\right)$$
$$+ \left(x^2 + 2x(\delta x) + (\delta x)^2\right) - (y + \delta y) = 0.$$

To make this look less complicated, rearrange the terms according to the degree to which δx and δy appear, as

$$\overbrace{\left(x^3 + y^3 + x^2 - y\right)}^{\text{degree 0}} + \overbrace{\left(3x^2(\delta x) + 3y^2(\delta y) + 2x(\delta x) - \delta y\right)}^{\text{degree 1}}$$

$$+ \overbrace{\left(3x(\delta x)^2 + 3y(\delta y)^2 + (\delta x)^2\right)}^{\text{degree 2}} + \overbrace{\left((\delta x)^3 + (\delta y)^3\right)}^{\text{degree 3}} = 0.$$

The first group of terms is just the left side of Equation P, so it is zero. Since you want to find the gradient of the chord, $\dfrac{\delta y}{\delta x}$, rewrite the other groups to show this fraction:

$$(0) + \left(3x^2 + 3y^2 \frac{\delta y}{\delta x} + 2x - \frac{\delta y}{\delta x}\right)\delta x + \left(3x + 3y\left(\frac{\delta y}{\delta x}\right)^2 + 1\right)(\delta x)^2 + \left(1 + \left(\frac{\delta y}{\delta x}\right)^3\right)(\delta x)^3 = 0.$$

There is now a common factor δx (which is non-zero), so divide by it to get

$$\left(3x^2 + 3y^2 \frac{\delta y}{\delta x} + 2x - \frac{\delta y}{\delta x}\right) + \left(3x + 3y\left(\frac{\delta y}{\delta x}\right)^2 + 1\right)\delta x + \left(1 + \left(\frac{\delta y}{\delta x}\right)^3\right)(\delta x)^2 = 0.$$

There is one last step, to see what happens as Q approaches P, when δx tends to 0. Then $\dfrac{\delta y}{\delta x}$ becomes $\dfrac{dy}{dx}$, so the equation becomes

$$\left(3x^2 + 3y^2 \frac{dy}{dx} + 2x - \frac{dy}{dx}\right) + \left(3x + 3y\left(\frac{dy}{dx}\right)^2 + 1\right)\times 0 + \left(1 + \left(\frac{dy}{dx}\right)^3\right)\times 0^2 = 0,$$

which is simply $3x^2 + 3y^2 \dfrac{dy}{dx} + 2x - \dfrac{dy}{dx} = 0$.

Now compare this with the original equation, Equation P. You can see that each term has been replaced by its derivative with respect to x. Thus x^3 has become $3x^2$, x^2 has become $2x$ and y has become $\dfrac{dy}{dx}$. The only term which calls for comment is the second, which is an application of the chain rule:

$$\frac{d}{dx}\left(y^3\right) = \frac{d}{dy}\left(y^3\right) \times \frac{dy}{dx} = 3y^2 \frac{dy}{dx}.$$

This is an example of a general rule:

To find $\dfrac{dy}{dx}$ from an implicit equation, differentiate each term with respect to x, using the chain rule to differentiate any function $f(y)$ as $f'(y)\dfrac{dy}{dx}$.

For the curve in Fig. 11.2, you can find the gradient by rearranging the differentiated equation as $\left(3x^2 + 2x\right) = \left(1 - 3y^2\right)\dfrac{dy}{dx}$, so

$$\frac{dy}{dx} = \frac{3x^2 + 2x}{1 - 3y^2}.$$

It is interesting to notice that $\dfrac{dy}{dx} = 0$ when $x = 0$ or $x = -\frac{2}{3}$. Fig. 11.2 shows that each of these values of x corresponds to three points on the curve: $x = 0$ at $(0,1)$, $(0,0)$ and $(0,-1)$, and $x = -\frac{2}{3}$ where $y^3 - y = -\frac{4}{27}$. This is a cubic equation whose roots can be found by numerical methods of the kind described in Chapter 8; they are 0.92, 0.15 and -1.07, correct to 2 decimal places.

Since equations in implicit form treat the x- and y-coordinates equally, you might also want to find $\dfrac{dx}{dy}$, which is $1 \Big/ \dfrac{dy}{dx}$:

$$\frac{dx}{dy} = \frac{1 - 3y^2}{3x^2 + 2x}.$$

The proof that $\dfrac{dx}{dy} = 1 \Big/ \dfrac{dy}{dx}$ is given in P3 Section 19.2.

The tangent to the curve is parallel to the y-axis when $\dfrac{dx}{dy} = 0$, which is when $y = \dfrac{1}{\sqrt{3}}$ or $-\dfrac{1}{\sqrt{3}}$. These points are labelled A and B in Fig. 11.2.

If you imagine the curve split into three pieces by making cuts at A and B, then each of these pieces defines y as a function of x (since for each x there is a unique y).

On each piece $\dfrac{dy}{dx}$ can be defined as the limit of $\dfrac{\delta y}{\delta x}$ in the usual way. If the curve is then stitched up again, you have a definition of $\dfrac{dy}{dx}$ at every point of the curve except at A and B, which are the points where the gradient of the tangent is not defined.

This process makes it possible to justify the rule in the box on page 144. Although the algebraic expression for y in terms of x is not known, the implicit equation defines y in terms of x on each piece of the curve; and when this y is substituted, the equation becomes an identity which is true for all relevant values of x. Any identity in x can be differentiated to give another identity. This produces an equation in which each term is differentiated with respect to x, as described by the rule.

Example 11.3.1

Show that $(1,2)$ is on the circle $x^2 + y^2 - 6x + 2y - 3 = 0$, and find the gradient there.

Substituting $x = 1$, $y = 2$ in the left side of the equation gives $1 + 4 - 6 + 4 - 3$, which is equal to 0.

Method 1   Differentiating term by term with respect to x gives

$$2x + 2y\frac{dy}{dx} - 6 + 2\frac{dy}{dx} - 0 = 0, \quad \text{that is} \quad x + y\frac{dy}{dx} - 3 + \frac{dy}{dx} = 0.$$

Setting $x = 1$, $y = 2$ gives $1 + 2\frac{dy}{dx} - 3 + \frac{dy}{dx} = 0$, so $\frac{dy}{dx} = \frac{2}{3}$ at this point.

Method 2   Using the method of Example 11.1.2, the equation can be written as $(x-3)^2 + (y+1)^2 = 13$. The centre of the circle is $(3,-1)$, so the gradient of the radius to $(1,2)$ is $\dfrac{2 - (-1)}{1 - 3} = -\dfrac{3}{2}$.

Since the tangent is perpendicular to the radius, the gradient of the tangent at $(1,2)$ is $-\dfrac{1}{-\frac{3}{2}} = \dfrac{2}{3}$.

Example 11.3.2

Find an expression for $\dfrac{dy}{dx}$ on the curve $3x^2 - 2y^3 = 1$.

Method 1   Differentiating term by term gives

$$6x - 6y^2\frac{dy}{dx} = 0, \quad \text{so} \quad \frac{dy}{dx} = \frac{x}{y^2}.$$

Method 2   This equation can be written explicitly as $y = \left(\frac{3}{2}x^2 - \frac{1}{2}\right)^{\frac{1}{3}}$, and by the chain rule

$$\frac{dy}{dx} = \frac{1}{3}\left(\frac{3}{2}x^2 - \frac{1}{2}\right)^{-\frac{2}{3}} \times 3x = x\left(y^3\right)^{-\frac{2}{3}} = \frac{x}{y^2}.$$

Example 11.3.3

Sketch the graph of $\cos x + \cos y = \frac{1}{2}$, and find the equation of the tangent at the point $\left(\frac{1}{2}\pi, \frac{1}{3}\pi\right)$.

Fig. 11.3 shows the part of the graph for which the values of both of x and y are between $-\pi$ and π. Since $\cos y \leqslant 1$, $\cos x \geqslant -\frac{1}{2}$, so $-\frac{2}{3}\pi \leqslant x \leqslant \frac{2}{3}\pi$. Similarly $-\frac{2}{3}\pi \leqslant y \leqslant \frac{2}{3}\pi$.

Because $\cos$ is an even function, the graph is symmetrical about both axes; and because interchanging x and y does not alter the equation,

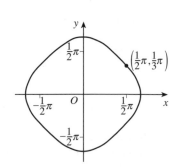

Fig. 11.3

the graph is also symmetrical about $y = x$, and hence about $y = -x$. Also, because the function cos has period 2π, this shape is repeated over the whole plane at intervals of 2π in both x- and y-directions.

Differentiating the equation gives

$$-\sin x + (-\sin y)\frac{dy}{dx} = 0, \quad \text{so} \quad \frac{dy}{dx} = -\frac{\sin x}{\sin y}.$$

At $\left(\frac{1}{2}\pi, \frac{1}{3}\pi\right)$ the gradient is $-\dfrac{1}{\frac{1}{2}\sqrt{3}} = -\dfrac{2}{\sqrt{3}}$, so the equation of the tangent is

$$y - \tfrac{1}{3}\pi = -\tfrac{2}{\sqrt{3}}\left(x - \tfrac{1}{2}\pi\right), \quad \text{or} \quad y + \tfrac{2}{\sqrt{3}}x = \tfrac{1}{3}\pi\left(1 + \sqrt{3}\right).$$

Exercise 11A

1 Each of the following equations represents a circle. Find the gradient of the tangent at the given point (i) by finding the coordinates of the centre as in method 2 of Example 11.3.1, and (ii) by differentiating the implicit equation.

(a) $x^2 + y^2 = 25$ $(-3,4)$

(b) $x^2 + y^2 + 4x - 6y = 24$ $(4,2)$

(c) $x^2 + y^2 - 6x + 8y = 0$ $(6,-8)$

(d) $x^2 + y^2 - 2x - 4y = 0$ $(0,0)$

2 Differentiate the implicit equation $y^2 = 4x$ to find the gradient at $(9,-6)$ on the curve.

3 Differentiate the implicit equation of the ellipse $3x^2 + 4y^2 = 16$ to find the equation of the tangent at the point $(2,-1)$.

4 Differentiate the implicit equation of the hyperbola $4x^2 - 3y^2 = 24$ to find the equation of the normal at the point $(3,-2)$. Find the y-coordinate of the point where the normal meets the curve again.

5 Consider the curve with equation $x^2 + 4y^2 = 1$.

(a) Find the coordinates of the points where the curve cuts the coordinate axes.

(b) Find the interval of possible values of x and y for points on the curve.

(c) Show that the curve is symmetrical about both the x- and y-axes.

(d) Differentiate the equation with respect to x, and show that $\dfrac{dy}{dx} = 0$ when $x = 0$. Interpret this geometrically.

(e) Repeat part (d) with the roles of x and y reversed.

(f) Use your results to sketch the curve.

6 Repeat Question 5, using the curve with equation $x^2 - y^2 = 1$. If there are parts of the question which have no answer, or are impossible, say why that is so.

7 Consider the curve $y^3 = (x-1)^2$.

    (a) Find the coordinates of the points where the curve crosses the axes.

    (b) Are there any values which either x or y cannot take?

    (c) Differentiate the equation $y^3 = (x-1)^2$ to find an expression for the gradient in terms of x and y. Find the gradient of the curve where it crosses the y-axis.

    (d) What happens to the gradient as x gets close to 1?

    (e) By making the substitution $x = 1 + X$, and examining the resulting equation between y and X, show that the curve is symmetrical about the line $x = 1$.

    (f) Sketch the curve. If you can, use a graphic calculator to check your results.

8 Use methods similar to those of Question 5 to sketch the curve $x^4 + y^4 = 1$. On the same diagram, sketch the curve $x^2 + y^2 = 1$.

9 (a) Show that the origin lies on the curve $e^x + e^y = 2$.

    (b) Differentiate the equation with respect to x, and explain why the gradient is always negative.

    (c) Find any restrictions that you can on the values of x and y, and sketch the curve.

10 (a) Show that if (a,b) lies on the curve $x^2 + y^3 = 2$, then so does $(-a,b)$. What can you deduce from this about the shape of the curve?

    (b) Differentiate $x^2 + y^3 = 2$ with respect to x, and deduce what you can about the gradient for negative and for positive values of x.

    (c) Show that there is a stationary point at $\left(0, \sqrt[3]{2}\right)$, and deduce its nature.

    (d) Sketch the curve.

11 Find the coordinates of the points at which the curve $y^5 + y = x^3 + x^2$ meets the coordinate axes, and find the gradients of the curve at each of these points.

12 Find the gradient of the curve $y^3 - 3y^2 + 2y = e^x + x - 1$ at the points where it crosses the y-axis.

11.4 Implicit equations including products

The implicit equations in Section 11.3 contained terms in x and terms in y, but there were no terms which involved both x and y. These more complicated terms can be differentiated using the product or quotient rule, sometimes in conjunction with the chain rule.

Example 11.4.1

Find the derivatives with respect to x of

(a) $y \sin x$,   (b) $y^3 \ln x$,   (c) $e^{x^2 y}$,   (d) $\cos \dfrac{x}{y}$.

    (a) By the product rule,

$$\frac{d}{dx} y \sin x = \frac{d}{dx} y \times \sin x + y \times \frac{d}{dx} \sin x = \frac{dy}{dx} \sin x + y \cos x.$$

(b) $\dfrac{d}{dx} y^3 \ln x = \dfrac{d}{dx} y^3 \times \ln x + y^3 \times \dfrac{d}{dx} \ln x = 3y^2 \dfrac{dy}{dx} \ln x + \dfrac{y^3}{x}$.

(c) Use the chain rule followed by the product rule.

$$\dfrac{d}{dx} e^{x^2 y} = e^{x^2 y} \times \dfrac{d}{dx} x^2 y = e^{x^2 y}\left(2xy + x^2 \dfrac{dy}{dx}\right).$$

(d) $\dfrac{d}{dx} \cos \dfrac{x}{y} = -\sin \dfrac{x}{y} \times \dfrac{1 \times y - x \times \dfrac{dy}{dx}}{y^2} = \dfrac{x \dfrac{dy}{dx} - y}{y^2} \sin \dfrac{x}{y}$.

Example 11.4.2
Find the gradient of $x^2 y^3 = 72$ at the point $(3,2)$.

Two methods are given. The first is direct. The second begins by taking logarithms; this makes expressions involving products of powers easier to handle.

Method 1   Differentiating with respect to x,

$$2xy^3 + x^2\left(3y^2 \dfrac{dy}{dx}\right) = 0.$$

At $(3,2)$, $2 \times 3 \times 8 + 9 \times 3 \times 4 \dfrac{dy}{dx} = 0$, so $\dfrac{dy}{dx} = -\dfrac{4}{9}$.

Method 2   Write the equation as $\ln\left(x^2 y^3\right) = \ln 72$. By the laws of logarithms,

$\ln\left(x^2 y^3\right) = \ln x^2 + \ln y^3 = 2\ln x + 3\ln y$, so the equation is $2\ln x + 3\ln y = \ln 72$.

Differentiating gives $\dfrac{2}{x} + \dfrac{3}{y} \dfrac{dy}{dx} = 0$, so $\dfrac{dy}{dx} = -\dfrac{2y}{3x}$. At $(3,2)$, $\dfrac{dy}{dx} = -\dfrac{4}{9}$.

Method 2 is sometimes called 'logarithmic differentiation'.

Example 11.4.3
The equation $x^2 - 6xy + 25y^2 = 16$ represents an ellipse with its centre at the origin. What ranges of values of x and y would you need in order to plot the whole of the curve on a computer screen?

Method 1   The problem is equivalent to finding the points where the tangent to the curve is parallel to one of the axes.

Differentiating gives

$$2x - 6\left(1 \times y + x \times \dfrac{dy}{dx}\right) + 50y \dfrac{dy}{dx} = 0, \quad \text{that is} \quad (x - 3y) + (25y - 3x)\dfrac{dy}{dx} = 0.$$

The tangent is parallel to the x-axis when $\dfrac{dy}{dx} = 0$, which is when $x = 3y$.

Substituting this into the equation of the ellipse gives

$$(3y)^2 - 6(3y)y + 25y^2 = 16, \quad 16y^2 = 16, \quad y = -1 \text{ or } 1.$$

The tangents are therefore parallel to the x-axis at $(-3,-1)$ and $(3,1)$.

The tangent is parallel to the y-axis when $\dfrac{dx}{dy} = 0$. Since $\dfrac{dx}{dy} = 1 \Big/ \dfrac{dy}{dx}$, this occurs when $25y = 3x$. Substituting $y = \frac{3}{25}x$ gives

$$x^2 - 6x\left(\tfrac{3}{25}x\right) + 25\left(\tfrac{3}{25}x\right)^2 = 16, \quad \tfrac{16}{25}x^2 = 16, \quad x = -5 \text{ or } 5.$$

The points of contact are $\left(-5,-\tfrac{3}{5}\right)$ and $\left(5,\tfrac{3}{5}\right)$.

To fit the curve on the screen you need $-5 \leqslant x \leqslant 5$ and $-1 \leqslant y \leqslant 1$. This is illustrated in Fig. 11.4.

Method 2   The equation can be written as a quadratic in x:

$$x^2 - 6yx + \left(25y^2 - 16\right) = 0.$$

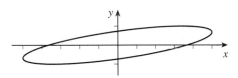

Fig. 11.4

The condition for this to give real values of x is

$$(6y)^2 - 4\left(25y^2 - 16\right) \geqslant 0 \quad \text{that is} \quad 64 - 64y^2 \geqslant 0, \quad -1 \leqslant y \leqslant 1.$$

Similarly, from the quadratic in y, which is $25y^2 - 6xy + \left(x^2 - 16\right) = 0$, you get the condition

$$(6x)^2 - 4 \times 25\left(x^2 - 16\right) \geqslant 0, \quad \text{that is} \quad 1600 - 64x^2 \geqslant 0, \quad -5 \leqslant x \leqslant 5.$$

Exercise 11B

1 Find the derivatives with respect to x of

(a) xy,

(b) xy^2,

(c) x^2y^2,

(d) $\dfrac{x^2}{y}$.

2 Find the derivatives with respect to x of

(a) $\sqrt{xy}$,

(b) $\sin\left(x^2y\right)$,

(c) $\ln(xy)$,

(d) e^{xy+y}.

3 Differentiate the implicit equations of the following curves to find the gradients at the point $(3,4)$.

(a) $xy = 12$

(b) $4x^2 - xy - y^2 = 8$

4   Find the gradient of each of the following curves at the point given.

(a) $x \sin y = \frac{1}{2}$ $\left(1, \frac{1}{6}\pi\right)$

(b) $ye^x = xy + y^2$ $(0,1)$

(c) $\ln(x + y) = -x$ $(0,1)$

(d) $\cos(xy) = \frac{1}{2}$ $\left(1, \frac{1}{3}\pi\right)$

5   Find the equation of the tangent to the curve $x^2 - 2xy + 2y^2 = 5$ at the point $(1,2)$.

6   Find the equation of the normal to the curve $2xy^2 - x^2y^3 = 1$ at the point $(1,1)$.

7   Find the points on the curve $4x^2 + 2xy - 3y^2 = 39$ at which the tangent is parallel to one of the axes.

8   (a)   Show that the curve $x^3 + y^3 = 3xy$ is symmetrical about the line $y = x$, and find the gradient of the curve at the point other than the origin for which $y = x$.

(b)   Show that, close to the origin, if y is very small compared with x, then the curve is approximately given by the equation $y = kx^2$. Give the value of k.

(c)   Find the coordinates of the points on the graph of $x^3 + y^3 = 3xy$ at which the tangent is parallel to one or other of the axes.

(d)   Suppose now that $|x|$ and $|y|$ are both very large. Explain why $x + y \approx k$, where k is a constant, and substitute $y = k - x$ into the equation of the curve. Show that, if this equation is to be approximately satisfied by a large value of $|x|$, then $k = -1$.

(e)   Sketch the curve.

9   (a)   Explain why all the points on the curve $\left(x^2 + y^2\right)^2 = x^2 - y^2$ lie in the region $x^2 \geqslant y^2$.

(b)   Find the coordinates of the points at which the tangent is either parallel to the x-axis or parallel to the y-axis.

(c)   By considering where the curve meets the circle $x^2 + y^2 = r^2$, show that $r^2 \leqslant 1$, so the curve is bounded.

(d)   Sketch the curve, which is called the *lemniscate of Bernoulli*.

Miscellaneous exercise 11

1   Find the equation of the normal at the point $(2,1)$ on the curve $x^3 + xy + y^3 = 11$, giving your answer in the form $ax + by + c = 0$.   (OCR)

2   A curve has implicit equation $x^2 - 2xy + 4y^2 = 12$.

(a)   Find an expression for $\dfrac{dy}{dx}$ in terms of y and x. Hence determine the coordinates of the points where the tangents to the curve are parallel to the x-axis.

(b)   Find the equation of the normal to the curve at the point $\left(2\sqrt{3}, \sqrt{3}\right)$.   (OCR)

3   A curve has equation $y^3 + 3xy + 2x^3 = 9$. Obtain the equation of the normal at the point $(2,-1)$.   (OCR)

4 A curve is defined implicitly by the equation $4y - x^2 + 2x^2 y = 4x$.

    (a) Use implicit differentiation to find $\dfrac{dy}{dx}$.

    (b) Find the coordinates of the turning points on the curve.   (OCR, adapted)

5* Show that the tangent to the ellipse $\dfrac{x^2}{a^2} + \dfrac{y^2}{b^2} = 1$ at the point $P(a\cos\theta, b\sin\theta)$ has equation $bx\cos\theta + ay\sin\theta = ab$.

    (a) The tangent to the ellipse at P meets the x-axis at Q and the y-axis at R. The mid-point of QR is M. Find a cartesian equation for the locus of M as θ varies.

    (b) The tangent to the ellipse at P meets the line $x = a$ at T. The origin is at O and A is the point $(-a, 0)$. Prove that OT is parallel to AP.   (OCR)

6 The equation of a curve is $x^2 + 4xy + 5y^2 = 9$. Show by differentiation that the maximum and minimum values of y occur at the intersections of $x + 2y = 0$ with the curve. Find the maximum and minimum values of y.   (OCR)

7* The curve C, whose equation is $x^2 + y^2 = e^{x+y} - 1$, passes through the origin O. Show that $\dfrac{dy}{dx} = -1$ at O. Find the value of $\dfrac{d^2 y}{dx^2}$ at O.   (OCR)

8* A curve C has equation $y = x + 2y^4$.

    (a) Find $\dfrac{dy}{dx}$ in terms of y.

    (b) Show that $\dfrac{d^2 y}{dx^2} = \dfrac{24y^2}{\left(1 - 8y^3\right)^3}$.

    (c) Write down the value of $\dfrac{dy}{dx}$ at the origin. Hence, by considering the sign of $\dfrac{d^2 y}{dx^2}$, draw a diagram to show the shape of C in the neighbourhood of the origin.   (OCR)

Revision exercise 2

1 Use a numerical method to find all the roots of the cubic equation $x^3 - 2x^2 - 2x + 2 = 0$, giving your answers correct to 2 decimal places.

2 The region R, bounded by the curve $y = 2x + \dfrac{1}{x^2}$, the x-axis and the lines $x = 1$ and $x = k$ is shaded in the figure.

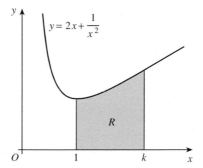

(a) Use integration to calculate the area of the region R when $k = 2$.

For a different value of k, the area of R is 10 square units.

(b) Show that k satisfies the equation $k^3 - 10k - 1 = 0$, and use a numerical method to find the value of k correct to 3 significant figures.

(OCR, adapted)

3 Differentiate the following with respect to x, and simplify your answers as much as possible.

(a) $x^2 \ln x$

(b) $\dfrac{\ln x}{x^2}$

(c) $\dfrac{e^x}{x^2 e^x + 1}$

(d) xe^{x^2}

4 A region R is bounded by part of the curve with equation $y = \sqrt{64 - x^3}$, the positive x-axis and the positive y-axis. Use the trapezium rule with 4 intervals to approximate to the area of R, giving your answer correct to 1 decimal place.

5 A chord of a circle which subtends an angle of θ at the centre cuts off a segment equal in area to $\frac{3}{8}$ of the whole circle.

Use a numerical method to find the value of θ correct to 3 significant figures.

6 Show graphically that there is a number α between π and $\frac{3}{2}\pi$ such that the tangent to $y = \sin x$ at $(\alpha, \sin \alpha)$ passes through the origin. Show that α is the smallest positive root of the equation $x = \tan x$.

Use a numerical method to find an approximate value for α, correct to 4 decimal places.

7 (a) Use the trapezium rule with 6 ordinates to calculate an approximation to

$$\int_0^1 \sqrt{4 - x^2}\, dx.$$ Give your answer to 4 decimal places.

(b) The graph of $y = \sqrt{4 - x^2}$ is a semicircle. Sketch the graph, and hence calculate the area exactly.

(c) Find to 1 decimal place the percentage error of your answer in part (a).

8 A curve has parametric equations $x = 3t^2 + 2t$, $y = 2t^2 + 3t$. Find the coordinates of the point where the tangent has gradient $\frac{3}{4}$.

9 Find the gradient at the point $(2,1)$ on the curve with equation $x^3 - 2xy + y^2 = 5$.

10 Find the equation of the tangent at the point P with parameter t to the curve with parametric equations $x = ct$, $y = \dfrac{c}{t}$, where c is a constant. Show that, if this tangent meets the x- and y-axes at X and Y, then P is the mid-point of XY.

11 A curve is defined parametrically by $x = \sqrt{3}\tan\theta$, $y = \sqrt{3}\cos\theta$, $0 \leqslant \theta \leqslant \pi$.

   (a) Find $\dfrac{dy}{dx}$ in terms of θ.

   (b) Find the equation of the tangent to the curve at the point where $\theta = \frac{1}{6}\pi$.   (OCR)

12 Find the coordinates of the points at which the tangent to the curve with equation $x^2 + 4xy + 5y^2 = 4$ is parallel to one of the axes.

13 Differentiate each of the following with respect to x.

   (a) $e^{2x}(2 + 3x)$ (b) $\dfrac{\cos x}{x^2}$ (c) $\dfrac{\sin 2x}{e^{2x}}$

   (d) $\dfrac{\tan 2x}{x}$ (e) $xe^{-x}\sin x$ (f) $\dfrac{e^{2x}\sin 2x}{4x^2}$

14 Let $f(x) = \dfrac{e^{-x}}{1 + x^2}$.

   (a) Find and simplify an expression for $f'(x)$.

   (b) Show that $f'(x) = 0$ when $x = -1$.

   (c) By considering the sign of $f'(x)$, show that the graph of $y = f(x)$ has a horizontal point of inflection when $x = -1$.

15 (a) Find the equation of the normal to the curve $xy + y^2 = 2x$ at the point $(1,1)$.

   (b) Find the coordinates of the point where the normal meets the curve again.

16 The curve C has parametric equation $x = \cos t$, $y = \cos 2t$ for $0 \leqslant t \leqslant \pi$.

   (a) Find the interval of values of t for which $\dfrac{dy}{dx}$ is negative.

   (b) Find the equation of the normal to C at the point for which $t = T$.

   (c) Find an equation satisfied by T if the normal passes through the point with coordinates $(1,1)$.

   (d) By putting $\cos T = X$, find a cubic equation satisfied by X, and write down one solution for X, and hence for T.

   (e) Find the other values of T, correct to 2 decimal places.

Practice examination 1 for P2

Time 1 hour 15 minutes

Answer all the questions.
The use of an electronic calculator is expected, where appropriate.

1 Solve the equation $|x-2|=|3-2x|$. [4]

2 (i) Show that the equation $x^3 - 3x - 10 = 0$ has a root between $x = 2$ and $x = 3$. [2]

(ii) Find an approximation, correct to 2 decimal places, to this root using an iteration based on the equation in the form

$$x = (3x+10)^{\frac{1}{3}}$$

and starting with $x_1 = 3$. [3]

3 The cubic polynomial $x^3 + x^2 + Ax + B$, where A and B are constants, is denoted by $f(x)$. When $f(x)$ is divided by $x-1$ the remainder is 4, and when $f(x)$ is divided by $x+2$ the remainder is 10. Prove that $x+3$ is a factor of $f(x)$. [6]

4 (a) Calculate the value of

$$\int_1^{13} \frac{1}{2x+1}\,dx,$$

giving your answer in an exact form, simplified as far as possible. [4]

(b) Use the trapezium rule, with 4 intervals of equal width, to estimate the value of

$$\int_1^{13} \frac{1}{2\sqrt{x}+1}\,dx.$$ [3]

5 The equation of a curve is $y = \sin x \sin 2x$.

(i) Show that $\dfrac{dy}{dx}$ may be written in the form $2\sin x\left(3\cos^2 x - 1\right)$. [4]

(ii) Hence show that the value of y at any stationary point on the curve is either 0 or $\pm\dfrac{4}{3\sqrt{3}}$. [4]

6 The amount, q units, of radioactivity present in a substance at time t seconds is given by the equation

$$q = 10\,e^{-\frac{1}{100}t}.$$

Calculate

(i) the amount of radioactivity present when $t = 5$, [1]

(ii) the value of t when the amount of radioactivity has halved from its value when $t = 0$, [4]

(iii) the rate of decrease in the amount of radioactivity when $t = 5$. [4]

7 (i)   Sketch the graphs of $y = \tan\theta$ and $y = \sec\theta$, in each case for $0 \leqslant \theta < \frac{1}{2}\pi$.   [2]

    (ii)   Prove that

$$\sec\theta - \tan\theta \equiv \frac{1}{\sec\theta + \tan\theta}.$$

                                                                        [2]

   (iii)   Deduce from parts (i) and (ii) that

$$0 < \sec\theta - \tan\theta \leqslant 1,$$

      for values of θ such that $0 \leqslant \theta < \frac{1}{2}\pi$. Explain your reasoning clearly.   [3]

   (iv)   Solve the equation

$$\sec\theta - \tan\theta = \frac{1}{2},$$

      for $0 \leqslant \theta < \frac{1}{2}\pi$.   [4]

Practice examination 2 for P2

Time 1 hour 15 minutes

Answer all the questions.
The use of an electronic calculator is expected, where appropriate.

1 Find the quotient and remainder when $2x^3 + x^2 + 3x + 1$ is divided by $x^2 + x + 2$. [4]

2 (i) Solve the inequality
$$|3-x| < 2.$$ [2]

   (ii) Hence solve the inequality
$$|3 - 2^y| < 2,$$
   expressing your answer in terms of logarithms where appropriate. [3]

3

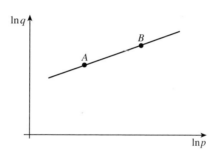

The variables p and q are related by an equation of the form
$$q = k p^z,$$
where k and z are constants. The diagram shows the graph of $\ln q$ against $\ln p$. The graph
is a straight line, and it passes through the points $A(1.61, 2.82)$ and $B(3.22, 3.62)$. Find the
values of k and z, giving the answers correct to 1 decimal place. [6]

4 (a) By first expressing $\sin^2 2x$ in terms of $\cos 4x$, find $\displaystyle\int_0^{\frac{1}{6}\pi} \sin^2 2x \, dx$. [4]

   (b) Find $\displaystyle\int_0^{\frac{1}{2}\pi} (\sin x + \cos x)^2 \, dx$. [4]

5

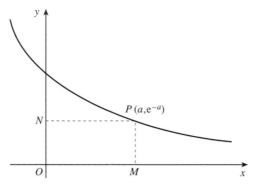

The diagram shows the graph of $y = e^{-x}$. The point P has coordinates $\left(a, e^{-a}\right)$, and the lines PM and PN are parallel to the axes.

(i)   Find $\displaystyle\int_0^a e^{-x}\, dx$ in terms of a.   [2]

The area of the rectangle $OMPN$ is one quarter of the area under the curve $y = e^{-x}$ from $x = 0$ to $x = a$.

(ii)   Show that $e^a = 4a + 1$.   [3]

(iii)   Use the iteration

$$a_{n+1} = \ln\left(4a_n + 1\right),$$

with $a_1 = 2$, to find the non-zero value of a satisfying the equation in part (ii). Give your answer correct to 1 decimal place.   [3]

6   Differentiate the following with respect to x, simplifying your answers.

(i) $x^2 e^{2x}$ [3]

(ii) $\dfrac{\ln 2x}{x}$ [3]

(iii) $\ln\left(\dfrac{x}{x+1}\right)$ [3]

7   The parametric equations of a curve are

$$x = 2\theta + \cos\theta, \quad y = \theta + \sin\theta,$$

where $0 \leqslant \theta \leqslant 2\pi$.

(i)   Find $\dfrac{dy}{dx}$ in terms of θ.   [2]

(ii)   Show that, at points on the curve where the gradient is $\frac{3}{4}$, the parameter θ satisfies an equation of the form

$$5\sin(\theta + \alpha) = 2,$$

where the value of α is to be stated.   [4]

(iii)   Solve the equation in part (ii) to find the two possible values of θ.   [4]

Unit P3

The subject content of unit P2 is a subset of the subject content of unit P3. This part of the book (pages 159–304) contains the subject content of unit P3 that *is not* common with unit P2. It is required for unit P3 but not for unit P2.

12 Vectors: lines in two and three dimensions

This chapter shows how to use vectors to describe lines in three dimensions. When you have completed it, you should

- know the form of the vector equation of a line
- be able to solve problems involving intersecting, parallel and skew lines
- be able to find the distance of a point from a line.

12.1 Vector equation of a line in two dimensions

Fig. 12.1 shows a line through a point A in the direction of a non-zero vector $\mathbf{p}$. If R is any point on the line, the displacement vector $\overrightarrow{AR}$ is a multiple of $\mathbf{p}$, so

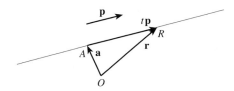

$$\mathbf{r} = \overrightarrow{OR} = \overrightarrow{OA} + \overrightarrow{AR} = \mathbf{a} + t\mathbf{p},$$

Fig. 12.1

where t is a scalar, and the alphabet convention (P1 Section 13.4) is used. The value of t measures the ratio of the displacement $\overrightarrow{AR}$ to $\mathbf{p}$, and takes a different value for each point R on the line.

> Points on a line through A in the direction of $\mathbf{p}$ have position vectors $\mathbf{r} = \mathbf{a} + t\mathbf{p}$, where t is a variable scalar. This is called the **vector equation** of the line.

The following examples show how vector equations can be used as an alternative to the cartesian equations with which you are familiar.

To illustrate alternative techniques the first is solved by using vectors in column form, and the second by using the basic unit vectors.

Example 12.1.1
Find a vector equation for the line through $(2,-1)$ with gradient $\frac{3}{4}$, and deduce its cartesian equation.

The position vector of the point $(2,-1)$ is $\begin{pmatrix} 2 \\ -1 \end{pmatrix}$. There are many vectors with gradient $\frac{3}{4}$, but the simplest is the vector which goes 4 units across the grid and 3 units up, that is $\begin{pmatrix} 4 \\ 3 \end{pmatrix}$. So an equation of the line is

$$\mathbf{r} = \begin{pmatrix} 2 \\ -1 \end{pmatrix} + t \begin{pmatrix} 4 \\ 3 \end{pmatrix}.$$

If R has coordinates (x, y), the position vector $\mathbf{r}$ is $\begin{pmatrix} x \\ y \end{pmatrix}$. This can be written

$$\begin{pmatrix} x \\ y \end{pmatrix} = \begin{pmatrix} 2 + 4t \\ -1 + 3t \end{pmatrix}.$$

This is equivalent to the two equations

$$x = 2 + 4t, \quad y = -1 + 3t,$$

which you will recognise as parametric equations for the line.

The cartesian equation is found by eliminating t:

$$3x - 4y = 3(2 + 4t) - 4(-1 + 3t) = 10.$$

You can check that $3x - 4y = 10$ has gradient $\frac{3}{4}$ and contains the point $(2, -1)$.

Example 12.1.2

Find a vector equation for the line through $(3, 1)$ parallel to the y-axis, and deduce its cartesian equation.

A vector parallel to the y-axis is $\mathbf{j}$, and the position vector of $(3, 1)$ is $3\mathbf{i} + \mathbf{j}$, so a vector equation of the line is

$$\mathbf{r} = (3\mathbf{i} + \mathbf{j}) + t\mathbf{j}.$$

Writing $\mathbf{r}$ as $x\mathbf{i} + y\mathbf{j}$, this is

$$x\mathbf{i} + y\mathbf{j} = (3\mathbf{i} + \mathbf{j}) + t\mathbf{j}.$$

This is equivalent to the two equations $x = 3$, $y = 1 + t$.

No elimination is necessary this time: the first equation does not involve t, so the cartesian equation is just $x = 3$.

Example 12.1.3

Find the points common to the pairs of lines

(a) $\mathbf{r} = \begin{pmatrix} 1 \\ 2 \end{pmatrix} + s \begin{pmatrix} 1 \\ 1 \end{pmatrix}$ and $\mathbf{r} = \begin{pmatrix} 3 \\ -2 \end{pmatrix} + t \begin{pmatrix} 1 \\ 4 \end{pmatrix}$, (b) $\mathbf{r} = \begin{pmatrix} 3 \\ 1 \end{pmatrix} + s \begin{pmatrix} 4 \\ -2 \end{pmatrix}$ and $\mathbf{r} = \begin{pmatrix} 1 \\ 2 \end{pmatrix} + t \begin{pmatrix} -6 \\ 3 \end{pmatrix}$.

Notice that different letters are used for the variable scalars on the two lines.

(a) Position vectors of points on the two lines can be written as

$$\mathbf{r} = \begin{pmatrix} 1 + s \\ 2 + s \end{pmatrix} \quad \text{and} \quad \mathbf{r} = \begin{pmatrix} 3 + t \\ -2 + 4t \end{pmatrix}.$$

If these are the same point they have the same position vectors, so

$$1 + s = 3 + t \quad \text{and} \quad 2 + s = -2 + 4t,$$

that is $\quad s - t = 2 \quad$ and $\quad s - 4t = -4$.

This is a pair of simultaneous equations for s and t, with solution $s = 4$, $t = 2$.

Substituting these values into the equation of one of the lines gives $\mathbf{r} = \begin{pmatrix} 5 \\ 6 \end{pmatrix}$. So the point common to the two lines has coordinates $(5,6)$.

(b) You can check for yourself that the procedure used in (a) leads to the equations

$$3 + 4s = 1 - 6t \quad \text{and} \quad 1 - 2s = 2 + 3t,$$

that is $2s + 3t = -1$ and $2s + 3t = -1$.

The two equations are the same! So there is really only one equation to solve, which has infinitely many solutions in s and t. If you take any value for s, say $s = 7$, and calculate the corresponding value $t = -5$, then you have a solution of both vector equations. You can easily check that $s = 7$, $t = -5$ gives the position vector $\begin{pmatrix} 31 \\ -13 \end{pmatrix}$ in both lines. (Try some other pairs of values for yourself.)

You can see that in (b) the direction vectors of the lines are $\begin{pmatrix} 4 \\ -2 \end{pmatrix} = 2\begin{pmatrix} 2 \\ -1 \end{pmatrix}$ and $\begin{pmatrix} -6 \\ 3 \end{pmatrix} = -3\begin{pmatrix} 2 \\ -1 \end{pmatrix}$. This means that the lines have the same direction, so they are either parallel or the same line. Also the position vectors of the given points on the two lines are $\begin{pmatrix} 3 \\ 1 \end{pmatrix}$ and $\begin{pmatrix} 1 \\ 2 \end{pmatrix}$, and $\begin{pmatrix} 3 \\ 1 \end{pmatrix} - \begin{pmatrix} 1 \\ 2 \end{pmatrix} = \begin{pmatrix} 2 \\ -1 \end{pmatrix}$; so the line joining these points is also in the same direction. The lines are therefore identical. This is illustrated in Fig. 12.2.

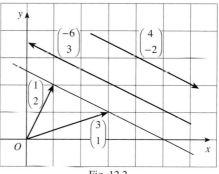

Fig. 12.2

The general result demonstrated in Example 12.1.3(b) is:

The lines with vector equations $\mathbf{r} = \mathbf{a} + s\mathbf{p}$ and $\mathbf{r} = \mathbf{b} + t\mathbf{q}$ have the same direction if $\mathbf{p}$ is a multiple of $\mathbf{q}$. If in addition $\mathbf{b} - \mathbf{a}$ is a multiple of $\mathbf{q}$, the lines are the same; otherwise the lines are parallel.

This shows that lines do not have unique vector equations. Two equations may represent the same line even though the vectors $\mathbf{a}$ and $\mathbf{b}$, and the vectors $\mathbf{p}$ and $\mathbf{q}$, are different.

Example 12.1.4

Show that the lines with vector equations $\mathbf{r} = 2\mathbf{i} - 3\mathbf{j} + s(-\mathbf{i} + 3\mathbf{j})$ and $\mathbf{r} = 4\mathbf{i} + t(2\mathbf{i} - 6\mathbf{j})$ are parallel, and find a vector equation for the parallel line through $(1,1)$.

The direction vectors of the two lines are $-\mathbf{i} + 3\mathbf{j}$ and $2\mathbf{i} - 6\mathbf{j}$.

As $2\mathbf{i} - 6\mathbf{j} = -2(-\mathbf{i} + 3\mathbf{j})$, $2\mathbf{i} - 6\mathbf{j}$ is a scalar multiple of $-\mathbf{i} + 3\mathbf{j}$, so the lines are in the same direction. But $4\mathbf{i} - (2\mathbf{i} - 3\mathbf{j}) = 2\mathbf{i} + 3\mathbf{j}$ is not a multiple of $-\mathbf{i} + 3\mathbf{j}$, so the lines are not the same. The lines are therefore parallel.

The position vector of $(1,1)$ is $\mathbf{i} + \mathbf{j}$, so an equation for the parallel line through $(1,1)$ is $\mathbf{r} = \mathbf{i} + \mathbf{j} + s(-\mathbf{i} + 3\mathbf{j})$. Or, alternatively, you could use $\mathbf{r} = \mathbf{i} + \mathbf{j} + t(2\mathbf{i} - 6\mathbf{j})$.

Example 12.1.5

Find a vector equation for the line with cartesian equation $2x + 5y = 1$ and use it to find where the line meets the circle with equation $x^2 + y^2 = 10$.

The gradient of the line is $-\frac{2}{5}$, so the direction vector could be taken as $\begin{pmatrix} 5 \\ -2 \end{pmatrix}$.

A point on the line is $(-2, 1)$, with position vector $\begin{pmatrix} -2 \\ 1 \end{pmatrix}$. So a possible vector

equation is $\mathbf{r} = \begin{pmatrix} -2 \\ 1 \end{pmatrix} + t \begin{pmatrix} 5 \\ -2 \end{pmatrix}$.

Writing $\mathbf{r}$ as $\begin{pmatrix} x \\ y \end{pmatrix}$, this equation becomes $\begin{pmatrix} x \\ y \end{pmatrix} = \begin{pmatrix} -2 \\ 1 \end{pmatrix} + t \begin{pmatrix} 5 \\ -2 \end{pmatrix}$, giving

$$x = -2 + 5t, \ y = 1 - 2t.$$

Substituting these values for x and y into the equation $x^2 + y^2 = 10$ gives $(-2 + 5t)^2 + (1 - 2t)^2 = 10$, which reduces to $29t^2 - 24t - 5 = 0$, or $(t - 1)(29t + 5) = 0$, giving $t = 1$ or $-\frac{5}{29}$.

Putting these values for t into the equation of the line gives the points with

position vectors $\begin{pmatrix} 3 \\ -1 \end{pmatrix}$ and $\begin{pmatrix} -2\frac{25}{29} \\ 1\frac{10}{29} \end{pmatrix}$.

So the line meets the circle at $(3, -1)$ and $\left(-2\frac{25}{29}, 1\frac{10}{29} \right)$.

Exercise 12A

1 Write down vector equations for the line through the given point in the specified direction. Then eliminate t to obtain the cartesian equation.

   (a) $(2, -3), \begin{pmatrix} 1 \\ 2 \end{pmatrix}$ (b) $(4, 1), \begin{pmatrix} -3 \\ 2 \end{pmatrix}$ (c) $(5, 7)$, parallel to the x-axis

   (d) $(0, 0), \begin{pmatrix} 2 \\ -1 \end{pmatrix}$ (e) $(a, b), \begin{pmatrix} 0 \\ 1 \end{pmatrix}$ (f) $(\cos\alpha, \sin\alpha), \begin{pmatrix} -\sin\alpha \\ \cos\alpha \end{pmatrix}$

2 Find vector equations for lines with the following cartesian equations.

   (a) $x = 2$ (b) $x + 3y = 7$ (c) $2x - 5y = 3$

3 Find the coordinates of the points common to the following pairs of lines, if any.

   (a) $\mathbf{r} = \begin{pmatrix} 2 \\ 0 \end{pmatrix} + s \begin{pmatrix} 5 \\ 3 \end{pmatrix}, \ \mathbf{r} = \begin{pmatrix} 3 \\ -1 \end{pmatrix} + t \begin{pmatrix} 1 \\ 1 \end{pmatrix}$ (b) $\mathbf{r} = \begin{pmatrix} 5 \\ 1 \end{pmatrix} + s \begin{pmatrix} -1 \\ 2 \end{pmatrix}, \ \mathbf{r} = \begin{pmatrix} 3 \\ -5 \end{pmatrix} + t \begin{pmatrix} 1 \\ 0 \end{pmatrix}$

   (c) $\mathbf{r} = \begin{pmatrix} 2 \\ -1 \end{pmatrix} + s \begin{pmatrix} 1 \\ -3 \end{pmatrix}, \ \mathbf{r} = \begin{pmatrix} 4 \\ 0 \end{pmatrix} + t \begin{pmatrix} -2 \\ 6 \end{pmatrix}$ (d) $\mathbf{r} = \begin{pmatrix} -1 \\ -4 \end{pmatrix} + s \begin{pmatrix} 3 \\ 4 \end{pmatrix}, \ \mathbf{r} = \begin{pmatrix} 11 \\ -1 \end{pmatrix} + t \begin{pmatrix} -4 \\ 3 \end{pmatrix}$

   (e) $\mathbf{r} = \begin{pmatrix} 7 \\ 1 \end{pmatrix} + s \begin{pmatrix} 6 \\ -4 \end{pmatrix}, \ \mathbf{r} = \begin{pmatrix} 10 \\ -1 \end{pmatrix} + t \begin{pmatrix} -9 \\ 6 \end{pmatrix}$ (f) $\mathbf{r} = \begin{pmatrix} 2 \\ 1 \end{pmatrix} + s \begin{pmatrix} 3 \\ 0 \end{pmatrix}, \ \mathbf{r} = \begin{pmatrix} -1 \\ 3 \end{pmatrix} + t \begin{pmatrix} 0 \\ -2 \end{pmatrix}$

4 Write down in parametric form the coordinates of any point on the line through $(2,-1)$ in the direction $\mathbf{i} + 3\mathbf{j}$. Use these to find the point where this line intersects the line $5y - 6x = 1$.

5 Find the coordinates of the point where the line with vector equation $\mathbf{r} = \begin{pmatrix} -3 \\ 4 \end{pmatrix} + t \begin{pmatrix} 2 \\ -1 \end{pmatrix}$ intersects the line with cartesian equation $2x + y = 7$.

6 Which of the following points lie on the line joining $(2,0)$ to $(4,3)$?

    (a) $(8,9)$ (b) $(12,13)$ (c) $(-4,-1)$ (d) $(-6,-12)$ (e) $\left(3\tfrac{1}{3},2\right)$

7 Find vector equations for the lines joining the following pairs of points.

    (a) $(3,7)$, $(5,4)$ (b) $(2,3)$, $(2,8)$ (c) $(-1,2)$, $(5,-1)$

    (d) $(-3,-4)$, $(5,8)$ (e) $(-2,7)$, $(4,7)$ (f) $(1,3)$, $(-4,-2)$

8 A quadrilateral $ABCD$ has vertices $A(4,-1)$, $B(-3,2)$, $C(-8,-5)$ and $D(4,-5)$.

    (a) Find vector equations for the diagonals, AC and BD, and find their point of intersection.

    (b) Find the points of intersection of BA produced and CD produced, and of CB produced and DA produced.

9 Show that the vectors $a\mathbf{i} + b\mathbf{j}$ and $-b\mathbf{i} + a\mathbf{j}$ are perpendicular to each other. Is this still true

    (a) if a is zero but b is not,   (b) if b is zero but a is not,

    (c) if both a and b are zero?

    Find a vector equation for the line through $(1,2)$ perpendicular to the line with vector equation $\mathbf{r} = 7\mathbf{i} + 2\mathbf{j} + t(3\mathbf{i} + 4\mathbf{j})$.

10 Find a vector in the direction of the line l with cartesian equation $3x - y - 8$. Write down a vector equation for the line through $P(1,5)$ which is perpendicular to l. Hence find the coordinates of the foot of the perpendicular from P to l.

11 Use the method of Question 10 to find the coordinates of the foot of the perpendicular from $(-3,-2)$ to $5x + 2y = 10$.

12 Find a vector equation for the line joining the points $(-1,1)$ and $(4,11)$. Use this to write parametric equations for any point on the line. Hence find the coordinates of the points where the line meets the parabola $y = x^2$.

13 Find the coordinates of the points where the line through $(-5,-1)$ in the direction $\begin{pmatrix} 2 \\ 3 \end{pmatrix}$ meets the circle $x^2 + y^2 = 65$.

12.2 Vector equation of a line in three dimensions

Everything that you have learnt about the vector equation of a line in two dimensions carries over into three dimensions in an obvious way. However there are two important differences between two dimensions and three dimensions.

- The idea of the gradient of a line does not carry over into three dimensions. However, you can still use a vector to describe the direction of a line. This is one of the main reasons why vectors are especially useful in three dimensions.
- In three dimensions lines which are not parallel may or may not meet. Non-parallel lines which do not meet are said to be **skew**. (Imagine two vapour trails made by aeroplanes flying at different heights in different directions.)

The following examples show some of the situations which can occur when working with lines in three dimensions.

Example 12.2.1

Points A and B have coordinates $(-5,3,4)$ and $(-2,9,1)$. The line AB meets the xy-plane at C. Find the coordinates of C.

The displacement vector $\overrightarrow{AB}$ is

$$\mathbf{b} - \mathbf{a} = \begin{pmatrix} -2 \\ 9 \\ 1 \end{pmatrix} - \begin{pmatrix} -5 \\ 3 \\ 4 \end{pmatrix} = \begin{pmatrix} 3 \\ 6 \\ -3 \end{pmatrix} = 3 \begin{pmatrix} 1 \\ 2 \\ -1 \end{pmatrix}.$$

So $\begin{pmatrix} 1 \\ 2 \\ -1 \end{pmatrix}$ can be taken as a direction vector for the line. A vector equation for the

line is therefore $\mathbf{r} = \begin{pmatrix} -5 \\ 3 \\ 4 \end{pmatrix} + t \begin{pmatrix} 1 \\ 2 \\ -1 \end{pmatrix}$, or $\mathbf{r} = \begin{pmatrix} -5+t \\ 3+2t \\ 4-t \end{pmatrix}$.

The xy-plane consists of the points with coordinates $(s,t,0)$. So C is the point on the line at which $z = 0$, so that $4 - t = 0$, $t = 4$. It therefore has position vector

$\mathbf{c} = \begin{pmatrix} -1 \\ 11 \\ 0 \end{pmatrix}$ and coordinates $(-1,11,0)$.

Example 12.2.2

Find the value of u for which the lines $\mathbf{r} = (\mathbf{j} - \mathbf{k}) + s(\mathbf{i} + 2\mathbf{j} + \mathbf{k})$ and $\mathbf{r} = (\mathbf{i} + 7\mathbf{j} - 4\mathbf{k}) + t(\mathbf{i} + u\mathbf{k})$ intersect.

Points on the lines can be written as $s\mathbf{i} + (1+2s)\mathbf{j} + (-1+s)\mathbf{k}$ and $(1+t)\mathbf{i} + 7\mathbf{j} + (-4+tu)\mathbf{k}$. If these are the same point, then

$$s = 1 + t, \quad 1 + 2s = 7, \quad \text{and} \quad -1 + s = -4 + tu.$$

The first two equations give $s = 3$ and $t = 2$. Putting these values into the third equation gives $-1 + 3 = -4 + 2u$, so $u = 3$.

You can easily check that, with these values, both equations give $\mathbf{r} = 3\mathbf{i} + 7\mathbf{j} + 2\mathbf{k}$, so the point of intersection has coordinates $(3,7,2)$.

Example 12.2.3

Determine whether the points with coordinates $(5,1,-6)$ and $(-7,5,9)$ lie on the line joining $A(1,2,-1)$ to $B(-3,3,4)$.

A vector in the direction of the line is

$$\overrightarrow{AB} = \mathbf{b} - \mathbf{a} = \begin{pmatrix} -3 \\ 3 \\ 4 \end{pmatrix} - \begin{pmatrix} 1 \\ 2 \\ -1 \end{pmatrix} = \begin{pmatrix} -4 \\ 1 \\ 5 \end{pmatrix}.$$

The equation of the line is $\mathbf{r} = \begin{pmatrix} 1 \\ 2 \\ -1 \end{pmatrix} + t\begin{pmatrix} -4 \\ 1 \\ 5 \end{pmatrix}$, which is $\begin{pmatrix} x \\ y \\ z \end{pmatrix} = \begin{pmatrix} 1 \\ 2 \\ -1 \end{pmatrix} + t\begin{pmatrix} -4 \\ 1 \\ 5 \end{pmatrix}$.

To find whether $(5,1,-6)$ lies on this line, substitute $x = 5$, $y = 1$ and $z = -6$ to get the vector equation

$$\begin{pmatrix} 5 \\ 1 \\ -6 \end{pmatrix} = \begin{pmatrix} 1 \\ 2 \\ 1 \end{pmatrix} + t\begin{pmatrix} -4 \\ 1 \\ 5 \end{pmatrix}, \quad \text{which is} \quad t\begin{pmatrix} -4 \\ 1 \\ 5 \end{pmatrix} = \begin{pmatrix} 5 \\ 1 \\ -6 \end{pmatrix} - \begin{pmatrix} 1 \\ 2 \\ -1 \end{pmatrix} = \begin{pmatrix} 4 \\ -1 \\ -5 \end{pmatrix}.$$

As $\begin{pmatrix} 4 \\ -1 \\ -5 \end{pmatrix}$ is a multiple of $\begin{pmatrix} -4 \\ 1 \\ 5 \end{pmatrix}$, this vector equation has a solution $(t = -1)$, so $(5,1,-6)$ lies on AB.

To find whether $(-7,5,9)$ lies on AB, try to solve the vector equation

$$\begin{pmatrix} -7 \\ 5 \\ 9 \end{pmatrix} = \begin{pmatrix} 1 \\ 2 \\ -1 \end{pmatrix} + t\begin{pmatrix} -4 \\ 1 \\ 5 \end{pmatrix}, \quad \text{which is} \quad t\begin{pmatrix} -4 \\ 1 \\ 5 \end{pmatrix} = \begin{pmatrix} -7 \\ 5 \\ 9 \end{pmatrix} - \begin{pmatrix} 1 \\ 2 \\ -1 \end{pmatrix} = \begin{pmatrix} -8 \\ 3 \\ 10 \end{pmatrix}.$$

As $\begin{pmatrix} -8 \\ 3 \\ 10 \end{pmatrix}$ is not a multiple of $\begin{pmatrix} -4 \\ 1 \\ 5 \end{pmatrix}$, this vector equation has no solution, so $(-7,5,9)$ does not lie on AB.

Example 12.2.4

Prove that the straight line with equation $\mathbf{r} = \begin{pmatrix} 1 \\ 2 \\ -3 \end{pmatrix} + t\begin{pmatrix} 2 \\ -1 \\ 4 \end{pmatrix}$ meets the line joining $(2,4,4)$ to $(3,3,5)$, and find the cosine of the angle between the lines.

The line joining $(2,4,4)$ and $(3,3,5)$ has direction $\begin{pmatrix} 3 \\ 3 \\ 5 \end{pmatrix} - \begin{pmatrix} 2 \\ 4 \\ 4 \end{pmatrix} = \begin{pmatrix} 1 \\ -1 \\ 1 \end{pmatrix}$, so it has equation $\mathbf{r} = \begin{pmatrix} 2 \\ 4 \\ 4 \end{pmatrix} + s\begin{pmatrix} 1 \\ -1 \\ 1 \end{pmatrix}$.

To prove that the lines intersect, you have to show that there is a point on one line which is the same as a point on the other. Suppose that the lines meet when the parameter of the first line is t and the parameter of the second line is s. Then

$$\begin{pmatrix} 1 \\ 2 \\ -3 \end{pmatrix} + t\begin{pmatrix} 2 \\ -1 \\ 4 \end{pmatrix} = \begin{pmatrix} 2 \\ 4 \\ 4 \end{pmatrix} + s\begin{pmatrix} 1 \\ -1 \\ 1 \end{pmatrix}; \quad \text{that is,} \quad \left.\begin{matrix} 2t - s = 1 \\ -t + s = 2 \\ 4t - s = 7 \end{matrix}\right\}.$$

Sets of three equations in two unknowns may not always have a solution. If there is a solution, the equations are said to be **consistent**.

These equations are consistent, with solution $t = 3$ and $s = 5$. Hence the lines intersect. (The point of intersection is $(7,-1,9)$.)

The angle between the lines is the angle between their direction vectors. Calling this angle θ and using the scalar product (see P1 Section 13.8),

$$\begin{pmatrix} 2 \\ -1 \\ 4 \end{pmatrix} \cdot \begin{pmatrix} 1 \\ -1 \\ 1 \end{pmatrix} = \left|\begin{pmatrix} 2 \\ -1 \\ 4 \end{pmatrix}\right|\left|\begin{pmatrix} 1 \\ -1 \\ 1 \end{pmatrix}\right| \cos\theta, \quad \text{giving} \quad \cos\theta = \frac{7}{\sqrt{21}\sqrt{3}} = \frac{\sqrt{7}}{3}.$$

Exercise 12B

1 Write down vector equations for the following straight lines which pass through the given points and lie in the given directions.

(a) $(1,2,3)$, $\begin{pmatrix} 0 \\ 1 \\ 2 \end{pmatrix}$ (b) $(0,0,0)$, $\begin{pmatrix} 0 \\ 0 \\ 1 \end{pmatrix}$ (c) $(2,-1,1)$, $\begin{pmatrix} 3 \\ -1 \\ 1 \end{pmatrix}$ (d) $(3,0,2)$, $\begin{pmatrix} 4 \\ -2 \\ 3 \end{pmatrix}$

2 Find vector equations for the lines joining the following pairs of points.

(a) $(2,-1,2)$, $(3,-1,4)$ (b) $(1,2,2)$, $(2,-2,2)$ (c) $(3,1,4)$, $(-1,2,3)$

3 Which of these equations of straight lines represent the same straight line as each other?

(a) $\mathbf{r} = \begin{pmatrix} 1 \\ 4 \\ 2 \end{pmatrix} + t\begin{pmatrix} 2 \\ -1 \\ 2 \end{pmatrix}$ (b) $\mathbf{r} = \begin{pmatrix} 3 \\ 3 \\ 4 \end{pmatrix} + t\begin{pmatrix} 2 \\ -1 \\ 2 \end{pmatrix}$

(c) $\mathbf{r} = 5\mathbf{i} + 2\mathbf{j} + 6\mathbf{k} + t(-2\mathbf{i} + \mathbf{j} - 2\mathbf{k})$ (d) $\mathbf{r} = \mathbf{i} + 4\mathbf{j} + 2\mathbf{k} + t(-2\mathbf{i} + \mathbf{j} - 2\mathbf{k})$

(e) $\mathbf{r} = -\mathbf{i} + 5\mathbf{j} + t(2\mathbf{i} - \mathbf{j} + 2\mathbf{k})$ (f) $\mathbf{r} = -\mathbf{i} + 5\mathbf{j} + t(-2\mathbf{i} + \mathbf{j} - 2\mathbf{k})$

4 Find whether or not the point $(-3,1,5)$ lies on each of the following lines.

(a) $\mathbf{r} = \begin{pmatrix} 1 \\ 3 \\ 1 \end{pmatrix} + t\begin{pmatrix} -2 \\ -1 \\ 2 \end{pmatrix}$ (b) $\mathbf{r} = \begin{pmatrix} 0 \\ 1 \\ 2 \end{pmatrix} + t\begin{pmatrix} 1 \\ 0 \\ 3 \end{pmatrix}$ (c) $\mathbf{r} = \begin{pmatrix} 1 \\ -2 \\ 4 \end{pmatrix} + t\begin{pmatrix} -4 \\ -3 \\ -1 \end{pmatrix}$

5 Determine whether each of the following sets of points lies on a straight line.

(a) $(1,2,-1)$, $(2,4,-3)$, $(4,8,-7)$ (b) $(5,2,-3)$, $(-1,6,-11)$, $(3,-2,4)$

6 Investigate whether or not it is possible to find numbers s and t which satisfy the following vector equations.

(a) $s\begin{pmatrix} 3 \\ 4 \\ 1 \end{pmatrix} + t\begin{pmatrix} 2 \\ -1 \\ 0 \end{pmatrix} = \begin{pmatrix} 0 \\ 11 \\ 2 \end{pmatrix}$ (b) $\begin{pmatrix} -1 \\ -2 \\ 3 \end{pmatrix} + s\begin{pmatrix} 1 \\ 2 \\ -1 \end{pmatrix} + t\begin{pmatrix} 3 \\ -1 \\ 1 \end{pmatrix} = \begin{pmatrix} 5 \\ 3 \\ 4 \end{pmatrix}$ (c) $s\begin{pmatrix} 1 \\ 2 \\ -3 \end{pmatrix} + t\begin{pmatrix} 5 \\ 1 \\ 1 \end{pmatrix} = \begin{pmatrix} 1 \\ -7 \\ 11 \end{pmatrix}$

7 Find the point of intersection, if any, of each of the following pairs of lines.

(a) $\mathbf{r} = \begin{pmatrix} 1 \\ 3 \\ 1 \end{pmatrix} + s \begin{pmatrix} -2 \\ -1 \\ 2 \end{pmatrix}$, $\mathbf{r} = \begin{pmatrix} 0 \\ -2 \\ 8 \end{pmatrix} + t \begin{pmatrix} 1 \\ -1 \\ 1 \end{pmatrix}$ (b) $\mathbf{r} = \begin{pmatrix} 1 \\ -1 \\ 2 \end{pmatrix} + s \begin{pmatrix} -1 \\ 2 \\ -1 \end{pmatrix}$, $\mathbf{r} = \begin{pmatrix} 1 \\ 3 \\ -1 \end{pmatrix} + t \begin{pmatrix} 2 \\ -8 \\ 5 \end{pmatrix}$

8 If $\mathbf{p} = 2\mathbf{i} - \mathbf{j} + 3\mathbf{k}$, $\mathbf{q} = 5\mathbf{i} + 2\mathbf{j}$ and $\mathbf{r} = 4\mathbf{i} + \mathbf{j} + \mathbf{k}$, find a set of numbers f, g and h such that $f\mathbf{p} + g\mathbf{q} + h\mathbf{r} = \mathbf{0}$. What does this tell you about the translations represented by $\mathbf{p}$, $\mathbf{q}$ and $\mathbf{r}$?

9 A and B are points with coordinates $(2,1,4)$ and $(5,-5,-2)$. Find the coordinates of the point C such that $\overrightarrow{AC} = \frac{2}{3}\overrightarrow{AB}$.

10 Four points A, B, C and D with position vectors $\mathbf{a}$, $\mathbf{b}$, $\mathbf{c}$ and $\mathbf{d}$ are vertices of a tetrahedron. The mid-points of BC, CA, AB, AD, BD, CD are denoted by P, Q, R, U, V, W. Find the position vectors of the mid-points of PU, QV and RW.

What do you notice about the answer? State your conclusion as a geometrical theorem.

11 If E and F are two points with position vectors $\mathbf{e}$ and $\mathbf{f}$, find the position vector of the point H such that $\overrightarrow{EH} = \frac{3}{4}\overrightarrow{EF}$.

With the notation of Question 10, express in terms of $\mathbf{a}$, $\mathbf{b}$, $\mathbf{c}$ and $\mathbf{d}$ the position vectors of G, the centroid of triangle ABC, and of H, the point on DG such that $DH:HG = 3:1$.

12 For each of the following sets of points A, B, C and D, determine whether the lines AB and CD are parallel, intersect each other, or are skew.

(a) $A(3,2,4)$, $B(-3,-7,-8)$, $C(0,1,3)$, $D(-2,5,9)$

(b) $A(3,1,0)$, $B(-3,1,3)$, $C(5,0,-1)$, $D(1,0,1)$

(c) $A(-5,-4,-3)$, $B(5,1,2)$, $C(-1,-3,0)$, $D(8,0,6)$

13 A student displays her birthday cards on strings which she has pinned to opposite walls of her room, whose floor measures 3 metres by 4 metres. Relative to one corner of the room, the coordinates of the ends of the first string are $(0,3.3,2.4)$ and $(3,1.3,1.9)$ in metre units. The coordinates of the ends of the second string are $(0.7,0,2.3)$ and $(1.5,4,1.5)$. Assuming that the strings are straight lines, find the difference in the heights of the two strings where one passes over the other.

12.3 The distance from a point to a line

If a point does not lie on a line, it is natural to ask how far from the line it is.

Before embarking on examples, it is worth noting that in the vector equation of a line

$$\mathbf{r} = \mathbf{a} + t\mathbf{p}$$

the vector $\mathbf{p}$ can be *any* vector in the direction of the line. In the case when $\mathbf{p}$ is a unit vector $\mathbf{u}$, the equation becomes $\mathbf{r} = \mathbf{a} + t\mathbf{u}$. In this case, $|t|$ is the distance along the line from A to the point with parameter t.

Example 12.3.1

Let A be the point with position vector $\mathbf{i} - \mathbf{j} + \mathbf{k}$. A line l through A is given by the vector equation $\mathbf{r} = \mathbf{i} - \mathbf{j} + \mathbf{k} + t\left(\frac{1}{3}\mathbf{i} + \frac{2}{3}\mathbf{j} - \frac{2}{3}\mathbf{k}\right)$.

(a) Verify that $\frac{1}{3}\mathbf{i} + \frac{2}{3}\mathbf{j} - \frac{2}{3}\mathbf{k}$ is a unit vector.

(b) Show that l intersects the line m with equation $\mathbf{r} = -3\mathbf{i} + 6\mathbf{k} + s(2\mathbf{i} + \mathbf{j} - 3\mathbf{k})$. Call this point of intersection B.

(c) Find the distance AB.

(a) $\left|\frac{1}{3}\mathbf{i} + \frac{2}{3}\mathbf{j} - \frac{2}{3}\mathbf{k}\right| = \sqrt{\left(\frac{1}{3}\right)^2 + \left(\frac{2}{3}\right)^2 + \left(-\frac{2}{3}\right)^2} = \sqrt{\frac{1}{9} + \frac{4}{9} + \frac{4}{9}} = 1$, so $\frac{1}{3}\mathbf{i} + \frac{2}{3}\mathbf{j} - \frac{2}{3}\mathbf{k}$ is a unit vector.

(b) If the point with parameter t on l coincides with the point with parameter s on m, then

$$\mathbf{i} - \mathbf{j} + \mathbf{k} + t\left(\frac{1}{3}\mathbf{i} + \frac{2}{3}\mathbf{j} - \frac{2}{3}\mathbf{k}\right) = -3\mathbf{i} + 6\mathbf{k} + s(2\mathbf{i} + \mathbf{j} - 3\mathbf{k}),$$

giving $\left(1 + \frac{1}{3}t\right)\mathbf{i} + \left(-1 + \frac{2}{3}t\right)\mathbf{j} + \left(1 - \frac{2}{3}t\right)\mathbf{k} = (-3 + 2s)\mathbf{i} + s\mathbf{j} + (6 - 3s)\mathbf{k}$,

leading to the equations

$$\frac{1}{3}t - 2s = -4,$$

$$\frac{2}{3}t - s = 1,$$

$$-\frac{2}{3}t + 3s = 5.$$

By adding the last two equations you can see that they have the solution $s = 3$, $t = 6$. You can easily check that this solution satisfies the first equation.

As the equations are consistent, the lines intersect.

(c) The equation of l is $\mathbf{r} = \mathbf{i} - \mathbf{j} + \mathbf{k} + t\left(\frac{1}{3}\mathbf{i} + \frac{2}{3}\mathbf{j} - \frac{2}{3}\mathbf{k}\right)$, which shows that the parameter t takes the value 0 at the point A.

Since the parameter of B on the line l is 6 and the vector $\frac{1}{3}\mathbf{i} + \frac{2}{3}\mathbf{j} - \frac{2}{3}\mathbf{k}$ is a unit vector, the distance from A to B is 6.

You could have found the coordinates of the point of intersection of l and m as $(3,3,-3)$, and then found the distance of $(3,3,-3)$ from A using the formula for the distance between two points.

If the parameter of the point of intersection had been negative, then you would have taken the modulus of the parameter to find the distance. So, if the parameter had been -6, taking the modulus would have given the distance as 6.

Finding the perpendicular distance of a point from a line is a little more complicated. Here are two examples to show the method. The first is in two dimensions, the second in three.

Example 12.3.2

Find the distance of the point Q with coordinates $(3,1)$ from the line l through the

origin with vector equation $\mathbf{r} = t\begin{pmatrix} \frac{5}{13} \\ \frac{12}{13} \end{pmatrix}$.

Let the perpendicular from Q to the line
l meet it at N, so you need to find QN
(see Fig. 12.3).

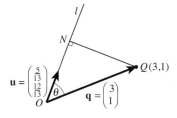

Notice that $\begin{pmatrix} \frac{5}{13} \\ \frac{12}{13} \end{pmatrix}$ is a unit vector, and

denote it by $\mathbf{u}$. Let θ be the angle
between OQ and l.

Fig. 12.3

In the right-angled triangle OQN, you can find the lengths of OQ and ON, and
then calculate the length QN by using Pythagoras' theorem.

$$OQ = \sqrt{3^2 + 1^2} = \sqrt{10}.$$

The length ON is called the projection of the vector $\overrightarrow{OQ}$ on the line l. It was
shown in P1 Section 13.10 that this is given by $\mathbf{q \cdot u}$. That is,

$$ON = OQ\cos\theta = OQ \times 1 \times \cos\theta = \mathbf{q \cdot u}, \text{ so } ON = \begin{pmatrix} 3 \\ 1 \end{pmatrix} \cdot \begin{pmatrix} \frac{5}{13} \\ \frac{12}{13} \end{pmatrix} = 3 \times \frac{5}{13} + 1 \times \frac{12}{13} = \frac{27}{13}.$$

Therefore,

$$QN^2 = OQ^2 - ON^2 = 10 - \left(\frac{27}{13}\right)^2 = 10 - \frac{729}{169} = \frac{961}{169},$$

giving $QN = \frac{31}{13}$.

The required distance is $2\frac{5}{13}$.

In three dimensions the principle is the same. In the next example the line does not pass
through the origin.

Example 12.3.3

Find the distance of the point Q with coordinates $(1,2,3)$ from the straight line with
equation $\mathbf{r} = 3\mathbf{i} + 4\mathbf{j} - 2\mathbf{k} + t(\mathbf{i} - 2\mathbf{j} + 2\mathbf{k})$.

Let A be the point $(3,4,-2)$, and N be the foot
of the perpendicular from Q to the line.

Focus on the triangle ANQ (see Fig. 12.4).

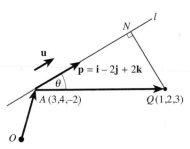

$$\overrightarrow{AQ} = \mathbf{q} - \mathbf{a} = (\mathbf{i} + 2\mathbf{j} + 3\mathbf{k}) - (3\mathbf{i} + 4\mathbf{j} - 2\mathbf{k})$$

$$= -2\mathbf{i} - 2\mathbf{j} + 5\mathbf{k}.$$

Fig. 12.4

So

$$AQ^2 = \left| -2\mathbf{i} - 2\mathbf{j} + 5\mathbf{k} \right|^2 = (-2)^2 + (-2)^2 + 5^2 = 33.$$

The length of the vector $\mathbf{p} = \mathbf{i} - 2\mathbf{j} + 2\mathbf{k}$ is $\sqrt{1^2 + (-2)^2 + 2^2} = 3$, so the unit vector in the direction of $\mathbf{p}$ is $\mathbf{u} = \frac{1}{3}\mathbf{i} - \frac{2}{3}\mathbf{j} + \frac{2}{3}\mathbf{k}$.

Then $AN = AQ \cos\theta = AQ \times 1 \times \cos\theta = (\mathbf{q} - \mathbf{a}).\mathbf{u}$, so

$$\begin{aligned}
AN &= (\mathbf{q} - \mathbf{a}).\mathbf{u} \\
&= (-2\mathbf{i} - 2\mathbf{j} + 5\mathbf{k}).\left(\tfrac{1}{3}\mathbf{i} - \tfrac{2}{3}\mathbf{j} + \tfrac{2}{3}\mathbf{k}\right) \\
&= (-2) \times \tfrac{1}{3} + (-2) \times \left(-\tfrac{2}{3}\right) + 5 \times \tfrac{2}{3} \\
&= -\tfrac{2}{3} + \tfrac{4}{3} + \tfrac{10}{3} = 4.
\end{aligned}$$

So $QN^2 = AQ^2 - AN^2 = 33 - 16 = 17$, giving $QN = \sqrt{17}$.

The required distance is $\sqrt{17}$.

Exercise 12C

1 Find the distance of $(3,4)$ from the straight line $3x + 4y = 0$.

2 Find the distance of $(1,0,0)$ from the line $\mathbf{r} = t(12\mathbf{i} - 3\mathbf{j} - 4\mathbf{k})$.

3 Find the distance of the point $(1,1,4)$ from the line $\mathbf{r} = \mathbf{i} - 2\mathbf{j} + \mathbf{k} + t(-2\mathbf{i} + \mathbf{j} + 2\mathbf{k})$.

4 Find the distance of the point $(1,\pm 6,-1)$ from the line with vector equation

$$\mathbf{r} = \begin{pmatrix} 1 \\ 7 \\ -1 \end{pmatrix} + t\begin{pmatrix} 3 \\ 12 \\ -4 \end{pmatrix}.$$

5 Find a vector equation of the line l containing the points $(1,3,1)$ and $(1,-3,-1)$. Find the perpendicular distance of the point with coordinates $(2,-1,1)$ from l.

Miscellaneous exercise 12

1 Two lines have equations $\mathbf{r} = \begin{pmatrix} 1 \\ 3 \\ 2 \end{pmatrix} + \lambda\begin{pmatrix} 4 \\ -2 \\ 1 \end{pmatrix}$ and $\mathbf{r} = \begin{pmatrix} 3 \\ 8 \\ 7 \end{pmatrix} + \mu\begin{pmatrix} 2 \\ -3 \\ -1 \end{pmatrix}$. Show that the lines intersect, and find the position vector of the point of intersection.   (OCR)

2 (a) Find a vector equation for the line joining $(1,1)$ and $(5,-1)$.

   (b) Another line has the vector equation $\mathbf{r} = \begin{pmatrix} 3 \\ 4 \end{pmatrix} + t\begin{pmatrix} 1 \\ 3 \end{pmatrix}$. Find the point of intersection of the two lines.

3 A tunnel is to be excavated through a hill. In order to define position, coordinates (x, y, z) are taken relative to an origin O such that x is the distance east from O, y is the distance north and z is the vertical distance upwards, with one unit equal to 100 m. The tunnel starts at point $A(2,3,5)$ and runs in a straight line in the direction $\begin{pmatrix} 1 \\ 1 \\ -0.5 \end{pmatrix}$.

(a) Write down the equation of the tunnel in the form $\mathbf{r} = \mathbf{u} + \lambda \mathbf{t}$.

(b) An old tunnel through the hill has equation $\mathbf{r} = \begin{pmatrix} 4 \\ 1 \\ 2 \end{pmatrix} + \mu \begin{pmatrix} 7 \\ 15 \\ 0 \end{pmatrix}$. Show that the point P on

the new tunnel where $x = 7\frac{1}{2}$ is directly above a point Q in the old tunnel. Find the vertical separation PQ of the tunnels at this point. (MEI)

4 Find the intersection of the lines $\mathbf{r} = \begin{pmatrix} -1 \\ 0 \end{pmatrix} + s \begin{pmatrix} \cos \alpha \\ \sin \alpha \end{pmatrix}$ and $\mathbf{r} = \begin{pmatrix} 1 \\ 0 \end{pmatrix} + t \begin{pmatrix} -\sin \alpha \\ \cos \alpha \end{pmatrix}$, giving

your answer in a simplified form. Interpret your answer geometrically.

5 An aeroplane climbs so that its position relative to the airport control tower t minutes after

take-off is given by the vector $\mathbf{r} = \begin{pmatrix} -1 \\ -2 \\ 0 \end{pmatrix} + t \begin{pmatrix} 4 \\ 5 \\ 0.6 \end{pmatrix}$, the units being kilometres. The x- and

y-axes point towards the east and the north respectively. Calculate the closest distance of the aeroplane from the airport control tower during this flight, giving your answer correct to 2 decimal places. To the nearest second, how many seconds after leaving the ground is the aeroplane at its closest to the airport control tower?

6 Lines l_1 and l_2 have vector equations $\mathbf{r} = \begin{pmatrix} 2 \\ -3 \\ 1 \end{pmatrix} + \lambda \begin{pmatrix} 5 \\ 1 \\ 2 \end{pmatrix}$ and $\mathbf{r} = \begin{pmatrix} m \\ 2 \\ 5 \end{pmatrix} + \mu \begin{pmatrix} 2 \\ 1 \\ -1 \end{pmatrix}$

respectively, where λ and μ are scalar parameters, and m is a constant.

(a) The point P has position vector $\begin{pmatrix} 7 \\ -2 \\ 1 \end{pmatrix}$ and point Q has position vector $\begin{pmatrix} 5 \\ 4 \\ 3 \end{pmatrix}$.

 (i) Determine the vector $\overrightarrow{PQ}$, and show that $\left| \overrightarrow{PQ} \right| = 2\sqrt{11}$.

 (ii) Verify that $\overrightarrow{PQ}$ is perpendicular to both l_1 and l_2.

 (iii) Show that P lies on l_1 and find the value of m for which Q lies on l_2. Write down the shortest distance between l_1 and l_2 in this case.

(b) Find the size of the acute angle between the lines l_1 and l_2, giving your answer correct to the nearest 0.1°.

(c) Determine the value of m, different from the value you found in part (a), for which l_1 and l_2 intersect. (OCR)

7 An aeroplane climbs so that its position relative to the airport control tower t minutes after take-off is given by the vector $\mathbf{r} = \begin{pmatrix} 1 \\ 2 \\ 0 \end{pmatrix} + t \begin{pmatrix} 4 \\ 5 \\ 0.6 \end{pmatrix}$, the units being kilometres. The x- and y-axes point towards the east and the north respectively.

(a) Find the position of the aeroplane when it reaches its cruising height of 9 km.

(b) With reference to (x, y) coordinates on the ground, the coastline has equation $x + 3y = 140$. How high is the aircraft flying as it crosses the coast?

(c) Calculate the speed of the aeroplane over the ground in kilometres per hour, and the bearing on which it is flying.

(d) Calculate the speed of the aeroplane through the air, and the angle to the horizontal at which it is climbing.

8 The line l has vector equation $\mathbf{r} = 2\mathbf{i} + s(\mathbf{i} + 3\mathbf{j} + 4\mathbf{k})$.

(a) (i) Show that the line l intersects the line with equation $\mathbf{r} = \mathbf{k} + t(\mathbf{i} + \mathbf{j} + \mathbf{k})$ and determine the position vector of the point of intersection.

(ii) Calculate the acute angle, to the nearest degree, between these lines.

(b) Find the position vectors of the points on l which are exactly $5\sqrt{10}$ units from the origin.

(c) Determine the position vector of the point on l which is closest to the point with position vector $6\mathbf{i} - \mathbf{j} + 3\mathbf{k}$.   (OCR)

9 Two aeroplanes take off simultaneously from different airports. As they climb, their positions relative to an air traffic control centre t minutes later are given by the vectors $\mathbf{r}_1 = \begin{pmatrix} 5 \\ -30 \\ 0 \end{pmatrix} + t \begin{pmatrix} 8 \\ 2 \\ 0.5 \end{pmatrix}$ and $\mathbf{r}_2 = \begin{pmatrix} 13 \\ 26 \\ 0 \end{pmatrix} + t \begin{pmatrix} 6 \\ -3 \\ 0.6 \end{pmatrix}$, the units being kilometres. Find the coordinates of the point on the ground over which both aeroplanes pass. Find also the difference in heights, and the difference in the times, when they pass over that point.

10 The centre line of an underground railway tunnel follows a line given by $\mathbf{r} = t \begin{pmatrix} 10 \\ 8 \\ -1 \end{pmatrix}$ for $0 \leqslant t \leqslant 40$, the units being metres. The centre line of another tunnel at present stops at the point with position vector $\begin{pmatrix} 200 \\ 100 \\ -25 \end{pmatrix}$ and it is proposed to extend this in a direction $\begin{pmatrix} 5 \\ 7 \\ u \end{pmatrix}$. The constant u has to be chosen so that, at the point where one tunnel passes over the other, there is at least 15 metres difference in depth between the centre lines of the two tunnels. What restriction does this impose on the value of u?

Another requirement is that the tunnel must not be inclined at more than 5° to the horizontal. What values of u satisfy both requirements?

13 Vectors: planes in three dimensions

This chapter uses vectors to investigate the geometry of planes. When you have completed it, you should

- be able to find the equation of a plane
- be able to find out whether or not a given line intersects a given plane
- be able to find the line of intersection of two planes, the angle between two planes, and the angle between a line and a plane
- be able to find the distance of a point from a plane.

13.1 The cartesian equation of a line in two dimensions

Although this chapter is about planes it is helpful to begin by looking at the equation of a line in two dimensions in another way.

In two dimensions, you can describe the direction of a line by a single number, its gradient. (The only exception is a line parallel to the y-axis.) But even in two dimensions it is often simpler to describe the direction of a line by a vector rather than by its gradient, as the following example shows. The second method suggests an approach which you can extend to three dimensions.

Example 13.1.1
Find the cartesian equation of the line through $(1,2)$ perpendicular to the non-zero vector $\begin{pmatrix} l \\ m \end{pmatrix}$.

Method 1   As $\begin{pmatrix} -m \\ l \end{pmatrix} \cdot \begin{pmatrix} l \\ m \end{pmatrix} = 0$, the vector $\begin{pmatrix} -m \\ l \end{pmatrix}$ is perpendicular to $\begin{pmatrix} l \\ m \end{pmatrix}$, so

$\mathbf{r} = \begin{pmatrix} x \\ y \end{pmatrix} = \begin{pmatrix} 1 \\ 2 \end{pmatrix} + t \begin{pmatrix} -m \\ l \end{pmatrix}$ is a vector equation for the line. Writing this as

$x = 1 - tm$ and $y = 2 + tl$ and eliminating t gives $l(x-1) + m(y-2) = 0$. This simplifies to

$$lx + my = l + 2m.$$

Look at the left side of this equation. The line perpendicular to the vector $\begin{pmatrix} l \\ m \end{pmatrix}$ has an equation of the form $lx + my = k$ where k is a constant. This suggests method 2.

Method 2   Let $\mathbf{n} = \begin{pmatrix} l \\ m \end{pmatrix}$ be the perpendicular to the line,

and $\mathbf{a}$ be the position vector of $(1,2)$. Let $\mathbf{r}$ be the position vector of any other point on the line. See Fig. 13.1.

Then $\mathbf{r} - \mathbf{a}$ is a vector in the direction of the line, and is therefore perpendicular to $\mathbf{n}$.

Therefore $(\mathbf{r} - \mathbf{a}) \cdot \mathbf{n} = 0$, or $\mathbf{r} \cdot \mathbf{n} = \mathbf{a} \cdot \mathbf{n}$.

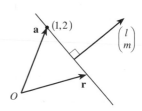

Fig. 13.1

If $\mathbf{n} = \begin{pmatrix} l \\ m \end{pmatrix}$, $\mathbf{r} = \begin{pmatrix} x \\ y \end{pmatrix}$ and $\mathbf{a} = \begin{pmatrix} 1 \\ 2 \end{pmatrix}$, then the equation $\mathbf{r} \cdot \mathbf{n} = \mathbf{a} \cdot \mathbf{n}$ becomes

$\begin{pmatrix} x \\ y \end{pmatrix} \cdot \begin{pmatrix} l \\ m \end{pmatrix} = \begin{pmatrix} 1 \\ 2 \end{pmatrix} \cdot \begin{pmatrix} l \\ m \end{pmatrix}$, that is $lx + my = l + 2m$.

Method 2 suggests what is generally the best way to find the cartesian equation of a plane.

13.2 The cartesian equation of a plane

What is a plane? Of course you know the answer to this – it is a flat surface. But this, on its own, will not help you to find its equation. You need to find a way to express the fact that the plane is a flat surface in a mathematical way.

One possibility is to use a property in three dimensions similar to that used in the previous section for a straight line in two dimensions.

On any smooth surface (such as a sphere) there is at each point a line perpendicular to the surface. The special property of a plane is that these perpendiculars are in the same direction at every point.

A vector in this direction is called a **normal** to the plane.

Fig. 13.2 shows a plane with a normal $\mathbf{n}$ drawn at three points. The normal to a plane is not unique. If $\mathbf{n}$ is normal to a plane, any multiple of $\mathbf{n}$ (except zero) is also normal to the plane.

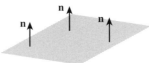

Fig. 13.2

Every vector lying in the plane is then perpendicular to $\mathbf{n}$. It is this property that enables you to find the equation of the plane.

Let $\mathbf{n}$ be a vector perpendicular to a plane and let $\mathbf{a}$ be the position vector of a point on the plane. Let $\mathbf{r}$ be the position vector of any other point on the plane, as in Fig. 13.3.

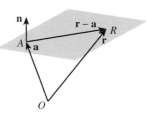

Then $\mathbf{r} - \mathbf{a}$ is a vector parallel to the plane, and is therefore perpendicular to $\mathbf{n}$. Therefore $(\mathbf{r} - \mathbf{a}) \cdot \mathbf{n} = 0$, or $\mathbf{r} \cdot \mathbf{n} = \mathbf{a} \cdot \mathbf{n}$.

Fig. 13.3

This equation, $\mathbf{r} \cdot \mathbf{n} = \mathbf{a} \cdot \mathbf{n}$, is one form of the equation of a plane.

> Points of a plane through A and perpendicular to the normal vector $\mathbf{n}$ have position vectors $\mathbf{r}$ which satisfy $\mathbf{r} \cdot \mathbf{n} = \mathbf{a} \cdot \mathbf{n}$.
> This is called the **normal equation** of the plane.

If you write $\mathbf{r} = \begin{pmatrix} x \\ y \\ z \end{pmatrix}$ and $\mathbf{n} = \begin{pmatrix} p \\ q \\ r \end{pmatrix}$, the equation $\mathbf{r} \cdot \mathbf{n} = \mathbf{a} \cdot \mathbf{n}$ becomes

$$px + qy + rz = \mathbf{a} \cdot \mathbf{n}.$$

This is the cartesian form of the equation of a plane. Notice that the right side is a constant, and the coefficients on the left side are the components of the normal vector. Thus, you can write down the equation of the plane directly if you know a vector normal to it and you know a point on it.

Points of a plane through A and perpendicular to the normal vector $\mathbf{n} = \begin{pmatrix} p \\ q \\ r \end{pmatrix}$

(or $\mathbf{n} = p\mathbf{i} + q\mathbf{j} + r\mathbf{k}$) have coordinates (x, y, z) which satisfy
$px + qy + rz = k$, where k is a constant determined by the coordinates of A.
This is called the **cartesian equation** of the plane.

Note that in two dimensions the direction of a line is determined either by a vector along the line or by a vector perpendicular to it.

In three dimensions the situation is different: the direction of a line is determined uniquely by a vector along the line, but there are many vectors perpendicular (normal) to it; the direction of a plane is determined uniquely by a vector perpendicular (normal) to it, but there are many different pairs of vectors parallel to the plane which can be used to determine it.

Example 13.2.1
Find the cartesian equation of the plane through the point $(1, 2, 3)$ with normal $\begin{pmatrix} 4 \\ 5 \\ 6 \end{pmatrix}$.

Method 1   The equation is $4x + 5y + 6z = k$, where k is a constant.

The constant has to be chosen so that the plane passes through $(1, 2, 3)$.

The constant is therefore $4 \times 1 + 5 \times 2 + 6 \times 3 = 32$,
so the equation is $4x + 5y + 6z = 32$.

Method 2   Using the equation $\mathbf{r} \cdot \mathbf{n} = \mathbf{a} \cdot \mathbf{n}$ gives $\begin{pmatrix} x \\ y \\ z \end{pmatrix} \cdot \begin{pmatrix} 4 \\ 5 \\ 6 \end{pmatrix} = \begin{pmatrix} 1 \\ 2 \\ 3 \end{pmatrix} \cdot \begin{pmatrix} 4 \\ 5 \\ 6 \end{pmatrix}$.

This is $4x + 5y + 6z = 1 \times 4 + 2 \times 5 + 3 \times 6$,
which is $4x + 5y + 6z = 32$.

Example 13.2.2

Find the coordinates of the point of intersection of the line $\mathbf{r} = \begin{pmatrix} 1 \\ 0 \\ 1 \end{pmatrix} + t \begin{pmatrix} 2 \\ 1 \\ -3 \end{pmatrix}$ with the plane $3x + 2y + 4z = 11$.

Rewriting the line in the form $\begin{pmatrix} x \\ y \\ z \end{pmatrix} = \begin{pmatrix} 1 \\ 0 \\ 1 \end{pmatrix} + t \begin{pmatrix} 2 \\ 1 \\ -3 \end{pmatrix}$ and taking components yields the equations $x = 1 + 2t$, $y = 0 + t$ and $z = 1 - 3t$. Substituting these into the equation of the plane gives

$$3(1 + 2t) + 2t + 4(1 - 3t) = 11, \quad \text{which gives} \quad t = -1.$$

So the line meets the plane at the point with parameter -1, namely $(-1, -1, 4)$.

Example 13.2.3

Find whether or not the lines   (a) $\mathbf{r} = \begin{pmatrix} 1 \\ 2 \\ -1 \end{pmatrix} + s \begin{pmatrix} 4 \\ 3 \\ 5 \end{pmatrix}$ and   (b) $\mathbf{r} = \begin{pmatrix} 3 \\ 2 \\ -3 \end{pmatrix} + t \begin{pmatrix} 1 \\ 2 \\ 0 \end{pmatrix}$, lie in the plane $2x - y - z = 1$.

(a) Writing $\mathbf{r}$ as $\begin{pmatrix} x \\ y \\ z \end{pmatrix}$ gives $x = 1 + 4s$, $y = 2 + 3s$ and $z = -1 + 5s$.

Substituting for x, y and z in the equation of the plane gives

$$2(1 + 4s) - (2 + 3s) - (-1 + 5s) = 1,$$

which simplifies to $2 + 8s - 2 - 3s + 1 - 5s = 1$, or $1 = 1$.

The equation $1 = 1$ is an identity. It is true for all values of s. So for all values of s the coordinates of the points of the line satisfy the equation of the plane. The line therefore lies in the plane.

(b) Using the same method as in part (a), $x = 3 + t$, $y = 2 + 2t$ and $z = -3$.

Substituting for x, y and z in the equation of the plane gives

$$2(3 + t) - (2 + 2t) - (-3) = 1,$$

which simplifies to $6 + 2t - 2 - 2t + 3 = 1$, or $7 = 1$.

No value of t can make $7 = 1$, so there are no values of t for which the coordinates of the points of the line satisfy the equation of the plane. The line therefore does not meet the plane.

In Example 13.2.3 the directions of both lines, $\begin{pmatrix} 4 \\ 3 \\ 5 \end{pmatrix}$ *and* $\begin{pmatrix} 1 \\ 2 \\ 0 \end{pmatrix}$, *are perpendicular to* $\begin{pmatrix} 2 \\ -1 \\ -1 \end{pmatrix}$,

the normal to the plane $2x - y - z = 1$*. So* $\begin{pmatrix} 4 \\ 3 \\ 5 \end{pmatrix}$ *and* $\begin{pmatrix} 1 \\ 2 \\ 0 \end{pmatrix}$ *are both parallel to the plane.*

In part (a) the point $(1, 2, -1)$ *lies in the plane, so the line lies in the plane. In part (b) the point* $(3, 2, -3)$ *does not lie in the plane, so the line is parallel to the plane.*

██████████████████████   **Exercise 13A**   ██████████████████████

1   Find the cartesian equation of the plane through $(1,1,1)$ normal to the vector $5\mathbf{i} - 8\mathbf{j} + 4\mathbf{k}$.

2   Find the cartesian equation of the plane through $(2,-1,1)$ normal to the vector $2\mathbf{i} - \mathbf{j} - \mathbf{k}$.

3   Find the coordinates of two points A and B on the plane $2x + 3y + 4z = 4$. Verify that the vector $\overrightarrow{AB}$ is perpendicular to the normal to the plane.

4   Verify that the line with equation $\mathbf{r} = 2\mathbf{i} + 4\mathbf{j} + \mathbf{k} + t(-4\mathbf{i} + 4\mathbf{j} - 5\mathbf{k})$ lies wholly in the plane with equation $3x - 2y + 4z = 2$.

5   Find the equation of the plane through $(1,2,-1)$ parallel to the plane $5x + y + 7z = 20$.

6   Find the equation of the line through $(4,2,-1)$ perpendicular to the plane $3x + 4y - z = 1$.

7   Verify that the plane with equation $x - 2y + 2z = 6$ is parallel to the plane with equation

$$\mathbf{r}\cdot\begin{pmatrix} 1 \\ -2 \\ 2 \end{pmatrix} = 4.$$ Find the perpendicular distance from the origin to each plane, and hence find

the perpendicular distance between the planes.

13.3 Some techniques for solving problems

Example 13.3.1
(a) Find the position vector of the foot A of the perpendicular from the point $(1,1,1)$ to the plane $x + 2y - 2z = 9$.
(b) Find the length of this perpendicular.

(a) The normal to the plane $x + 2y - 2z = 9$ has direction $\mathbf{i} + 2\mathbf{j} - 2\mathbf{k}$, so the equation of the line through $(1,1,1)$ perpendicular to the plane is
$\mathbf{r} = \mathbf{i} + \mathbf{j} + \mathbf{k} + t(\mathbf{i} + 2\mathbf{j} - 2\mathbf{k})$. That is, $\mathbf{r} = (1+t)\mathbf{i} + (1+2t)\mathbf{j} + (1-2t)\mathbf{k}$.

This meets the plane $x + 2y - 2z = 9$ where $(1+t) + 2(1+2t) - 2(1-2t) = 9$, that is where $9t = 8$, or $t = \frac{8}{9}$.

The position vector of A is $\mathbf{i} + \mathbf{j} + \mathbf{k} + \frac{8}{9}(\mathbf{i} + 2\mathbf{j} - 2\mathbf{k}) = \frac{17}{9}\mathbf{i} + \frac{25}{9}\mathbf{j} - \frac{7}{9}\mathbf{k}$.

(b) The coordinates of the point A are $\left(\frac{17}{9}, \frac{25}{9}, -\frac{7}{9}\right)$. You could find the distance of A from $(1,1,1)$ by using the distance formula.

But there is a quicker way.

The length of the vector from $(1,1,1)$ to the plane is $\frac{8}{9}$ of the length of the vector

$\mathbf{i} + 2\mathbf{j} - 2\mathbf{k}$, which is $\sqrt{1^2 + 2^2 + (-2)^2} = 3$.

So the perpendicular distance is $\frac{8}{9} \times 3 = \frac{8}{3}$.

Example 13.3.2

Find the acute angle between the line
$\mathbf{r} = \mathbf{i} + 3\mathbf{j} + 5\mathbf{k} + t(2\mathbf{i} + 4\mathbf{j} + \mathbf{k})$ and the plane $x - y + z = 0$.

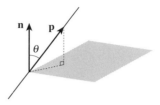

As shown in Fig. 13.4, take $\mathbf{n}$ to be a vector normal to the plane, and $\mathbf{p}$ a vector along the line. You can find the angle θ from the scalar product $\mathbf{n} \cdot \mathbf{p}$, and the angle between the line and the plane is then $\frac{1}{2}\pi - \theta$.

Fig. 13.4

The line has direction vector $2\mathbf{i} + 4\mathbf{j} + \mathbf{k}$. The normal to the plane is $\mathbf{i} - \mathbf{j} + \mathbf{k}$. So the angle θ between them is given by

$$(2\mathbf{i} + 4\mathbf{j} + \mathbf{k}) \cdot (\mathbf{i} - \mathbf{j} + \mathbf{k}) = |2\mathbf{i} + 4\mathbf{j} + \mathbf{k}| \times |\mathbf{i} - \mathbf{j} + \mathbf{k}| \times \cos\theta, \text{ that is}$$

$$\cos\theta = -\frac{1}{\sqrt{21}\sqrt{3}} = -\frac{1}{3\sqrt{7}}.$$

There is a problem. As $\cos\theta$ is negative, angle θ is obtuse, so Fig. 13.4 can't be right. The relation between $\mathbf{p}$ and $\mathbf{n}$ is correctly shown in Fig. 13.5. The required angle is $\theta - \frac{1}{2}\pi$, which is

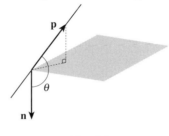

$$\cos^{-1}\left(-\frac{1}{3\sqrt{7}}\right) - \frac{1}{2}\pi = \left(\pi - \cos^{-1}\left(\frac{1}{3\sqrt{7}}\right)\right) - \frac{1}{2}\pi$$

$$= \frac{1}{2}\pi - \cos^{-1}\left(\frac{1}{3\sqrt{7}}\right)$$

$$= \sin^{-1}\frac{1}{3\sqrt{7}}.$$

Fig. 13.5

Example 13.3.3

A pyramid of height 3 units stands symmetrically on a rectangular base $ABCD$ with $AB = 2$ units and $BC = 4$ units. Find the angle between two adjacent slanting faces.

The strategy is to find the angle between the planes by finding the angle between the normals to the planes.

Let V be the vertex of the pyramid. Take the origin at the centre of the base, and the x- and y-axes parallel to CB and AB, as in Fig. 13.6. Then the coordinates of A, B, C and V are respectively $(2,-1,0)$, $(2,1,0)$, $(-2,1,0)$ and $(0,0,3)$.

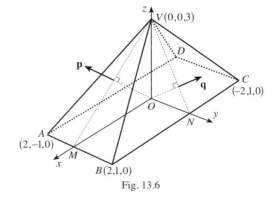

The normal vector $\mathbf{p}$ to the face VAB is in the direction of the perpendicular from O to VM, where M is the midpoint of AB with coordinates $(2,0,0)$.

Fig. 13.6

By symmetry, $\mathbf{p}$ has no y-component, and it is perpendicular to

$$\overrightarrow{MV} = \begin{pmatrix} 0 \\ 0 \\ 3 \end{pmatrix} - \begin{pmatrix} 2 \\ 0 \\ 0 \end{pmatrix} = \begin{pmatrix} -2 \\ 0 \\ 3 \end{pmatrix}, \text{ so } \mathbf{p} \text{ can be taken as } \begin{pmatrix} 3 \\ 0 \\ 2 \end{pmatrix}.$$

Similarly the normal vector $\mathbf{q}$ to the face VBC has no x-component and is

perpendicular to $\overrightarrow{NV} = \begin{pmatrix} 0 \\ -1 \\ 3 \end{pmatrix}$, so take $\mathbf{q} = \begin{pmatrix} 0 \\ 3 \\ 1 \end{pmatrix}$.

Let the angle between $\mathbf{p}$ and $\mathbf{q}$ be $\theta°$. Then

$$\left| \begin{pmatrix} 3 \\ 0 \\ 2 \end{pmatrix} \right| \left| \begin{pmatrix} 0 \\ 3 \\ 1 \end{pmatrix} \right| \cos\theta° = \begin{pmatrix} 3 \\ 0 \\ 2 \end{pmatrix} \cdot \begin{pmatrix} 0 \\ 3 \\ 1 \end{pmatrix},$$

so $\quad \cos\theta° = \dfrac{2}{\sqrt{13}\sqrt{10}} = \dfrac{2}{\sqrt{130}} \quad$ and $\quad \theta \approx 79.9.$

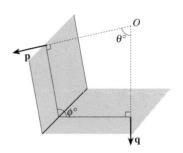

Fig. 13.7 illustrates the relation between the angle $\theta°$ between $\mathbf{p}$ and $\mathbf{q}$, and the angle $\phi°$ between the faces of the pyramid, as viewed from inside the pyramid. This shows that $\phi = 180 - \theta \approx 100.1$.

The faces of the pyramid are at $100.1°$ to each other.

Fig. 13.7

Example 13.3.4

Find the cartesian equation of the plane through $A(1,2,1)$, $B(2,-1,-4)$ and $C(1,0,-1)$.

You can tackle this problem in several ways. Two are given here.

Method 1   The equation of a plane is of the form $ax + by + cz = d$, where a, b and c are not all 0. Substitute the coordinates of the points in the equation. You then get the three equations

$$\begin{aligned} a + 2b + c &= d, \\ 2a - b - 4c &= d, \\ a - c &= d. \end{aligned}$$

If you subtract the third equation from the first, you get $2b + 2c = 0$, giving $c = -b$. This reduces the equations to

$$\begin{aligned} a + b &= d, \\ 2a + 3b &= d. \end{aligned}$$

You can now solve for a and b in terms of d, giving $a = 2d$ and $b = -d$. Using $c = -b$ means that $c = d$, so the equation $ax + by + cz = d$ becomes

$$2dx - dy + dz = d.$$

Note that if $d = 0$, then $a = b = c = 0$, which is not allowed. So $d \neq 0$ and you can divide by d to give $2x - y + z = 1$ as the equation of the plane.

Method 2   Let $\begin{pmatrix} p \\ q \\ r \end{pmatrix}$ be normal to the plane. As $\mathbf{b} - \mathbf{a} = \begin{pmatrix} 1 \\ -3 \\ -5 \end{pmatrix}$ and $\mathbf{c} - \mathbf{a} = \begin{pmatrix} 0 \\ -2 \\ -2 \end{pmatrix}$ are

vectors parallel to the plane, they are perpendicular to the normal, so both scalar

products $\begin{pmatrix} 1 \\ -3 \\ -5 \end{pmatrix} \cdot \begin{pmatrix} p \\ q \\ r \end{pmatrix}$ and $\begin{pmatrix} 0 \\ -2 \\ -2 \end{pmatrix} \cdot \begin{pmatrix} p \\ q \\ r \end{pmatrix}$ are zero. Therefore

$$\left. \begin{matrix} p - 3q - 5r = 0 \\ -2q - 2r = 0 \end{matrix} \right\}, \quad \text{so} \quad \left. \begin{matrix} p - 3q - 5r = 0 \\ q + r = 0 \end{matrix} \right\}.$$

This is a pair of simultaneous equations in three unknowns. The best that can be done is to say that

$$\left. \begin{matrix} p - 3q = 5r \\ q = -r \end{matrix} \right\},$$

and to solve for p and q in terms of r.

Substituting $q = -r$ in the first equation shows that $p - 3(-r) = 5r$, giving $p = 2r$.

Thus $\begin{pmatrix} p \\ q \\ r \end{pmatrix} = \begin{pmatrix} 2r \\ -r \\ r \end{pmatrix} = r \begin{pmatrix} 2 \\ -1 \\ 1 \end{pmatrix}$ is normal to the plane for all r, except $r = 0$. Since you

need only one normal, put $r = 1$, giving $\begin{pmatrix} 2 \\ -1 \\ 1 \end{pmatrix}$ as the normal to the plane.

Therefore the equation is $2x - y + z = k$; and since $(1,2,1)$ lies on the plane, the equation is $2x - y + z = 2 \times 1 - 1 \times 2 + 1 \times 1$, which is $2x - y + z = 1$.

It is good practice to verify that the other points also lie on this plane.

Method 2 of Example 13.3.4 is probably the best way to find the equation of a plane through three points, but it can be shortened further by a piece of theory.

13.4 Finding a common perpendicular

In Example 13.3.4 you had to find a vector which was perpendicular to both of two given vectors. This situation occurs quite frequently when you tackle problems concerned with lines and planes.

Suppose that the vectors $\begin{pmatrix} l \\ m \\ n \end{pmatrix}$ and $\begin{pmatrix} p \\ q \\ r \end{pmatrix}$ are both non-zero and non-parallel,

and that $\begin{pmatrix} x \\ y \\ z \end{pmatrix}$ is perpendicular to both of them.

Then the scalar products $\begin{pmatrix} l \\ m \\ n \end{pmatrix} \cdot \begin{pmatrix} x \\ y \\ z \end{pmatrix} = \begin{pmatrix} p \\ q \\ r \end{pmatrix} \cdot \begin{pmatrix} x \\ y \\ z \end{pmatrix} = 0$, so $\left. \begin{matrix} lx + my + nz = 0 \\ px + qy + rz = 0 \end{matrix} \right\}.$

This is a set of two equations in three unknowns.

You can verify by substitution that $\begin{pmatrix} mr - nq \\ np - lr \\ lq - mp \end{pmatrix}$ (and any multiple of it) is a solution.

$$\begin{pmatrix} l \\ m \\ n \end{pmatrix} \cdot \begin{pmatrix} x \\ y \\ z \end{pmatrix} = \begin{pmatrix} l \\ m \\ n \end{pmatrix} \cdot \begin{pmatrix} mr - nq \\ np - lr \\ lq - mp \end{pmatrix} = lmr - lnq + mnp - mlr + nlq - nmp = 0 \text{, and}$$

$$\begin{pmatrix} p \\ q \\ r \end{pmatrix} \cdot \begin{pmatrix} x \\ y \\ z \end{pmatrix} = \begin{pmatrix} p \\ q \\ r \end{pmatrix} \cdot \begin{pmatrix} mr - nq \\ np - lr \\ lq - mp \end{pmatrix} = pmr - pnq + qnp - qlr + rlq - rmp = 0.$$

This is true for all vectors $\begin{pmatrix} l \\ m \\ n \end{pmatrix}$ and $\begin{pmatrix} p \\ q \\ r \end{pmatrix}$. It is also true that, provided $\begin{pmatrix} l \\ m \\ n \end{pmatrix}$ and $\begin{pmatrix} p \\ q \\ r \end{pmatrix}$ are non-zero and non-parallel, $\begin{pmatrix} mr - nq \\ np - lr \\ lq - mp \end{pmatrix}$ cannot be zero.

If you go on to Further Mathematics you will recognise this new vector as the vector product. It is easy then to show that this new vector is non-zero if the original vectors are non-zero and non-parallel, and vice versa.

If $\begin{pmatrix} l \\ m \\ n \end{pmatrix}$ and $\begin{pmatrix} p \\ q \\ r \end{pmatrix}$ are non-zero, non-parallel vectors, then $\begin{pmatrix} mr - nq \\ np - lr \\ lq - mp \end{pmatrix}$ is non-zero and perpendicular to both of them.

At first sight this result looks hard to remember, but it isn't too bad if you can see where the individual terms come from.

To find the first component of the common perpendicular, start by blocking out the first components of the given vectors, as shown in Fig. 13.8. Then take the products of the remaining components as indicated by the arrows, and subtract them, getting $mr - nq$.

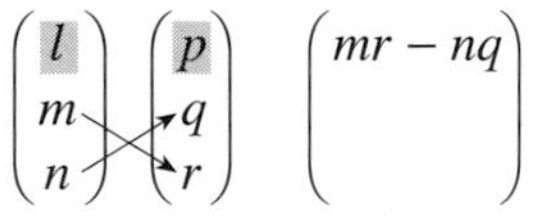

Fig. 13.8

To get the second component, block out the second components of the given vectors, as shown in Fig. 13.9. Then take the products of the remaining components indicated by the arrows, and subtract them, getting $np - lr$.

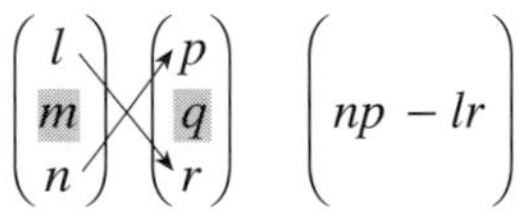

Fig. 13.9

Finally, get the third component by blocking out the third components of the given vectors, as shown in Fig. 13.10. Then take the products of the remaining components indicated by the arrows, and subtract them, getting $lq - mp$.

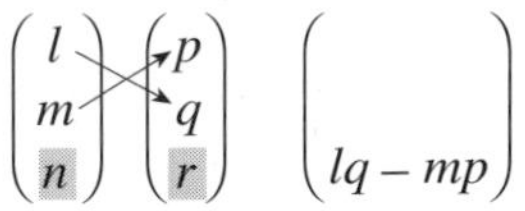

$$\begin{pmatrix} l \\ m \\ n \end{pmatrix} \begin{pmatrix} p \\ q \\ r \end{pmatrix} \quad \begin{pmatrix} \\ \\ lq - mp \end{pmatrix}$$

Fig. 13.10

In each component, note carefully which is the positive term, and which the negative.

Example 13.4.1

Find a common perpendicular to $\begin{pmatrix} 1 \\ 2 \\ 3 \end{pmatrix}$ and $\begin{pmatrix} 7 \\ 8 \\ 9 \end{pmatrix}$.

Using the formula in the box on page 183, a common perpendicular is

$$\begin{pmatrix} 2 \times 9 - 3 \times 8 \\ 3 \times 7 - 1 \times 9 \\ 1 \times 8 - 2 \times 7 \end{pmatrix} = \begin{pmatrix} -6 \\ 12 \\ -6 \end{pmatrix}.$$

Since any multiple of this vector is also perpendicular to both vectors, $\begin{pmatrix} -1 \\ 2 \\ -1 \end{pmatrix}$ is a simpler common perpendicular.

To check that you have a correct answer, calculate the scalar product of your vector with each of the original vectors. As $(-1) \times 1 + 2 \times 2 + (-1) \times 3 = -1 + 4 - 3 = 0$ and

$(-1) \times 7 + 2 \times 8 + (-1) \times 9 = -7 + 16 - 9 = 0$, $\begin{pmatrix} -1 \\ 2 \\ -1 \end{pmatrix}$ is perpendicular to both $\begin{pmatrix} 1 \\ 2 \\ 3 \end{pmatrix}$ and $\begin{pmatrix} 7 \\ 8 \\ 9 \end{pmatrix}$.

The next example shows method 2 of Example 13.3.4 again.

Example 13.4.2

Find the cartesian equation of the plane through $A(1,2,1)$, $B(2,-1,-4)$ and $C(1,0,-1)$.

As $\begin{pmatrix} 1 \\ -3 \\ -5 \end{pmatrix}$ and $\begin{pmatrix} 0 \\ -2 \\ -2 \end{pmatrix}$ are vectors parallel to the plane, the normal is perpendicular to both of them. So, using the result in the box on page 183,

$$\begin{pmatrix} (-3) \times (-2) - (-5) \times (-2) \\ (-5) \times 0 - 1 \times (-2) \\ 1 \times (-2) - (-3) \times 0 \end{pmatrix} = \begin{pmatrix} 6 - 10 \\ 0 + 2 \\ -2 + 0 \end{pmatrix} = \begin{pmatrix} -4 \\ 2 \\ -2 \end{pmatrix}$$

is in the direction of the normal.

Notice that the components of this vector have the common factor 2, and that $\begin{pmatrix} -4 \\ 2 \\ -2 \end{pmatrix} = -2 \begin{pmatrix} 2 \\ -1 \\ 1 \end{pmatrix}$. So the simpler vector $\begin{pmatrix} 2 \\ -1 \\ 1 \end{pmatrix}$ is also normal to the plane.

Therefore the equation is $2x - y + z = k$; and since $(1,2,1)$ lies on the plane, the equation is $2x - y + z = 2 \times 1 - 2 + 1 = 1$, which is $2x - y + z = 1$.

Example 13.4.3

The line l passes through $A(3,-4,\pm6)$ and has direction vector $\mathbf{p} = -2\mathbf{i} + 5\mathbf{j} + \mathbf{k}$. Find the equation of the plane through the origin O which contains l.

Let $\mathbf{n}$ be a normal to the plane. Then, as l lies in the plane, $\mathbf{n}$ is perpendicular to $\mathbf{p}$.

The line OA, with direction vector $3\mathbf{i} - 4\mathbf{j} - 6\mathbf{k}$, also lies in the plane, so it is also perpendicular to $\mathbf{n}$.

So $\mathbf{n}$ is perpendicular to $-2\mathbf{i} + 5\mathbf{j} + \mathbf{k}$ and $3\mathbf{i} - 4\mathbf{j} - 6\mathbf{k}$. Using the result in the box, a common perpendicular is

$$\left(5 \times (-6) - 1 \times (-4)\right)\mathbf{i} + \left(1 \times 3 - (-2) \times (-6)\right)\mathbf{j} + \left((-2) \times (-4) - 5 \times 3\right)\mathbf{k}$$
$$= (-30 + 4)\mathbf{i} + (3 - 12)\mathbf{j} + (8 - 15)\mathbf{k} = -26\mathbf{i} - 9\mathbf{j} - 7\mathbf{k}.$$

Don't forget to check that the relevant scalar products are both zero.

The equation of the plane through the origin with normal $-26\mathbf{i} - 9\mathbf{j} - 7\mathbf{k}$ is

$$-26x - 9y - 7z = 0, \quad \text{or} \quad 26x + 9y + 7z = 0.$$

Exercise 13B

1 In each part find vectors perpendicular to both of the given vectors.

(a) $\begin{pmatrix} 8 \\ -3 \\ 1 \end{pmatrix}$ and $\begin{pmatrix} 7 \\ -2 \\ 0 \end{pmatrix}$ (b) $\begin{pmatrix} 4 \\ -1 \\ -3 \end{pmatrix}$ and $\begin{pmatrix} 3 \\ 5 \\ 1 \end{pmatrix}$ (c) $\begin{pmatrix} 2 \\ 0 \\ 1 \end{pmatrix}$ and $\begin{pmatrix} 1 \\ 0 \\ 2 \end{pmatrix}$

2 In each part find vectors perpendicular to both of the given vectors.

(a) $(3\mathbf{i} - 2\mathbf{k})$ and $2\mathbf{k}$ (b) $5\mathbf{k}$ and $(\mathbf{i} + 2\mathbf{j} - 3\mathbf{k})$ (c) $(\mathbf{i} + \mathbf{j})$ and $(\mathbf{i} - \mathbf{j})$

3 Find a vector perpendicular to both $\mathbf{i} + 2\mathbf{j} - \mathbf{k}$ and $3\mathbf{i} - \mathbf{j} + \mathbf{k}$. Hence find the cartesian equation of the plane parallel to both $\mathbf{i} + 2\mathbf{j} - \mathbf{k}$ and $3\mathbf{i} - \mathbf{j} + \mathbf{k}$ which passes through the point $(2, 0, -3)$.

4 Find a vector perpendicular to both $\begin{pmatrix} -1 \\ 1 \\ 0 \end{pmatrix}$ and $\begin{pmatrix} 0 \\ 1 \\ -1 \end{pmatrix}$.

Hence find the cartesian equation of the plane parallel to both $\begin{pmatrix} -1 \\ 1 \\ 0 \end{pmatrix}$ and $\begin{pmatrix} 0 \\ 1 \\ -1 \end{pmatrix}$ which passes through the point $(1, -1, -3)$.

5 Find the cartesian equations of the planes through the given points.

(a) $(1,0,0), (0,0,0), (0,1,0)$ (b) $(1,-1,0), (0,1,-1), (-1,0,1)$

(c) $(1,2,3), (2,-1,2), (3,1,-1)$ (d) $(4,-1,2), (0,0,3), (-1,2,0)$

6 Find the coordinates of the foot of the perpendicular from the point $(2,-3,6)$ to the plane $2x - 3y + 6z = 0$. Hence find the perpendicular distance of the point from the plane.

7 Find the perpendicular distance of the point $(3,1,-2)$ from the plane $2x + y - 2z = 8$.

8 Find the equation of the plane through $(1,2,-1)$ parallel to the plane $5x + y + 7z = 20$.

9 Find the equation of the line through $(4,2,-1)$ perpendicular to the plane $3x + 4y - z = 1$.

10 A cave has a planar roof passing through the points $(0,0,-19), (5,0,-20)$ and $(0,5,-22)$. A tunnel is being bored through the rock from the point $(0,3,4)$ in the direction $-\mathbf{i} + 2\mathbf{j} - 20\mathbf{k}$. Find the angle between the tunnel and the cave roof in degrees, correct to the nearest degree.

11 Find whether or not the four points $(1,5,4), (2,0,3), (3,-5,0)$ and $(0,10,6)$ lie in a plane.

12 Find a vector equation of the line of intersection of the planes $x + 3y - 6z = 2$ and $2x + 7y - 3z = 7$. (Hint: put $z = 0$ to find a point on the line of intersection.)

13 Find a vector equation of the line through $(4,2,-3)$, parallel to the line of intersection of the planes $3x - 2y = 6$ and $4x + 2z = 7$.

14 Find the equation of the plane through $(3,-2,4)$ and $(2,-1,3)$ which is parallel to the line joining $(1,1,1)$ to $(2,3,5)$.

15 Show that the lines $\mathbf{r} = 3\mathbf{i} + 2\mathbf{j} + \mathbf{k} + s(-\mathbf{i} + 2\mathbf{j} + \mathbf{k})$ and $\mathbf{r} = 3\mathbf{i} + 9\mathbf{j} + 2\mathbf{k} + t(2\mathbf{i} + 3\mathbf{j} - \mathbf{k})$ lie in the same plane. Find the cartesian equation of this plane.

16 Find the cartesian equation of the plane which passes through the point $(1,2,3)$ and contains the line of intersection of the planes $2x - y + z = 4$ and $x + y + z = 4$.

17 Let l_1 denote the line passing through the points $A(2,-1,1)$ and $B(0,5,-7)$, and l_2 denote the line passing through the points $C(1,-1,1)$ and $D(1,-4,5)$.
  (a) Write down a vector equation of the line l_1 and a vector equation of the line l_2.
  (b) Show that the lines l_1 and l_2 intersect, and determine the point of intersection.
  (c) Calculate the acute angle between the lines l_1 and l_2.
  (d) Determine a vector perpendicular to both lines l_1 and l_2 and hence show that the cartesian equation of the plane containing l_1 and l_2 is $4y + 3z = -1$.

Miscellaneous exercise 13

1 Find the cartesian equation of the plane through $(1,3,-7)$, $(2,-5,-3)$ and $(-5,7,2)$.

2 Two planes are defined by the equations $x + 2y + z = 4$ and $2x - 3y = 6$.
  (a) Find the acute angle between them.
  (b) Find a vector equation of their line of intersection.

3 Two planes with vector equations $\mathbf{r} \cdot \begin{pmatrix} 3 \\ 1 \\ 1 \end{pmatrix} = 2$ and $\mathbf{r} \cdot \begin{pmatrix} 2 \\ 5 \\ -1 \end{pmatrix} = 15$ intersect in the line L.

    (a) Find a direction for the line L.

    (b) Show that the point $(1,2,-3)$ lies in both planes, and write down a vector equation for the line L.   (OCR)

4 Find the perpendicular distance of the point (p,q,r) from the plane $ax + by + cz = d$.

5 Find the equation of the plane through $(1,2,-4)$ perpendicular to the line joining $(3,1,-1)$ to $(1,4,7)$.

6 Prove that the planes $2x - 3y + z = 4$, $x + 4y - z = 7$ and $3x - 10y + 3z = 1$ meet in a line.

7 The straight line L_1 with vector equation $\mathbf{r} = \mathbf{a} + t\mathbf{b}$ cuts the plane $2x - 3y + z = 6$ at right angles, at the point $(5,1,-1)$.

    (a) Explain why suitable choices for $\mathbf{a}$ and $\mathbf{b}$ would be $\mathbf{a} = 5\mathbf{i} + \mathbf{j} - \mathbf{k}$ and $\mathbf{b} = 2\mathbf{i} - 3\mathbf{j} + \mathbf{k}$.
Another straight line, L_2, has vector equation $\mathbf{r} = s(\mathbf{i} + 3\mathbf{j} + 2\mathbf{k})$.

    (b) (i) Find the angle between the directions of L_1 and L_2, giving your answer to the nearest degree.
        (ii) Verify that L_2 cuts the plane $2x - 3y + z = 6$ at the point $(-1.2, -3.6, -2.4)$.
        (iii) Prove that L_1 and L_2 do not meet.   (OCR)

8 The line l_1 passes through the point P with position vector $2\mathbf{i} + \mathbf{j} - \mathbf{k}$ and has direction vector $\mathbf{i} - \mathbf{j}$. The line l_2 passes through the point Q with position vector $5\mathbf{i} - 2\mathbf{j} - \mathbf{k}$ and has direction vector $\mathbf{j} + 2\mathbf{k}$.

    (a) Write down equations for l_1 and l_2 in the form $\mathbf{r} = \mathbf{a} + t\mathbf{b}$.

    (b) Show that Q lies on l_1.

    (c) Find either the acute angle or the obtuse angle between l_1 and l_2.

    (d) Show that the vector $\mathbf{n} = 2\mathbf{i} + 2\mathbf{j} - \mathbf{k}$ is perpendicular to both l_1 and l_2.

    (e) Find the cartesian equation for the plane containing l_1 and l_2.   (OCR)

9 Find the cartesian equation of the plane which passes through the point $(3,-4,1)$ and which is parallel to the plane containing the point $(1,2,-1)$ and the line $\mathbf{r} = t \begin{pmatrix} 1 \\ 1 \\ 1 \end{pmatrix}$.

10 The line l_1 passes through the point A, whose position vector is $\mathbf{i} - \mathbf{j} - 5\mathbf{k}$, and is parallel to the vector $\mathbf{i} - \mathbf{j} - 4\mathbf{k}$. The line l_2 passes through the point B, whose position vector is $2\mathbf{i} - 9\mathbf{j} - 14\mathbf{k}$, and is parallel to the vector $2\mathbf{i} + 5\mathbf{j} + 6\mathbf{k}$. The point P on l_1 and the point Q on l_2 are such that PQ is perpendicular to both l_1 and l_2.

    (a) Find the length of PQ.

    (b) Find a vector perpendicular to the plane Π which contains PQ and l_2.

    (c) Find the perpendicular distance from A to Π.   (OCR)

14 The binomial expansion

The binomial theorem tells you how to expand $(1+x)^n$ when n is a positive integer. This chapter extends this result to all rational values of n. When you have completed it, you should

- be able to expand $(1+x)^n$ in ascending powers of x
- know that the expansion is valid for $|x| < 1$
- understand how to use expansions to find approximations
- know how to extend the method to expand powers of more general expressions.

14.1 Generalising the binomial theorem

You learnt in P1 Chapter 9 how to expand $(x+y)^n$ by the binomial theorem, when n is a positive integer. If you replace x by 1 and y by x, this expansion becomes

$$(1+x)^n = 1^n + \frac{n}{1} \times 1^{n-1} x + \frac{n(n-1)}{1 \times 2} \times 1^{n-2} x^2 + \frac{n(n-1)(n-2)}{1 \times 2 \times 3} \times 1^{n-3} x^3 + \dots .$$

You can remove all the powers of 1, and write this more simply as

$$(1+x)^n = 1 + \frac{n}{1} x + \frac{n(n-1)}{1 \times 2} x^2 + \frac{n(n-1)(n-2)}{1 \times 2 \times 3} x^3 + \dots . \qquad \text{Equation A}$$

Notice that in this form the terms are written in ascending powers of x (see Section 1.1).

This chapter tackles the question 'Can you still use this expansion when n is not a positive integer?'

Before trying to answer this, you should notice something important about the terms. If n is a positive integer, then the form of the coefficients ensures that no terms have powers higher than x^n. For example, if $n = 5$ the coefficient of x^6 is

$$\frac{5 \times 4 \times 3 \times 2 \times 1 \times 0}{1 \times 2 \times 3 \times 4 \times 5 \times 6} = 0,$$

and all the coefficients which follow it are zero. But this only happens when n is a positive integer. For example, if $n = 4\frac{1}{2}$ the coefficients of x^4, x^5 and x^6 are

$$\frac{4\frac{1}{2} \times 3\frac{1}{2} \times 2\frac{1}{2} \times 1\frac{1}{2}}{1 \times 2 \times 3 \times 4}, \quad \frac{4\frac{1}{2} \times 3\frac{1}{2} \times 2\frac{1}{2} \times 1\frac{1}{2} \times \frac{1}{2}}{1 \times 2 \times 3 \times 4 \times 5} \quad \text{and} \quad \frac{4\frac{1}{2} \times 3\frac{1}{2} \times 2\frac{1}{2} \times 1\frac{1}{2} \times \frac{1}{2} \times \left(-\frac{1}{2}\right)}{1 \times 2 \times 3 \times 4 \times 5 \times 6} .$$

Whichever coefficient you consider, you never get a factor of 0. So, if n is not a positive integer, the expansion never stops.

The case $n = -1$

You know from P1 Section 14.3 that the sum to infinity of the geometric series

$1 + r + r^2 + r^3 + \dots$ is $\dfrac{1}{1-r}$, provided that $|r| < 1$.

Replacing r by $-x$ now gives

$$\frac{1}{1-(-x)} = (1+x)^{-1} = 1 - x + x^2 - x^3 + \dots .$$

Now try using Equation A with $n = -1$. This gives

$$(1+x)^{-1} = 1 + \frac{(-1)}{1}x + \frac{(-1)(-2)}{1 \times 2}x^2 + \frac{(-1)(-2)(-3)}{1 \times 2 \times 3}x^3 + \dots ,$$

which simplifies to

$$(1+x)^{-1} = 1 - x + x^2 - x^3 + \dots .$$

So far so good: Equation A works when $n = -1$.

The case $n = \frac{1}{2}$

If $(1+x)^{\frac{1}{2}}$ can be expanded in the form $A + Bx + Cx^2 + Dx^3 + \dots$, then you want to find $A, B, C, D, \dots$ so that

$$\left(A + Bx + Cx^2 + Dx^3 + \dots\right)^2 \equiv 1 + x .$$

This needs to be true for $x = 0$, so $A^2 = 1$. Since $(1+x)^{\frac{1}{2}}$ is the positive square root, this means that $A = 1$.

Then

$$\begin{aligned}
\left(1 + Bx + Cx^2 + Dx^3 + \dots\right)^2 &\equiv \left(1 + Bx + Cx^2 + Dx^3 + \dots\right)\left(1 + Bx + Cx^2 + Dx^3 + \dots\right) \\
&\equiv 1 + (B + B)x + \left(C + B^2 + C\right)x^2 \\
&\quad + (D + BC + CB + D)x^3 + \dots \\
&\equiv 1 + (2B)x + \left(2C + B^2\right)x^2 + (2D + 2BC)x^3 + \dots .
\end{aligned}$$

So $1 + x \equiv 1 + (2B)x + \left(2C + B^2\right)x^2 + (2D + 2BC)x^3 + \dots .$

Since this is an identity, you can equate coefficients of each power of x in turn:

$$2B = 1, \qquad \text{so } B = \tfrac{1}{2},$$

$$2C + B^2 = 0, \quad \text{so } C = -\tfrac{1}{8},$$

$$2D + 2BC = 0, \quad \text{so } D = \tfrac{1}{16}.$$

So it appears that $\left(1 + \tfrac{1}{2}x - \tfrac{1}{8}x^2 + \tfrac{1}{16}x^3 + \dots\right)^2 \equiv 1 + x$, and

$$(1+x)^{\frac{1}{2}} = 1 + \tfrac{1}{2}x - \tfrac{1}{8}x^2 + \tfrac{1}{16}x^3 + \dots .$$

Using Equation A with $n = \frac{1}{2}$ gives

$$(1+x)^{\frac{1}{2}} = 1 + \frac{\frac{1}{2}}{1}x + \frac{\frac{1}{2}\left(-\frac{1}{2}\right)}{1 \times 2}x^2 + \frac{\frac{1}{2}\left(-\frac{1}{2}\right)\left(-\frac{3}{2}\right)}{1 \times 2 \times 3}x^3 + \ldots$$

$$= 1 + \frac{1}{2}x - \frac{1}{8}x^2 + \frac{1}{16}x^3 + \ldots.$$

So Equation A seems to work when $n = \frac{1}{2}$, at least for the first few terms.

The general case

In fact Equation A works for all rational powers of n, positive or negative. There is, however, an important restriction. You will remember, from P1 Section 14.3, that the series $1 + r + r^2 + r^3 + \ldots$ only converges to $\dfrac{1}{1-r}$ if $|r| < 1$. A similar condition applies to the binomial expansion of $(1+x)^n$ for any value of n which is not a positive integer.

Binomial expansion   When n is rational, but not a positive integer, and $|x| < 1$,

$$(1+x)^n = 1 + \frac{n}{1}x + \frac{n(n-1)}{1 \times 2}x^2 + \frac{n(n-1)(n-2)}{1 \times 2 \times 3}x^3 + \ldots.$$

This is sometimes called the **binomial series**.

Example 14.1.1
Find the expansion of $(1+x)^{-2}$ in ascending powers of x up to the term in x^4.

Putting $n = -2$ in the formula for $(1+x)^n$,

$$(1+x)^{-2} = 1 + \frac{(-2)}{1}x + \frac{(-2)(-3)}{1 \times 2}x^2 + \frac{(-2)(-3)(-4)}{1 \times 2 \times 3}x^3 + \frac{(-2)(-3)(-4)(-5)}{1 \times 2 \times 3 \times 4}x^4 + \ldots$$

$$= 1 - 2x + 3x^2 - 4x^3 + 5x^4 + \ldots.$$

The required expansion is $1 - 2x + 3x^2 - 4x^3 + 5x^4$.

Example 14.1.2
Find the expansion of $(1+3x)^{\frac{3}{2}}$ in ascending powers of x up to and including the term in x^3. For what values of x is the expansion valid?

Putting $n = \frac{3}{2}$ in the formula for $(1+x)^n$, and writing $3x$ in place of x,

$$(1+3x)^{\frac{3}{2}} = 1 + \frac{\frac{3}{2}}{1}(3x) + \frac{\left(\frac{3}{2}\right)\left(\frac{1}{2}\right)}{1 \times 2}(3x)^2 + \frac{\left(\frac{3}{2}\right)\left(\frac{1}{2}\right)\left(-\frac{1}{2}\right)}{1 \times 2 \times 3}(3x)^3 + \ldots$$

$$= 1 + \frac{9}{2}x + \frac{27}{8}x^2 - \frac{27}{16}x^3 + \ldots.$$

The required expansion is $1 + \frac{9}{2}x + \frac{27}{8}x^2 - \frac{27}{16}x^3$.

The expansion $(1+x)^n$ is valid for $|x|<1$, so this expansion is valid for $|3x|<1$, that is for $|x|<\frac{1}{3}$.

You should notice one other point. When n is a positive integer, the coefficient $\dfrac{n(n-1)\ldots(n-(r-1))}{1\times 2\times\ldots\times r}$ can be written more concisely using factorials, as $\dfrac{n!}{r!(n-r)!}$.
You can't use this notation when n is not a positive integer, since $n!$ is only defined when n is a positive integer or zero. However, r is always an integer, so you can still if you like write the coefficient as $\dfrac{n(n-1)\ldots(n-(r-1))}{r!}$.

14.2 Approximations

One use of binomial expansions is to find numerical approximations to square roots, cube roots and other calculations. If $|x|$ is much smaller than 1, the power $|x^2|$ is very small, $|x^3|$ is smaller still, and you soon reach a power which, for all intents and purposes, can be neglected. So the sum of the first few terms of the expansion is a very close approximation to $(1+x)^n$.

Example 14.2.1
Find the expansion of $(1-2x)^{\frac{1}{2}}$ in ascending powers of x up to and including the term in x^3. By choosing a suitable value for x, find an approximation for $\sqrt{2}$.

$$(1-2x)^{\frac{1}{2}} = 1 + \frac{\frac{1}{2}}{1}(-2x) + \frac{\frac{1}{2}\left(-\frac{1}{2}\right)}{1\times 2}(-2x)^2 + \frac{\frac{1}{2}\left(-\frac{1}{2}\right)\left(-\frac{3}{2}\right)}{1\times 2\times 3}(-2x)^3 + \ldots$$
$$= 1 - x - \frac{1}{2}x^2 - \frac{1}{2}x^3 + \ldots.$$

Choosing a suitable value for x needs a bit of ingenuity. It is no use simply taking x so that $1-2x=2$, which would give $x=-\frac{1}{2}$, since this is not nearly small enough for the terms in x^4, x^5, … to be neglected. The trick is to find a value of x so that $1-2x$ has the form $2\times$ a perfect square. A good choice is to take $x=0.01$, so that $1-2\times 0.01 = 0.98$, which is 2×0.7^2.

So put $x=0.01$ in the expansion. This gives

$$0.98^{\frac{1}{2}} = 1 - 0.01 - \frac{1}{2}\times 0.01^2 - \frac{1}{2}\times 0.01^3 - \ldots,$$

so $\quad 0.7\sqrt{2} \approx 1 - 0.01 - 0.000\,05 - 0.000\,000\,5 = 0.989\,949\,5.$

Therefore $\frac{7}{10}\sqrt{2} \approx 0.989\,949\,5$, giving $\sqrt{2} \approx 1.414\,214$.

14.3 Expanding other expressions

The binomial series can also be used to expand powers of expressions more complicated than $1+x$ or $1+ax$. If you can rewrite an expression as $Y(1+Z)^n$ where Y and Z are expressions involving x, then you can expand $(1+Z)^n$, substitute the appropriate expression for Z in the result and then multiply through by the expression for Y.

Example 14.3.1

Find the binomial expansion of $\left(4-3x^2\right)^{\frac{1}{2}}$ up to and including the term in x^4.

$4-3x^2$ is not of the required form, but you can write it as $4\left(1-\frac{3}{4}x^2\right)$. So, using the factor rule for indices,

$$\left(4-3x^2\right)^{\frac{1}{2}} = 4^{\frac{1}{2}}\left(1-\tfrac{3}{4}x^2\right)^{\frac{1}{2}} = 2\left(1-\tfrac{3}{4}x^2\right)^{\frac{1}{2}}.$$

Then $\left(1-\tfrac{3}{4}x^2\right)^{\frac{1}{2}} = 1 + \dfrac{\frac{1}{2}}{1}\left(-\tfrac{3}{4}x^2\right) + \dfrac{\left(\frac{1}{2}\right)\left(-\frac{1}{2}\right)}{1\times 2}\left(-\tfrac{3}{4}x^2\right)^2 + \dots .$

Therefore $\left(4-3x^2\right)^{\frac{1}{2}} = 2\left(1-\tfrac{3}{4}x^2\right)^{\frac{1}{2}} = 2 - \tfrac{3}{4}x^2 - \tfrac{9}{64}x^4 + \dots$ and the required

expansion is $2 - \tfrac{3}{4}x^2 - \tfrac{9}{64}x^4$.

Example 14.3.2

Expand $\dfrac{5+x}{2-x+x^2}$ in ascending powers of x up to the term in x^3.

Write $\dfrac{5+x}{2-x+x^2} = \dfrac{5+x}{2\left(1-\frac{1}{2}x+\frac{1}{2}x^2\right)} = \tfrac{1}{2}(5+x)\left(1-\tfrac{1}{2}\left(x-x^2\right)\right)^{-1}.$

Now $(1-u)^{-1} = 1 + \dfrac{(-1)}{1}(-u) + \dfrac{(-1)(-2)}{1\times 2}(-u)^2 + \dfrac{(-1)(-2)(-3)}{1\times 2\times 3}(-u)^3 + \dots$

$$= 1 + u + u^2 + u^3 + \dots .$$

Writing $\tfrac{1}{2}\left(x-x^2\right)$ in place of u,

$$\left(1-\tfrac{1}{2}\left(x-x^2\right)\right)^{-1} = 1 + \left(\tfrac{1}{2}\left(x-x^2\right)\right) + \left(\tfrac{1}{2}\left(x-x^2\right)\right)^2 + \left(\tfrac{1}{2}\left(x-x^2\right)\right)^3 + \dots .$$

Collecting together the terms on the right, and ignoring any powers higher than x^3,

$$\left(1-\tfrac{1}{2}\left(x-x^2\right)\right)^{-1} = 1 + \tfrac{1}{2}\left(x-x^2\right) + \tfrac{1}{4}\left(x^2-2x^3+\dots\right) + \tfrac{1}{8}\left(x^3+\dots\right) + \dots$$

$$= 1 + \tfrac{1}{2}x - \tfrac{1}{4}x^2 - \tfrac{3}{8}x^3 + \dots .$$

Therefore, multiplying by $\tfrac{1}{2}(5+x)$,

$$\dfrac{5+x}{2-x+x^2} = \tfrac{1}{2}(5+x)\left(1+\tfrac{1}{2}x-\tfrac{1}{4}x^2-\tfrac{3}{8}x^3+\dots\right)$$

$$= \tfrac{1}{2}\left(5+\tfrac{5}{2}x-\tfrac{5}{4}x^2-\tfrac{15}{8}x^3+\dots+x+\tfrac{1}{2}x^2-\tfrac{1}{4}x^3+\dots\right)$$

$$= \tfrac{1}{2}\left(5+\tfrac{7}{2}x-\tfrac{3}{4}x^2-\tfrac{17}{8}x^3+\dots\right)$$

$$= \tfrac{5}{2}+\tfrac{7}{4}x-\tfrac{3}{8}x^2-\tfrac{17}{16}x^3+\dots .$$

The required expansion is $\frac{5}{2} + \frac{7}{4}x - \frac{3}{8}x^2 - \frac{17}{16}x^3$.

If an algebraic expression has a denominator which factorises, like

$$\frac{5+x}{2-x-x^2} = \frac{5+x}{(2+x)(1-x)},$$

there is a simpler way of expanding it, by first splitting it into partial fractions. This technique is explained in Chapter 15.

Exercise 14

1 Expand the following in ascending powers of x up to and including the term in x^2.

    (a) $(1+x)^{-3}$ (b) $(1+x)^{-5}$ (c) $(1-x)^{-4}$ (d) $(1-x)^{-6}$

2 Find the expansion of the following in ascending powers of x up to and including the term in x^2.

    (a) $(1+4x)^{-1}$ (b) $(1-2x)^{-3}$ (c) $(1-3x)^{-4}$ (d) $\left(1+\frac{1}{7}x\right)^{-2}$

3 Find the coefficient of x^3 in the expansions of the following.

    (a) $(1-x)^{-7}$ (b) $(1+2x)^{-1}$ (c) $(1+3x)^{-3}$ (d) $(1-4x)^{-2}$

    (e) $\left(1-\frac{1}{3}x\right)^{-6}$ (f) $(1+ax)^{-4}$ (g) $(1-bx)^{-4}$ (h) $(1-cx)^{-n}$

4 Find the expansion of the following in ascending powers of x up to and including the term in x^2.

    (a) $(1+x)^{\frac{1}{3}}$ (b) $(1+x)^{\frac{3}{4}}$ (c) $(1-x)^{\frac{3}{2}}$ (d) $(1-x)^{-\frac{1}{2}}$

5 Find the expansion of the following in ascending powers of x up to and including the term in x^2.

    (a) $(1+4x)^{\frac{1}{2}}$ (b) $(1+3x)^{-\frac{1}{3}}$ (c) $(1-6x)^{\frac{4}{3}}$ (d) $\left(1-\frac{1}{2}x\right)^{-\frac{1}{4}}$

6 Find the coefficient of x^3 in the expansions of the following.

    (a) $(1+2x)^{\frac{3}{2}}$ (b) $(1-5x)^{-\frac{1}{2}}$ (c) $\left(1+\frac{3}{2}x\right)^{\frac{1}{3}}$ (d) $(1-4x)^{\frac{3}{4}}$

    (e) $(1-7x)^{-\frac{1}{7}}$ (f) $\left(1+\sqrt{2}x\right)^{\frac{1}{2}}$ (g) $(1+ax)^{\frac{3}{2}}$ (h) $(1-bx)^{-\frac{1}{2}n}$

7 Show that, for small x, $\sqrt{1+\frac{1}{4}x} \approx 1+\frac{1}{8}x-\frac{1}{128}x^2$. Deduce the first three terms in the expansions of the following.

    (a) $\sqrt{1-\frac{1}{4}x}$ (b) $\sqrt{1+\frac{1}{4}x^2}$ (c) $\sqrt{4+x}$ (d) $\sqrt{36+9x}$

8 Show that $\dfrac{1}{\left(1-\frac{3}{2}x\right)^2} \approx 1+3x+\frac{27}{4}x^2+\frac{27}{2}x^3$ and state the interval of values of x for which the expansion is valid. Deduce the first four terms in the expansions of the following.

    (a) $\dfrac{4}{\left(1-\frac{3}{2}x\right)^2}$ (b) $\dfrac{1}{(2-3x)^2}$

9 Find the first four terms in the expansion of each of the following in ascending powers of x. State the interval of values of x for which each expansion is valid.

(a) $\sqrt{1-6x}$ (b) $\dfrac{1}{1+5x}$ (c) $\dfrac{1}{\sqrt[3]{1+9x}}$ (d) $\dfrac{1}{(1-2x)^4}$

(e) $\sqrt{1+2x^2}$ (f) $\sqrt[3]{8-16x}$ (g) $\dfrac{10}{\left(1+\frac{1}{5}x\right)^2}$ (h) $\dfrac{2}{2-x}$

(i) $\dfrac{1}{(2+x)^3}$ (j) $\dfrac{4x}{\sqrt{4+x^3}}$ (k) $\sqrt[4]{1+8x}$ (l) $\dfrac{12}{\left(\sqrt{3}-x\right)^4}$

10 Expand $\sqrt{1+8x}$ in ascending powers of x up to and including the term in x^3. By giving a suitable value to x, find an approximation for $\sqrt{1.08}$. Deduce approximations for

(a) $\sqrt{108}$,   (b) $\sqrt{3}$.

11 Expand $\sqrt[3]{1+4x}$ in ascending powers of x up to and including the term in x^2.

(a) By putting $x=0.01$, determine an approximation for $\sqrt[3]{130}$.

(b) By putting $x=-0.000\,25$, determine an approximation for $\sqrt[3]{999}$.

12 Given that the coefficient of x^3 in the expansion of $\dfrac{1}{(1+ax)^3}$ is -2160, find a.

13 Find the coefficient of x^2 in the expansion of $\dfrac{(1-2x)^2}{(1+x)^2}$.

14 Find the first three terms in the expansion in ascending powers of x of $\dfrac{\sqrt{1+2x}}{\sqrt{1-4x}}$. State the values of x for which the expansion is valid. By substituting $x=0.01$ in your expansion, find an approximation for $\sqrt{17}$.

15 Given that terms involving x^4 and higher powers may be ignored and that
$$\dfrac{1}{(1+ax)^3}-\dfrac{1}{(1+3x)^4}=bx^2+cx^3,$$ find the values of a, b and c.

16 Find the expansion of $\dfrac{1}{1-\left(x+x^2\right)}$ in ascending powers of x up to and including the term in x^4. By substitution of a suitable value of x, find the approximation, correct to 12 decimal places, of $\dfrac{1}{0.998\,999}$.

17 Find the first three terms in the expansion in ascending powers of x of

(a) $\dfrac{8}{\left(2+x-x^2\right)^2}$,   (b) $\dfrac{1+2x}{\left(1-x+2x^2\right)^3}$.

18 Given that the expansion of $(1+ax)^n$ is $1-2x+\frac{7}{3}x^2+kx^3+\dots$, find the value of k.

1 Find the series expansion of $(1 + 2x)^{\frac{5}{2}}$ up to and including the term in x^3, simplifying the coefficients. (OCR)

2 Expand $(1 - 4x)^{\frac{1}{2}}$ as a series of ascending powers of x, where $|x| < \frac{1}{4}$, up to and including the term in x^3, expressing the coefficients in their simplest form. (OCR)

3 Expand $(1 + 2x)^{-3}$ as a series of ascending powers of x, where $|x| < \frac{1}{2}$, up to and including the term in x^3, expressing the coefficients in their simplest form. (OCR)

4 Expand $\dfrac{1}{\left(1 + 2x^2\right)^2}$ as a series in ascending powers of x, up to and including the term in

x^6, giving the coefficients in their simplest form. (OCR)

5 Obtain the first three terms in the expansion, in ascending powers of x, of $(4 + x)^{\frac{1}{2}}$. State the set of values of x for which the expansion is valid. (OCR)

6 If x is small compared with a, expand $\dfrac{a^3}{\left(a^2 + x^2\right)^{\frac{3}{2}}}$ in ascending powers of $\dfrac{x}{a}$ up to and including the term in $\dfrac{x^4}{a^4}$. (OCR)

7 Given that $|x| < 1$, expand $\sqrt{1 + x}$ as a series of ascending powers of x, up to and including the term in x^2. Show that, if x is small, then $(2 - x)\sqrt{1 + x} \approx a + bx^2$, where the values of a and b are to be stated. (OCR)

8 Expand $(1 - x)^{-2}$ as a series of ascending powers of x, given that $|x| < 1$. Hence express

$\dfrac{1 + x}{(1 - x)^2}$ in the form $1 + 3x + ax^2 + bx^3 + \ldots$, where the values of a and b are to be stated.

(OCR)

9 Obtain the first three terms in the expansion, in ascending powers of x, of $(8 + 3x)^{\frac{2}{3}}$, stating the set of values of x for which the expansion is valid. (OCR)

10 Write down the first four terms of the series expansion in ascending powers of x of $(1 - x)^{\frac{1}{3}}$, simplifying the coefficients. By taking $x = 0.1$, use your answer to show

that $\sqrt[3]{900} \approx \dfrac{15\,641}{1620}$. (OCR)

11 Give the binomial expansion, for small x, of $(1 + x)^{\frac{1}{4}}$ up to and including the term in x^2, and simplify the coefficients. By putting $x = \frac{1}{16}$ in your expression, show that

$\sqrt[4]{17} \approx \dfrac{8317}{4096}$. (OCR)

12 Expand $\dfrac{2 + \left(1 + \frac{1}{2}x\right)^6}{2 + 3x}$ in ascending powers of x up to and including the term in x^2.

(OCR)

13 Show that $26\left(1-\dfrac{1}{26^2}\right)^{\frac{1}{2}} = n\sqrt{3}$, where n is an integer whose value is to be found. Given that $|x| < 1$, expand $(1-x)^{\frac{1}{2}}$ as a series of ascending powers of x, up to and including the term in x^2, simplifying the coefficients. By using the first *two* terms of the expansion of $26\left(1-\dfrac{1}{26^2}\right)^{\frac{1}{2}}$, obtain an approximate value for $\sqrt{3}$ in the form $\dfrac{p}{q}$, where p and q are integers.

                                                        (OCR)

14 Show that, for small values of x, $(1+x)^{\frac{1}{3}} \approx 1 + \frac{1}{3}x - \frac{1}{9}x^2$. Sketch on the same axes (with the aid of a graphic calculator if possible) the graphs of $y = (1+x)^{\frac{1}{3}}$, $y = 1 + \frac{1}{3}x$ and $y = 1 + \frac{1}{3}x - \frac{1}{9}x^2$.

Compare the graphs for values of x such that

    (a) $-3 < x < 3$,   (b) $-1 < x < 1$,   (c) $-0.2 < x < 0.2$.

15 Show that the expansion of $(1+4x)^{-2}$ in ascending powers of x is $1 - 8x + 48x^2 - \ldots$ and state the set of values of x for which the expansion is valid. Compare, for suitable values of x, the graphs of $y = (1+4x)^{-2}$, $y = 1 - 8x$ and $y = 1 - 8x + 48x^2$.

16 Given that $1 \equiv (1+x)^2\left(A + Bx + Cx^2 + Dx^3 + \ldots\right)$, equate coefficients of powers of x to find the values of A, B, C and D. Hence state the first four terms of the expansions in ascending powers of x of

    (a) $(1+x)^{-2}$,   (b) $\left(1-x^2\right)^{-2}$,   (c) $\left(1+2x^2\right)^{-2}$.

17 Given that $1 \equiv \left(1 + x + x^2\right)\left(A + Bx + Cx^2 + Dx^3 + Ex^4 + \ldots\right)$, equate coefficients of powers of x to find the values of A, B, C, D and E. Hence

    (a)   find the value of $\dfrac{1}{1.000\,300\,09}$ correct to 16 decimal places;

    (b)   show that $\dfrac{1}{\left(1 + x + x^2\right)\left(1 + 2x + 4x^2\right)} \approx 1 - 3x + 2x^2 + 9x^3 - 27x^4$ for small values of x.

18 Expand $\dfrac{1}{(1-x)^2}$, where $|x| < 1$, in ascending powers of x up to and including the term in x^3. You should simplify the coefficients. By putting $x = 10^{-4}$ in your expansion, find $\dfrac{1}{0.9999^2}$ correct to twelve decimal places.   (OCR)

19 Expand $(1+x)^{-\frac{1}{4}}$ in ascending powers of x as far as the term in x^2, simplifying the coefficients. Prove that $\frac{3}{2}\left(1 + \dfrac{1}{80}\right)^{-\frac{1}{4}} = 5^{\frac{1}{4}}$ and, using your expansion of $(1+x)^{-\frac{1}{4}}$ with $x = \frac{1}{80}$, find an approximate value for $5^{\frac{1}{4}}$, giving five places of decimals in your answer.

                                                        (OCR)

20 Write down and simplify the series expansion of $\dfrac{1}{\sqrt{1+x}}$, where $|x|<1$, up to and including the term in x^3. Show that using just these terms of the series with $x = 0.4$ gives a value for $\dfrac{1}{\sqrt{1.4}}$ which differs from the true value by less than 0.7%. By replacing x by z^2 in your series and then integrating, show that $\displaystyle\int_0^{0.2} \dfrac{1}{\sqrt{1+z^2}}\,dz \approx 0.1987$.

(OCR, adapted)

21* Show that the coefficient of x^n in the series expansion of $(1+2x)^{-2}$ is $(-1)^n(n+1)2^n$.

22* Show that the coefficient of x^n in the series expansion of $(1-x)^{-\frac{1}{2}}$ is $\dfrac{(2n)!}{2^{2n}(n!)^2}$.

23 Find the first three terms in the expansion in ascending powers of x of $\sqrt{\dfrac{1+2x}{1-x}}$. By putting $x = 0.02$ in your expansion, find an approximation for $\sqrt{13}$.

24 Find the first five terms in the series expansion of $\dfrac{1}{1+2x}$. Use the expansion to find an approximation to $\displaystyle\int_{-0.2}^{0.1} \dfrac{1}{1+2x}\,dx$. By also evaluating the integral exactly, find an approximation for $\ln 2$.

25 Find the first three terms in the expansion in ascending powers of x of $\dfrac{3+4x+x^2}{\sqrt[3]{1+\frac{1}{2}x}}$. Hence find an approximation to $\displaystyle\int_{-0.5}^{0.5} \dfrac{3+4x+x^2}{\sqrt[3]{1+\frac{1}{2}x}}\,dx$.

26 Find the first three terms in the expansion in ascending powers of x of
$$\dfrac{1}{\left(1-2x^2\right)^2\left(1+3x^2\right)^2}.$$

The diagram shows the graph of $y = 3 - 52x^2$ and part of the graph of
$$y = \dfrac{1}{\left(1-2x^2\right)^2\left(1+3x^2\right)^2}.$$

Use your expansion to find an approximation to the area of the region shaded in the diagram.

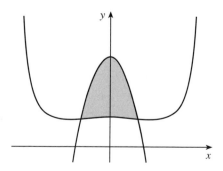

15 Rational functions

This chapter is about rational functions, which are fractions in which the numerator and the denominator are both polynomials. When you have completed the chapter, you should

- be able to simplify rational functions by cancelling
- be able to add, subtract, multiply and divide rational functions
- be able to split simple rational functions into their partial fractions
- be able to use partial fractions to integrate some rational functions and to find binomial expansions.

15.1 Simplifying rational functions

In Chapter 1 you saw that in many ways polynomials behave like integers. Similarly rational functions (also called 'algebraic fractions') have many properties in common with ordinary fractions. For example, just as you can cancel a fraction like $\frac{10}{15}$ to get $\frac{2}{3}$, you can cancel a rational function, but it is a little more complicated. Since you cancel $\frac{10}{15}$ in your head, it is worth looking to see what is actually happening:

$$\frac{10}{15} = \frac{2 \times 5}{3 \times 5} = \frac{2}{3}.$$

Start by factorising the numerator and denominator. Then you can divide the numerator and denominator by any common factor (5 in the example) to get the simplified fraction.

The same process is used to simplify rational functions. However, you need to realise that you cannot cancel common factors of single terms. For example, you can't cancel the 2s which appear in the numerator and denominator of $\frac{x-2}{2x-1}$. A fraction bar acts like a bracket, so $\frac{x-2}{2x-1}$ must be thought of as $\frac{(x-2)}{(2x-1)}$. Since $x-2$ and $2x-1$ have no common factor, no cancellation is possible.

Example 15.1.1

Simplify   (a) $\dfrac{x-2}{2x-4}$,   (b) $\dfrac{2x-3}{6x^2-x-12}$,   (c) $\dfrac{3x^2-8x+4}{6x^2-7x+2}$.

(a) $\dfrac{x-2}{2x-4} = \dfrac{(x-2)}{2(x-2)} = \dfrac{1}{2}$.

(b) $\dfrac{2x-3}{6x^2-x-12} = \dfrac{(2x-3)}{(2x-3)(3x+4)} = \dfrac{1}{3x+4}$.

(c) $\dfrac{3x^2-8x+4}{6x^2-7x+2} = \dfrac{(x-2)(3x-2)}{(2x-1)(3x-2)} = \dfrac{x-2}{2x-1}$.

You cannot cancel the last answer any further. If you have factorised fully, you can only cancel factors if they are exactly the same, or if one is the negative of the other. For example, you could cancel $\dfrac{2x-3}{3-2x}$ as $\dfrac{-(3-2x)}{3-2x} = -1$.

You can check these simplifications by putting x equal to a particular value, say $x = 0$ (provided your chosen value of x does not make the denominator equal to 0). In part (a), putting $x = 0$ in the original expression gives $\dfrac{-2}{-4} = \dfrac{1}{2}$, which is the same as the simplified version. In part (b), the original becomes $\dfrac{-3}{-12} = \dfrac{1}{4}$ and the answer becomes $\dfrac{1}{4}$. In part (c), the original is $\dfrac{4}{2} = 2$ and the answer is $\dfrac{-2}{-1} = 2$.

15.2 Adding and subtracting rational functions

In cancelling and simplifying rational functions, you worked in the same way as in normal arithmetic. To add and subtract rational functions you also follow the same principles as arithmetic, but you need to take special care of signs.

In arithmetic, to calculate $\dfrac{11}{15} - \dfrac{7}{20}$ you start by finding the lowest common multiple (LCM) of 15 and 20. You can easily see that the LCM is 60. But if you couldn't spot it, you would factorise the denominators,

$$\frac{11}{15} - \frac{7}{20} = \frac{11}{3 \times 5} - \frac{7}{2 \times 2 \times 5},$$

from which you can work out that the LCM is $2 \times 2 \times 3 \times 5 = 60$. Now

$$\frac{11}{15} - \frac{7}{20} = \frac{11 \times 4}{15 \times 4} - \frac{7 \times 3}{20 \times 3} = \frac{44}{60} - \frac{21}{60} = \frac{44-21}{60} = \frac{23}{60}.$$

Example 15.2.1

Express as single fractions in their simplest forms   (a) $\dfrac{1}{x} - \dfrac{2}{3}$,   (b) $\dfrac{3}{x+2} - \dfrac{6}{2x-1}$.

(a) The LCM of x and 3 is $3x$. So $\dfrac{1}{x} - \dfrac{2}{3} = \dfrac{1 \times 3}{3x} - \dfrac{2 \times x}{3x} = \dfrac{3}{3x} - \dfrac{2x}{3x} = \dfrac{3-2x}{3x}.$

(b) The LCM of $x+2$ and $2x-1$ is $(x+2)(2x-1)$. Subtracting in the usual way,

$$\frac{3}{x+2} - \frac{6}{2x-1} = \frac{3(2x-1)}{(x+2)(2x-1)} - \frac{6(x+2)}{(x+2)(2x-1)}$$

$$= \frac{6x-3-6x-12}{(x+2)(2x-1)} = \frac{-15}{(x+2)(2x-1)}.$$

Notice the sign change which gives $-6x - 12$ in the second step of part (b).

Don't forget to check mentally by substituting a numerical value for x which makes the calculations easy. Try $x = 1$ for part (a) and $x = 0$ for part (b).

Example 15.2.2

Express $\dfrac{31x-8}{2x^2+3x-2}-\dfrac{14}{x+2}$ as a single fraction in its lowest terms.

The first step, as before, is to find the LCM. The solution goes:

$$\frac{31x-8}{2x^2+3x-2}-\frac{14}{x+2}=\frac{31x-8}{(2x-1)(x+2)}-\frac{14}{x+2}=\frac{31x-8-14(2x-1)}{(2x-1)(x+2)}$$

$$=\frac{31x-8-28x+14}{(2x-1)(x+2)}=\frac{3x+6}{(2x-1)(x+2)}$$

$$=\frac{3(x+2)}{(2x-1)(x+2)}=\frac{3}{2x-1}.$$

15.3 Multiplying and dividing rational functions

Normal arithmetic methods also apply to multiplication and division of rational functions.

Division is defined as the reverse of multiplication in the sense that

$$a\times k=b \iff b\div k=a \qquad \text{(provided that } k\neq 0\text{)}.$$

Thus $\dfrac{4}{15}\div\dfrac{3}{20}$ is the number (or fraction) x, such that $\dfrac{3}{20}x=\dfrac{4}{15}$.

Multiply both sides of the equation by the inverse (reciprocal) of $\dfrac{3}{20}$, which is $\dfrac{20}{3}$:

$$\frac{20}{3}\times\left(\frac{3}{20}x\right)=\frac{20}{3}\times\frac{4}{15} \iff \left(\frac{20}{3}\times\frac{3}{20}\right)x=\frac{20}{3}\times\frac{4}{15} \iff 1x=\frac{20}{3}\times\frac{4}{15}$$

$$\iff x=\frac{20}{3}\times\frac{4}{15}.$$

This justifies the method for dividing by fractions in arithmetic ('turn it upside down and multiply'), which you may have used before. Then

$$x=\frac{20}{3}\times\frac{4}{15}=\frac{2\times2\times5}{3}\times\frac{2\times2}{3\times5}=\frac{16}{9}.$$

In general, if a, b, c and d are integers, and $x=\dfrac{c}{d}\div\dfrac{a}{b}$, then $\dfrac{a}{b}x=\dfrac{c}{d}$.

Multiplying by the inverse of $\dfrac{a}{b}$, which is $\dfrac{b}{a}$, gives $\dfrac{b}{a}\times\dfrac{a}{b}x=\dfrac{b}{a}\times\dfrac{c}{d} \iff x=\dfrac{b}{a}\times\dfrac{c}{d}$.

Example 15.3.1

Simplify the rational functions   (a) $\dfrac{2}{x}\times\dfrac{x^2-2x}{x-2}$,   (b) $\dfrac{x-2}{x^2-4x+3}\div\dfrac{x}{2x^2-7x+3}$.

(a) $\dfrac{2}{x}\times\dfrac{x^2-2x}{x-2}=\dfrac{2}{x}\times\dfrac{x(x-2)}{x-2}=\dfrac{2x(x-2)}{x(x-2)}=2.$

(b) $\dfrac{x-2}{x^2-4x+3} \div \dfrac{x}{2x^2-7x+3} = \dfrac{x-2}{x^2-4x+3} \times \dfrac{2x^2-7x+3}{x}$

$$= \dfrac{x-2}{(x-1)(x-3)} \times \dfrac{(2x-1)(x-3)}{x}$$

$$= \dfrac{(x-2)(2x-1)}{x(x-1)}.$$

Exercise 15A

1 Simplify

(a) $\dfrac{4x-8}{2}$,

(b) $\dfrac{9x+6}{3}$,

(c) $\dfrac{2x^2-6x+12}{2}$,

(d) $\dfrac{6}{18x+12}$,

(e) $\dfrac{(2x+6)(2x-4)}{4}$,

(f) $\dfrac{x}{x^3+x^2+x}$.

2 Simplify

(a) $\dfrac{5x+15}{x+3}$,

(b) $\dfrac{x+1}{4x+4}$,

(c) $\dfrac{2x+5}{5+2x}$,

(d) $\dfrac{3x-7}{7-3x}$,

(e) $\dfrac{(2x+8)(3x+6)}{(2x+4)(3x+12)}$,

(f) $\dfrac{2x^2-6x+10}{3x^2-9x+15}$.

3 Simplify

(a) $\dfrac{x^2+5x+4}{x+1}$,

(b) $\dfrac{x-2}{x^2+5x-14}$,

(c) $\dfrac{6x^2+4x}{4x^2+2x}$,

(d) $\dfrac{x^2+5x-6}{x^2-4x+3}$,

(e) $\dfrac{2x^2+5x-12}{2x^2-11x+12}$,

(f) $\dfrac{8x^2-6x-20}{2+5x-3x^2}$.

4 Simplify

(a) $\dfrac{2x}{3} - \dfrac{x}{4}$,

(b) $\dfrac{5x}{2} + \dfrac{x}{3} - \dfrac{3x}{4}$,

(c) $\dfrac{x+2}{3} + \dfrac{x+1}{4}$,

(d) $\dfrac{2x+1}{5} - \dfrac{x+2}{3}$,

(e) $\dfrac{(x+1)(x+3)}{2} - \dfrac{(x+2)^2}{4}$,

(f) $3x+4 - \dfrac{2(x+3)}{5}$.

5 Simplify

(a) $\dfrac{2}{x} + \dfrac{3}{4}$,

(b) $\dfrac{1}{2x} + \dfrac{2}{x}$,

(c) $\dfrac{5}{4x} - \dfrac{2}{3x}$,

(d) $\dfrac{x+3}{2x} + \dfrac{x-4}{x}$,

(e) $\dfrac{3x-1}{x} - \dfrac{x+1}{2}$,

(f) $\dfrac{x+1}{x} + \dfrac{x+1}{x^2}$.

6 Simplify

(a) $\dfrac{2}{x+1} + \dfrac{4}{x+3}$,

(b) $\dfrac{5}{x-2} + \dfrac{3}{2x+1}$,

(c) $\dfrac{4}{x+3} - \dfrac{2}{x+4}$,

(d) $\dfrac{7}{x-3} - \dfrac{2}{x+1}$,

(e) $\dfrac{4}{2x+3} + \dfrac{5}{3x+1}$,

(f) $\dfrac{6}{2x+1} - \dfrac{2}{5x-3}$.

7 Simplify

(a) $\dfrac{5}{3x-1} - \dfrac{2}{2x+1}$,

(b) $\dfrac{6}{4x+1} - \dfrac{3}{2x}$,

(c) $\dfrac{3x}{x+2} + \dfrac{5x}{x+1}$,

(d) $\dfrac{8x}{2x-1} - \dfrac{x}{x+2}$,

(e) $\dfrac{x+1}{x+2} + \dfrac{x+2}{x+1}$,

(f) $\dfrac{2x+1}{x+4} - \dfrac{x-5}{x-2}$.

8 Simplify

(a) $\dfrac{2x+3}{(x+1)(x+3)} + \dfrac{2}{x+3}$,

(b) $\dfrac{5x}{x^2+x-2} + \dfrac{1}{x+2}$,

(c) $\dfrac{5}{x-3} + \dfrac{x+2}{x^2-3x}$,

(d) $\dfrac{8}{x^2-4} - \dfrac{4}{x-2}$,

(e) $\dfrac{13-3x}{x^2-2x-3} + \dfrac{4}{x+1}$,

(f) $\dfrac{11x+27}{2x^2+11x-6} - \dfrac{3}{x+6}$.

9 Simplify

(a) $\dfrac{4x+6}{x-4} \times \dfrac{3x-12}{2x+3}$,

(b) $\dfrac{x^2-4}{x+2} \times \dfrac{3x}{x-2}$,

(c) $\dfrac{x^2+9x+20}{x+3} \times \dfrac{3}{x+4}$,

(d) $\dfrac{x^2+3x+2}{x^2+4x+4} \times \dfrac{x^2+5x+6}{x^2+2x+1}$,

(e) $\dfrac{4x+12}{2x+2} \times \dfrac{x^2+2x+1}{x^2+6x+9}$,

(f) $\dfrac{4x^2-9}{9x^2-4} \times \dfrac{9x^2-12x+4}{4x^2-12x+9}$.

10 Simplify

(a) $\dfrac{x+2}{2x+3} \div \dfrac{2x+4}{8x+12}$,

(b) $\dfrac{x}{5-2x} \div \dfrac{3x}{2x-5}$,

(c) $\dfrac{1}{x^2+6x+6} \div \dfrac{1}{x^2+8x+16}$,

(d) $\dfrac{5x-1}{2x^2+x-3} \div \dfrac{1}{2x^2+7x+6}$,

(e) $\dfrac{x^2+5x-6}{x^2-5x+4} \div \dfrac{x^2+9x+18}{x^2-x-12}$,

(f) $\dfrac{-2x^2+7x-6}{8x^2-10x-3} \div \dfrac{7x-x^2-10}{5+19x-4x^2}$.

11 Given that $\dfrac{a}{x-2} + \dfrac{b}{x+c} \equiv \dfrac{15x}{x^2+2x-8}$, find the values of the constants a, b and c.

12 Given that $\dfrac{(x+2)f(x)}{(x+3)(x^2-x-6)} \equiv 1$, find $f(x)$ in its simplest form.

13 Given that $P(x) \equiv \dfrac{5}{x+4}$ and $Q(x) \equiv \dfrac{2}{x-3}$,

(a) find $2P(x) + 3Q(x)$ in simplified form,

(b) find $R(x)$, where $R(x) + 4Q(x) \equiv 3P(x)$.

14 Simplify

(a) $\dfrac{2}{x^3-3x^2+2x} + \dfrac{1}{x^3-6x^2+11x-6}$,

(b) $\dfrac{5}{2x+1} - \dfrac{4}{3x-1} - \dfrac{7x-10}{6x^2+x-1}$.

15.4 Partial fractions with simple denominators

Sometimes you need to reverse the process of adding or subtracting rational functions.

For example, instead of adding $\dfrac{3}{2x-1}$ and $\dfrac{2}{x-2}$ to get $\dfrac{7x-8}{(2x-1)(x-2)}$, you might need

to find the fractions which, when added together, give $\dfrac{7x-8}{(2x-1)(x-2)}$.

This process is called **splitting into partial fractions**. Suppose you need to calculate

$$\int \frac{7x-8}{(2x-1)(x-2)}\,dx\,.$$

You cannot do this as it stands. However, if you rewrite the integrand using partial fractions, you can integrate it, as follows:

$$\int \frac{7x-8}{(2x-1)(x-2)}\,dx = \int \left(\frac{3}{2x-1}+\frac{2}{x-2}\right)dx = \int \frac{3}{2x-1}\,dx + \int \frac{2}{x-2}\,dx$$

$$= \tfrac{3}{2}\ln|2x-1| + 2\ln|x-2| + k.$$

You may need to refresh your memory from Section 4.5.

In this case, to split $\dfrac{7x-8}{(2x-1)(x-2)}$ into partial fractions start by supposing that you can write it in the form

$$\frac{7x-8}{(2x-1)(x-2)} \equiv \frac{A}{2x-1}+\frac{B}{x-2},$$

where the identity sign $\equiv$ means that the two sides are equal for all values of x for which they are defined: here, all values except $x=\tfrac{1}{2}$ and $x=2$, where the denominators become 0.

From this point, there are two methods you can use.

Equating coefficients method

Expressing the right side as a single fraction,

$$\frac{7x-8}{(2x-1)(x-2)} \equiv \frac{A(x-2)+B(2x-1)}{(2x-1)(x-2)}\,.$$

Multiplying both sides of the identity by $(2x-1)(x-2)$,

$$7x-8 \equiv A(x-2)+B(2x-1).$$

You can now find A and B by the method of equating coefficients (Section 1.3).

Equating coefficients of x^1 : $\quad 7 = \quad A+2B.$
Equating coefficients of x^0 : $\quad -8 = -2A - \ B.$

Solving these two equations simultaneously gives $A=3$, $B=2$.

Substitution method

To find A, multiply both sides of the identity $\dfrac{7x-8}{(2x-1)(x-2)} \equiv \dfrac{A}{2x-1} + \dfrac{B}{x-2}$ by $2x-1$ to obtain

$$\frac{7x-8}{x-2} \equiv A + B\frac{2x-1}{x-2}.$$

Putting $x = \tfrac{1}{2}$ gives $A = \dfrac{\tfrac{7}{2}-8}{\tfrac{1}{2}-2} = \dfrac{-\tfrac{9}{2}}{-\tfrac{3}{2}} = 3.$

Similarly, to find B, multiply the identity by $x-2$ to get $\dfrac{7x-8}{2x-1} \equiv A\dfrac{x-2}{2x-1} + B.$

Putting $x = 2$ gives $B = \dfrac{7 \times 2 - 8}{2 \times 2 - 1} = \dfrac{6}{3} = 2.$

By either method $\dfrac{7x-8}{(2x-1)(x-2)} \equiv \dfrac{3}{2x-1} + \dfrac{2}{x-2}.$

There are three important points to make about these solutions.

- Always check your answer by testing it with a simple value of x. For example, putting $x = 0$ gives -4 for the left side, and $-3 - 1 = -4$ for the right.
- In the substitution method, the values $x = \tfrac{1}{2}$ and $x = 2$ are chosen because they give simple equations for A and B. You could have chosen other values for x, but the equations for A and B would have been more complicated. Try it and see!
- You may be worried that, at the beginning of the example, the values $x = \tfrac{1}{2}$ and $x = 2$ were excluded, but they are then used in the substitution method. The fact is that $\dfrac{7x-8}{x-2} \equiv 3 + 2\dfrac{2x-1}{x-2}$ is true for all x except $x = 2$, including $x = \tfrac{1}{2}$. So it is all right to use $x = \tfrac{1}{2}$ to find A, since there is no need to exclude $x = \tfrac{1}{2}$ in the identity $\dfrac{7x-8}{x-2} \equiv A + B\dfrac{2x-1}{x-2}$. But the partial fraction form $\dfrac{A}{2x-1} + \dfrac{B}{x-2} \equiv \dfrac{7x-8}{(2x-1)(x-2)}$, has no meaning when $x = \tfrac{1}{2}$ (or when $x = 2$).

An expression of the form $\dfrac{ax+b}{(px+q)(rx+s)}$ can be split into partial fractions of the form $\dfrac{A}{px+q} + \dfrac{B}{rx+s}.$

Example 15.4.1

Split $\dfrac{13x - 6}{3x^2 - 2x}$ into partial fractions.

Rewrite $\dfrac{13x - 6}{3x^2 - 2x}$ in the form $\dfrac{13x - 6}{x(3x - 2)}$, and then put $\dfrac{13x - 6}{x(3x - 2)} \equiv \dfrac{A}{x} + \dfrac{B}{3x - 2}$.

Using the equating coefficients method, $\dfrac{13x - 6}{x(3x - 2)} \equiv \dfrac{A(3x - 2) + Bx}{x(3x - 2)}$, so
$A(3x - 2) + Bx \equiv 13x - 6$.

$$\text{Equating coefficients of } x^1 : \quad 3A + B = 13.$$
$$\text{Equating coefficients of } x^0 : \quad -2A \quad\quad = -6.$$

Solving these two equations simultaneously gives $A = 3$, $B = 4$.

Therefore $\dfrac{13x - 6}{x(3x - 2)} \equiv \dfrac{3}{x} + \dfrac{4}{3x - 2}$.

Example 15.4.2

Split $\dfrac{12x}{(x + 1)(2x + 3)(x - 3)}$ into partial fractions.

Put $\dfrac{12x}{(x + 1)(2x + 3)(x - 3)} \equiv \dfrac{A}{x + 1} + \dfrac{B}{2x + 3} + \dfrac{C}{x - 3}$.

Use the substitution method: multiplying by $x + 1$ gives

$$\dfrac{12x}{(2x + 3)(x - 3)} \equiv A + B\dfrac{x + 1}{2x + 3} + C\dfrac{x + 1}{x - 3}.$$

Putting $x = -1$, $A = \dfrac{12 \times (-1)}{(2 \times (-1) + 3)((-1) - 3)} = \dfrac{-12}{1 \times (-4)} = 3$.

Similarly, after multiplying by $2x + 3$ and putting $x = -\frac{3}{2}$, you get $B = -8$; multiplying by $x - 3$ and putting $x = 3$ gives $C = 1$.

Therefore $\dfrac{12x}{(x + 1)(2x + 3)(x - 3)} \equiv \dfrac{3}{x + 1} - \dfrac{8}{2x + 3} + \dfrac{1}{x - 3}$.

If you try to use the equating coefficients method in this example, you get three simultaneous equations, with three unknowns, to solve. In this case, the substitution method is easier.

Example 15.4.3

Calculate the value of $\displaystyle\int_1^4 \frac{1}{x(x-5)}\,dx$.

Write $\dfrac{1}{x(x-5)}$ in partial fraction form, as $\dfrac{1}{x(x-5)} \equiv \dfrac{A}{x} + \dfrac{B}{x-5}$.

Either method gives $A = -\frac15$ and $B = \frac15$, so $\dfrac{1}{x(x-5)} \equiv \dfrac{-\frac15}{x} + \dfrac{\frac15}{x-5}$.

Therefore $\displaystyle\int_1^4 \frac{1}{x(x-5)}\,dx = \int_1^4 \left(-\frac{\frac15}{x} + \frac{\frac15}{x-5} \right)dx.$

Remembering that $x-5$ is negative over the interval of integration, rewrite this integral as $\displaystyle\int_1^4 \left(-\frac{\frac15}{x} - \frac{\frac15}{5-x} \right)dx$ (this method was used in Example 4.5.1). Then

$$\int_1^4 \left(-\frac{\frac15}{x} - \frac{\frac15}{5-x} \right)dx = \left[-\tfrac15 \ln x + \tfrac15 \ln(5-x) \right]_1^4$$

$$= \left(-\tfrac15 \ln 4 + \tfrac15 \ln 1 \right) - \left(-\tfrac15 \ln 1 + \tfrac15 \ln 4 \right)$$

$$= -\tfrac15 \ln 4 + \tfrac15 \ln 1 + \tfrac15 \ln 1 - \tfrac15 \ln 4$$

$$= \tfrac25 \ln 1 - \tfrac25 \ln 4 = -\tfrac25 \ln 4.$$

You may find it helpful to look up Sections 4.4 and 4.5 on definite integrals involving logarithms.

Exercise 15B

1 Split the following into partial fractions.

(a) $\dfrac{2x+8}{(x+5)(x+3)}$
(b) $\dfrac{10x+8}{(x-1)(x+5)}$
(c) $\dfrac{x}{(x-4)(x-5)}$
(d) $\dfrac{28}{(2x-1)(x+3)}$

2 Split the following into partial fractions.

(a) $\dfrac{8x+1}{x^2+x-2}$
(b) $\dfrac{25}{x^2-3x-4}$
(c) $\dfrac{10x-6}{x^2-9}$
(d) $\dfrac{3}{2x^2+x}$

3 Split into partial fractions

(a) $\dfrac{35-5x}{(x+2)(x-1)(x-3)}$,
(b) $\dfrac{8x^2}{(x+1)(x-1)(x+3)}$,
(c) $\dfrac{15x^2-28x-72}{x^3-2x^2-24x}$.

4 Find

(a) $\displaystyle\int \frac{7x-1}{(x-1)(x-3)}\,dx$,

(b) $\displaystyle\int \frac{4}{x^2-4}\,dx$,

(c) $\displaystyle\int \frac{15x+35}{2x^2+5x}\,dx$,

(d) $\displaystyle\int \frac{x-8}{6x^2-x-1}\,dx$.

5 Evaluate the following, expressing each answer in a form involving a single logarithm.

(a) $\displaystyle\int_2^{10} \frac{2x+5}{(x-1)(x+6)}\,dx$

(b) $\displaystyle\int_0^3 \frac{3x+5}{(x+1)(x+2)}\,dx$

(c) $\displaystyle\int_4^5 \frac{6x}{x^2-9}\,dx$

(d) $\displaystyle\int_1^{\frac{3}{2}} \frac{4x-18}{4x^2+4x-3}\,dx$

6 Split $\dfrac{2-x}{(1+x)(1-2x)}$ into partial fractions and hence find the binomial expansion of

$\dfrac{2-x}{(1+x)(1-2x)}$ up to and including the term in x^3.

7 Split $\dfrac{3}{8x^2+6x+1}$ into partial fractions and hence find the binomial expansion of

$\dfrac{3}{8x^2+6x+1}$ up to and including the term in x^3. State the values of x for which the expansion is valid.

8 Split $\dfrac{4ax-a^2}{x^2+ax-2a^2}$ into partial fractions.

9 Find the exact value of $\displaystyle\int_{2\sqrt{5}}^{3\sqrt{5}} \frac{4\sqrt{5}}{x^2-5}\,dx$.

15.5 Partial fractions with a repeated factor

You will have noticed in the examples of the last section that when the denominator has two factors there are two partial fractions, with unknowns A and B. When the denominator has three factors, there are three fractions, with unknowns A, B and C. The equating coefficients method shows why, since you can find two unknowns by equating coefficients of x^0 and x^1, and three unknowns by equating coefficients of x^0, x^1 and x^2.

So you would expect $\dfrac{3x^2+6x+2}{(2x+3)(x+1)^2}$ to split into three fractions. Two of these must be

$\dfrac{A}{2x+3}$ and $\dfrac{B}{(x+1)^2}$. The third fraction is $\dfrac{C}{x+1}$. So write

$$\frac{3x^2+6x+2}{(2x+3)(x+1)^2} \equiv \frac{A}{2x+3} + \frac{B}{(x+1)^2} + \frac{C}{x+1}.$$

Then multiplying the identity by $2x+3$ gives $\dfrac{3x^2+6x+2}{(x+1)^2} \equiv A + \dfrac{B(2x+3)}{(x+1)^2} + \dfrac{C(2x+3)}{x+1}$.

Putting $x = -\frac{3}{2}$ gives $A = \dfrac{3\times\left(-\frac{3}{2}\right)^2 + 6\times\left(-\frac{3}{2}\right)+2}{\left(-\frac{3}{2}+1\right)^2} = \dfrac{\frac{27}{4}-9+2}{\left(-\frac{1}{2}\right)^2} = \dfrac{-\frac{1}{4}}{\frac{1}{4}} = -1.$

You might next try multiplying the identity by $x+1$, which gives

$$\dfrac{3x^2+6x+2}{(2x+3)(x+1)} \equiv \dfrac{A(x+1)}{2x+3} + \dfrac{B}{x+1} + C.$$

But you cannot put $x = -1$ because neither side of the identity is defined for $x = -1$.

However, you can multiply the original identity by $(x+1)^2$ to get

$$\dfrac{3x^2+6x+2}{2x+3} \equiv \dfrac{A(x+1)^2}{2x+3} + B + C(x+1).$$

Putting $x = -1$ now gives $B = \dfrac{3\times(-1)^2 + 6\times(-1)+2}{\left(2\times(-1)+3\right)} = \dfrac{3-6+2}{1} = -1.$

Thus $\dfrac{3x^2+6x+2}{(2x+3)(x+1)^2} \equiv \dfrac{-1}{2x+3} + \dfrac{-1}{(x+1)^2} + \dfrac{C}{x+1}.$

Here are two ways to find C. The first uses substitution and the second uses algebra.

Substitution method

There is no other especially convenient value to give x, but putting $x = 0$ in the original identity gives $\dfrac{3\times(0)^2 + 6\times 0 + 2}{(2\times 0 + 3)(0+1)^2} = \dfrac{A}{2\times 0 + 3} + \dfrac{B}{(0+1)^2} + \dfrac{C}{0+1}$, or $\frac{2}{3} = \frac{1}{3}A + B + C$. Using the values $A = -1$ and $B = -1$, which you know, leads to $C = 2$.

Thus $\dfrac{3x^2+6x+2}{(2x+3)(x+1)^2} \equiv \dfrac{-1}{2x+3} + \dfrac{-1}{(x+1)^2} + \dfrac{2}{x+1}.$

Algebraic method

Write $\dfrac{3x^2+6x+2}{(2x+3)(x+1)^2} \equiv \dfrac{-1}{2x+3} + \dfrac{-1}{(x+1)^2} + \dfrac{C}{x+1}$ as

$$\dfrac{C}{x+1} \equiv \dfrac{3x^2+6x+2}{(2x+3)(x+1)^2} + \dfrac{1}{2x+3} + \dfrac{1}{(x+1)^2} \equiv \dfrac{3x^2+6x+2+(x+1)^2+2x+3}{(2x+3)(x+1)^2}$$

$$\equiv \dfrac{3x^2+6x+2+x^2+2x+1+2x+3}{(2x+3)(x+1)^2} \equiv \dfrac{4x^2+10x+6}{(2x+3)(x+1)^2}$$

$$\equiv \dfrac{2(2x+3)(x+1)}{(2x+3)(x+1)^2} \equiv \dfrac{2}{x+1}.$$

Therefore $C = 2$, as before, and $\dfrac{3x^2 + 6x + 2}{(2x+3)(x+1)^2} \equiv \dfrac{-1}{2x+3} + \dfrac{-1}{(x+1)^2} + \dfrac{2}{x+1}$.

The key to finding partial fractions is to start with the correct form involving A, B and C. If you do not have that form, you will not be able to find the partial fractions.

> An expression of the form $\dfrac{ax^2 + bx + c}{(px+q)(rx+s)^2}$ can be split into partial
>
> fractions of the form $\dfrac{A}{px+q} + \dfrac{B}{(rx+s)^2} + \dfrac{C}{rx+s}$.

Example 15.5.1

Express $\dfrac{x^2 - 7x - 6}{x^2(x-3)}$ in partial fractions.

Write $\dfrac{x^2 - 7x - 6}{x^2(x-3)}$ in the form $\dfrac{x^2 - 7x - 6}{x^2(x-3)} \equiv \dfrac{A}{x^2} + \dfrac{B}{x} + \dfrac{C}{x-3}$.

You are advised to work through the details of this example for yourself.

Multiplying by x^2 gives $\dfrac{x^2 - 7x - 6}{x-3} \equiv A + Bx + \dfrac{Cx^2}{x-3}$: putting $x = 0$ leads to $A = 2$.

Multiplying by $x - 3$ gives $\dfrac{x^2 - 7x - 6}{x^2} \equiv \dfrac{A(x-3)}{x^2} + \dfrac{B(x-3)}{x} + C$; putting $x = 3$ leads to $C = -2$.

Therefore $\dfrac{x^2 - 7x - 6}{x^2(x-3)} \equiv \dfrac{2}{x^2} + \dfrac{B}{x} - \dfrac{2}{x-3}$.

Using the substitution method as on page 208, putting $x = 1$ leads to $B = 3$.

Thus $\dfrac{x^2 - 7x - 6}{x^2(x-3)} \equiv \dfrac{2}{x^2} + \dfrac{3}{x} - \dfrac{2}{x-3}$.

Example 15.5.2

Express $\dfrac{9 + 4x^2}{(1-2x)^2(2+x)}$ in partial fractions, and hence find the binomial expansion of

$\dfrac{9 + 4x^2}{(1-2x)^2(2+x)}$ up to and including the term in x^3. State the values of x for which the expansion is valid.

Write $\dfrac{9 + 4x^2}{(1-2x)^2(2+x)} \equiv \dfrac{A}{(1-2x)^2} + \dfrac{B}{1-2x} + \dfrac{C}{2+x}$.

Multiplying both sides by $(1-2x)^2$ gives $\dfrac{9+4x^2}{2+x} \equiv A + B(1-2x) + \dfrac{C(1-2x)^2}{2+x}$.

Putting $x = \frac{1}{2}$ leads to $A = 4$.

Multiplying both sides by $2 + x$ gives $\dfrac{9+4x^2}{(1-2x)^2} \equiv \dfrac{A(2+x)}{(1-2x)^2} + \dfrac{B(2+x)}{1-2x} + C$.

Putting $x = -2$ leads to $C = 1$.

Therefore $\dfrac{9+4x^2}{(1-2x)^2(2+x)} \equiv \dfrac{4}{(1-2x)^2} + \dfrac{B}{1-2x} + \dfrac{1}{2+x}$.

Using the algebraic method as on page 208,

$$\dfrac{B}{1-2x} \equiv \dfrac{9+4x^2}{(1-2x)^2(2+x)} - \dfrac{4}{(1-2x)^2} - \dfrac{1}{2+x} \equiv \dfrac{9+4x^2-4(2+x)-(1-2x)^2}{(1-2x)^2(2+x)}$$

$$\equiv \dfrac{9+4x^2-8-4x-1+4x-4x^2}{(1-2x)^2(2+x)} \equiv \dfrac{0}{(1-2x)^2(2+x)} \equiv 0.$$

Thus $\dfrac{9+4x^2}{(1-2x)^2(2+x)} \equiv \dfrac{4}{(1-2x)^2} + \dfrac{1}{2+x} \equiv 4(1-2x)^{-2} + (2+x)^{-1}$.

Using the binomial expansion,

$$4(1-2x)^{-2} = 4\left(1 + \dfrac{(-2)}{1}(-2x) + \dfrac{(-2)(-3)}{1\times 2}(-2x)^2 + \dfrac{(-2)(-3)(-4)}{1\times 2\times 3}(-2x)^3 + \ldots\right)$$

$$= 4\left(1 + 4x + 12x^2 + 32x^3 + \ldots\right) = 4 + 16x + 48x^2 + 128x^3 + \ldots,$$

$$(2+x)^{-1} = 2^{-1}\left(1 + \tfrac{1}{2}x\right)^{-1} = \tfrac{1}{2}\left(1 + \tfrac{1}{2}x\right)^{-1}$$

$$= \tfrac{1}{2}\left(1 + \dfrac{(-1)}{1}\left(\tfrac{1}{2}x\right) + \dfrac{(-1)(-2)}{1\times 2}\left(\tfrac{1}{2}x\right)^2 + \dfrac{(-1)(-2)(-3)}{1\times 2\times 3}\left(\tfrac{1}{2}x\right)^3 + \ldots\right)$$

$$= \tfrac{1}{2}\left(1 - \tfrac{1}{2}x + \tfrac{1}{4}x^2 - \tfrac{1}{8}x^3 + \ldots\right) = \tfrac{1}{2} - \tfrac{1}{4}x + \tfrac{1}{8}x^2 - \tfrac{1}{16}x^3 + \ldots.$$

Therefore

$$4(1-2x)^{-2} + (2+x)^{-1}$$

$$= 4 + 16x + 48x^2 + 128x^3 + \ldots + \tfrac{1}{2} - \tfrac{1}{4}x + \tfrac{1}{8}x^2 - \tfrac{1}{16}x^3 + \ldots$$

$$= \tfrac{9}{2} + \tfrac{63}{4}x + \tfrac{385}{8}x^2 + \tfrac{2047}{16}x^3 + \ldots,$$

so the required expansion is $\tfrac{9}{2} + \tfrac{63}{4}x + \tfrac{385}{8}x^2 + \tfrac{2047}{16}x^3$.

The expansion of $(1-2x)^{-2}$ is valid when $|2x| < 1$, that is when $|x| < \frac{1}{2}$.

The expansion of $\left(1 + \tfrac{1}{2}x\right)^{-1}$ is valid when $\left|\tfrac{1}{2}x\right| < 1$, that is when $|x| < 2$.

For the final result to hold you require both $|x| < \frac{1}{2}$ and $|x| < 2$, that is $|x| < \frac{1}{2}$.

Exercise 15C

1 Split into partial fractions

(a) $\dfrac{4}{(x-1)(x-3)^2}$,

(b) $\dfrac{6x^2+11x-8}{(x+2)^2(x-1)}$,

(c) $\dfrac{6}{x^3-4x^2+4x}$,

(d) $\dfrac{8-7x}{2x^3+3x^2-1}$.

2 Find

(a) $\displaystyle\int \frac{6x^2+27x+25}{(x+2)^2(x+1)}\,dx$,

(b) $\displaystyle\int \frac{97x+35}{(2x-3)(5x+2)^2}\,dx$.

3 Show that $\displaystyle\int_1^6 \frac{16}{x^2(x+4)}\,dx = \frac{10}{3}-\ln 3$.

4 Find the exact value of $\displaystyle\int_2^3 \frac{x(x+14)}{2x^3-3x^2+1}\,dx$.

5 Obtain the series expansion of $\dfrac{1}{(1+2x)^2(1-x)}$ up to and including the term in x^2 by

(a) multiplying the expansion of $(1+2x)^{-2}$ by the expansion of $(1-x)^{-1}$,

(b) splitting $\dfrac{1}{(1+2x)^2(1-x)}$ into partial fractions and finding the expansion of each fraction.

15.6 Partial fractions when the denominator includes a quadratic factor

Neither of the types considered so far, $\dfrac{ax+b}{(px+q)(rx+s)}$ and $\dfrac{ax^2+bx+c}{(px+q)(rx+s)^2}$, includes rational expressions such as $\dfrac{4x+6}{(x-1)(x^2+9)}$, where the quadratic factor x^2+9 in the denominator does not factorise.

In this case, you would certainly expect to write

$$\frac{4x+6}{(x-1)(x^2+9)} \equiv \frac{A}{x-1} + \frac{\text{something}}{x^2+9}.$$

But what is the 'something'?

Begin with the easy bit, and use the substitution method to find A.

Multiply by $x-1$, and put $x=1$ in the subsequent identity.

$$\frac{4x+6}{x^2+9} \equiv A + \frac{(\text{something})(x-1)}{x^2+9} \text{ with } x=1 \text{ gives } A=1.$$

Therefore $\dfrac{4x+6}{(x-1)(x^2+9)} \equiv \dfrac{1}{x-1} + \dfrac{\text{something}}{x^2+9}.$

You can now find the 'something' by the algebraic method, since

$$\frac{\text{something}}{x^2+9} \equiv \frac{4x+6}{(x-1)(x^2+9)} - \frac{1}{x-1} \equiv \frac{4x+6-(x^2+9)}{(x-1)(x^2+9)}$$

$$\equiv \frac{-x^2+4x-3}{(x-1)(x^2+9)} \equiv \frac{-(x-1)(x-3)}{(x-1)(x^2+9)} \equiv \frac{-x+3}{x^2+9}.$$

So the 'something' has the form $Bx+C$.

> An expression of the form $\dfrac{ax^2+bx+c}{(px+q)(rx^2+s)}$, where r and s have the same sign, can be split into partial fractions of the form $\dfrac{A}{px+q} + \dfrac{Bx+C}{rx^2+s}$.

Here are two other techniques which you can use.

If you write $\dfrac{4x+6}{(x-1)(x^2+9)} \equiv \dfrac{A}{x-1} + \dfrac{Bx+C}{x^2+9}$ initially, you can combine the partial fractions to get a numerator which is a quadratic; so you can equate the coefficients of x^0, x^1 and x^2 to get three equations, and solve them for A, B and C. This is the equating coefficients method.

Or you can multiply by $x-1$ and put $x=1$ as before to get $A=1$, and then equate coefficients to find B and C. This is a mixture of the substitution method and the equating coefficients method.

Example 15.6.1 shows these three techniques. Use whichever technique you find easiest.

The 'substitution and algebraic' method used above has the advantage that it is self-checking. If the fraction just before the final result does not cancel, you have made a mistake. If it cancels, there is probably no mistake.

The equating coefficients method is in many ways the most straightforward, but involves solving three equations for A, B and C.

Example 15.6.1

Split $\dfrac{5x-6}{(x+2)(x^2+4)}$ into partial fractions.

Substitution and algebraic method

Write $\dfrac{5x-6}{(x+2)(x^2+4)} \equiv \dfrac{A}{x+2} + \dfrac{Bx+C}{x^2+4}$.

Multiply by $x+2$ to get $\dfrac{5x-6}{x^2+4} \equiv A + \dfrac{(Bx+C)(x+2)}{x^2+4}$, and put $x=-2$. Then

$\dfrac{5\times(-2)-6}{(-2)^2+4} = A$, so $A=-2$.

Then $\dfrac{Bx+C}{x^2+4} \equiv \dfrac{5x-6}{(x+2)(x^2+4)} - \dfrac{-2}{x+2} \equiv \dfrac{5x-6-(-2)(x^2+4)}{(x+2)(x^2+4)}$

$\equiv \dfrac{2x^2+5x+2}{(x+2)(x^2+4)} \equiv \dfrac{(x+2)(2x+1)}{(x+2)(x^2+4)} \equiv \dfrac{2x+1}{x^2+4}$.

So $\dfrac{5x-6}{(x+2)(x^2+4)} \equiv \dfrac{-2}{x+2} + \dfrac{2x+1}{x^2+4}$.

Equating coefficients method

Write $\dfrac{5x-6}{(x+2)(x^2+4)} \equiv \dfrac{A}{x+2} + \dfrac{Bx+C}{x^2+4}$.

Multiply by $(x+2)(x^2+4)$ to get $5x-6 \equiv A(x^2+4) + (Bx+C)(x+2)$, which you can write as

$$5x-6 \equiv (A+B)x^2 + (2B+C)x + 4A + 2C.$$

The three equations which you get by equating coefficients of x^0, x^1 and x^2 are

$$4A+2C=-6, \quad 2B+C=5 \quad \text{and} \quad A+B=0.$$

To solve these equations, put $A=-B$ in the first equation to get $-4B+2C=-6$, or $-2B+C=-3$. Solving this and $2B+C=5$ gives $B=2$, $C=1$, so $A=-2$.

So $\dfrac{5x-6}{(x+2)(x^2+4)} \equiv \dfrac{-2}{x+2} + \dfrac{2x+1}{x^2+4}$.

Substitution and equating coefficients method

Follow the argument in the substitution and algebraic method to get $A=-2$.

Then, combining the fractions in $\dfrac{5x-6}{(x+2)(x^2+4)} \equiv \dfrac{-2}{x+2} + \dfrac{Bx+C}{x^2+4}$ and equating

the numerators gives $5x-6 \equiv -2(x^2+4) + (Bx+C)(x+2)$,

which you can write as

$$5x-6 \equiv (-2+B)x^2 + (2B+C)x - 8 + 2C.$$

Equating the coefficients of x^2 gives $-2 + B = 0$, so $B = 2$.

Equating the coefficients of x^0 gives $-6 = -8 + 2C$, so $C = 1$.

Checking the x-coefficient: on the left side 5; on the right $2B + C = 2 \times 2 + 1 = 5$.

Therefore $\dfrac{5x - 6}{(x + 2)(x^2 + 4)} \equiv \dfrac{-2}{x + 2} + \dfrac{2x + 1}{x^2 + 4}$.

Notice that in the 'substitution and equating coefficients' method the coefficients of x^2 and x^0 are used to get the values of B and C. This is because they give the simplest equations. You will usually find that the highest and lowest powers give the simplest equations in these situations. However, you should be aware that equating the coefficients of x^0 has the same result as putting $x = 0$; you get no extra information.

Example 15.6.2

Expand $\dfrac{1 + x}{(1 - x)(1 + x^2)}$ in ascending powers of x as far as the term in x^5.

Using the substitution and algebraic method, write $\dfrac{1 + x}{(1 - x)(1 + x^2)} \equiv \dfrac{A}{1 - x} + \dfrac{Bx + C}{1 + x^2}$.

Multiplying by $1 - x$ and putting $x = 1$ gives $A = 1$.

Simplifying $\dfrac{1 + x}{(1 - x)(1 + x^2)} - \dfrac{1}{1 - x}$ to find the other fraction on the right gives

$$\frac{1 + x}{(1 - x)(1 + x^2)} - \frac{1}{1 - x} \equiv \frac{1 + x - (1 + x^2)}{(1 - x)(1 + x^2)}$$

$$\equiv \frac{x - x^2}{(1 - x)(1 + x^2)}$$

$$\equiv \frac{x(1 - x)}{(1 - x)(1 + x^2)}$$

$$\equiv \frac{x}{1 + x^2}.$$

Therefore $\dfrac{1 + x}{(1 - x)(1 + x^2)} \equiv \dfrac{1}{1 - x} + \dfrac{x}{1 + x^2}$.

Use the binomial theorem on $\dfrac{1 + x}{(1 - x)(1 + x^2)}$ in the form

$$\frac{1 + x}{(1 - x)(1 + x^2)} \equiv (1 - x)^{-1} + x(1 + x^2)^{-1}.$$

This gives

$$(1-x)^{-1} + x(1+x^2)^{-1} = 1 + \frac{(-1)}{1}(-x) + \frac{(-1)(-2)}{1 \times 2}(-x)^2$$

$$+ \frac{(-1)(-2)(-3)}{1 \times 2 \times 3}(-x)^3 + \frac{(-1)(-2)(-3)(-4)}{1 \times 2 \times 3 \times 4}(-x)^4$$

$$+ \frac{(-1)(-2)(-3)(-4)(-5)}{1 \times 2 \times 3 \times 4 \times 5}(-x)^5 + \dots$$

$$+ x\left(1 + \frac{(-1)}{1}x^2 + \frac{(-1)(-2)}{1 \times 2}(x^2)^2 + \dots\right)$$

$$= 1 + x + x^2 + x^3 + x^4 + x^5 + \dots + x(1 - x^2 + x^4 - \dots)$$

$$= 1 + 2x + x^2 + x^4 + 2x^5 + \dots .$$

The required expansion is $1 + 2x + x^2 + x^4 + 2x^5$.

Example 15.6.3
Use partial fractions to differentiate $\dfrac{x^2 + x}{(x-4)(x^2+4)}$ with respect to x.

Putting $\dfrac{x^2 + x}{(x-4)(x^2+4)}$ into partial fractions gives

$$\frac{x^2 + x}{(x-4)(x^2+4)} \equiv \frac{1}{x-4} + \frac{1}{x^2+4}.$$

Then $\dfrac{d}{dx}\left(\dfrac{x^2 + x}{(x-4)(x^2+4)}\right) = \dfrac{d}{dx}\left(\dfrac{1}{x-4} + \dfrac{1}{x^2+4}\right)$

$$= \frac{d}{dx}(x-4)^{-1} + \frac{d}{dx}(x^2+4)^{-1}$$

$$= -1 \times (x-4)^{-2} + (-1) \times (x^2+4)^{-2} \times 2x$$

$$= -\frac{1}{(x-4)^2} - \frac{2x}{(x^2+4)^2}.$$

Exercise 15D

1 Express each of the following in partial fractions.

(a) $\dfrac{x^2 + x}{(x-1)(x^2+1)}$

(b) $\dfrac{4 - x}{(x+1)(x^2+4)}$

(c) $\dfrac{2x^2 + x - 2}{(x+3)(x^2+4)}$

(d) $\dfrac{2x^2 + 11x - 8}{(2x-3)(x^2+1)}$

(e) $\dfrac{x^2 - 3x + 14}{(3x+2)(x^2+16)}$

(f) $\dfrac{3 + 17x^2}{(1+4x)(4+x^2)}$

(g) $\dfrac{6 - 5x}{(1+2x)(4+x^2)}$

(h) $\dfrac{x^2 + 18}{x(x^2+9)}$

(i) $\dfrac{17 - 25x}{(x+4)(2x^2+7)}$

2 Expand the following expressions in ascending powers of x as far as the term in x^3.

(a) $\dfrac{1 - 2x - x^2}{(1 + x)(1 + x^2)}$ (b) $\dfrac{x^2 - 8x}{(2x + 1)(x^2 + 4)}$ (c) $\dfrac{75 - 2x - 3x^2}{(1 + 3x)(25 + x^2)}$

3 Use partial fractions to differentiate the following functions with respect to x.

(a) $\dfrac{3x^2 - 2x - 1}{(x + 2)(x^2 + 1)}$ (b) $\dfrac{3x^2 + 4}{x(x^2 + 4)}$ (c) $\dfrac{x^2 - 4x}{(x + 4)(x^2 + 16)}$

15.7 Improper fractions

So far all the rational functions you have seen have been 'proper' fractions. That is, the degree of the numerator has been less than the degree of the denominator.

Rational functions such as $\dfrac{x^2 - 3x + 5}{(x + 1)(x - 2)}$ and $\dfrac{x^3}{x^2 + 1}$, in which the degree of the numerator is greater than or equal to the degree of the denominator, are called **improper fractions**.

Improper fractions may not be the most convenient form to work with; it is often better to express them in a different way.

- You cannot find $\displaystyle\int \dfrac{6x}{x - 1}\,dx$ with the integrand in its present form, but it is straightforward to integrate the same expression as $\displaystyle\int \left(6 + \dfrac{6}{x - 1} \right) dx$. You get

$$\int \left(6 + \dfrac{6}{x - 1} \right) dx = 6x + 6\ln|x - 1| + k.$$

- If you wish to sketch the graph of $y = \dfrac{6x}{x - 1}$, although you can see immediately that the graph passes through the origin, other features are much clearer in the form $y = 6 + \dfrac{6}{x - 1}$.

You may find it helpful to think about an analogy between improper fractions in arithmetic and improper fractions in algebra. Sometimes in arithmetic it is more useful to think of the number $\frac{25}{6}$ in that form; at other times it is better in the form $4\frac{1}{6}$. The same is true in algebra, and you need to be able to change from one form to the other.

In Section 1.4, you learned how to divide one polynomial by another polynomial to get a quotient and a remainder.

For example, when you divide the polynomial $a(x)$ by the polynomial $b(x)$ you will get a quotient $q(x)$ and a remainder $r(x)$ defined by

$$a(x) \equiv b(x)q(x) + r(x)$$

where the degree of the remainder $r(x)$ is less than the degree of the divisor $b(x)$.

If you divide this equation by $b(x)$, you get

$$\frac{a(x)}{b(x)} \equiv \frac{b(x)q(x) + r(x)}{b(x)} \equiv q(x) + \frac{r(x)}{b(x)}.$$

Therefore, if you divide $x^2 - 3x + 5$ by $(x+1)(x-2)$ to get a number A and a remainder of the form $Px + Q$, it is equivalent to saying that

$$x^2 - 3x + 5 \equiv A(x+1)(x-2) + Px + Q$$

and $\quad \dfrac{x^2 - 3x + 5}{(x+1)(x-2)} \equiv A + \dfrac{Px + Q}{(x+1)(x-2)}.$

This form will be called **divided out form**.

An analogous statement in arithmetic is 25 *divided by* 6 *is* 4 *with remainder* 1; *that is,* $25 = 4 \times 6 + 1$, *or* $\frac{25}{6} = 4 + \frac{1}{6} = 4\frac{1}{6}$.

In the examples which follow the degree of the numerator is equal to the degree of the denominator; in this case the quotient $q(x)$ is just a constant.

Example 15.7.1

Split $\dfrac{2x^2 + 4x - 3}{(x+1)(2x-3)}$ into partial fractions.

You have a choice between going immediately into partial fraction form or putting the expression into divided out form first. Either way, you can then use one of the standard methods to find the coefficients.

Method 1 This method goes straight into partial fraction form and then uses the substitution technique.

Write $\dfrac{2x^2 + 4x - 3}{(x+1)(2x-3)} \equiv A + \dfrac{B}{x+1} + \dfrac{C}{2x-3}.$

Multiplying by $x+1$ gives $\dfrac{2x^2 + 4x - 3}{2x-3} \equiv A(x+1) + B + \dfrac{C(x+1)}{2x-3}.$

Putting $x = -1$ gives $\dfrac{2 \times (-1)^2 + 4 \times (-1) - 3}{2 \times (-1) - 3} = B$, so $B = 1$.

Multiplying instead by $2x - 3$ gives $\dfrac{2x^2 + 4x - 3}{x+1} \equiv A(2x-3) + \dfrac{B(2x-3)}{x+1} + C.$

Putting $x = \frac{3}{2}$ gives $\dfrac{2 \times \left(\frac{3}{2}\right)^2 + 4 \times \left(\frac{3}{2}\right) - 3}{\frac{3}{2} + 1} = C$, so $C = 3$.

Therefore $\dfrac{2x^2 + 4x - 3}{(x+1)(2x-3)} \equiv A + \dfrac{1}{x+1} + \dfrac{3}{2x-3}.$

You can now substitute any other value of x to find A. The simplest is $x = 0$, which gives $\dfrac{-3}{1 \times (-3)} = A + \dfrac{1}{1} + \dfrac{3}{-3}$, so $A = 1$.

Therefore $\dfrac{2x^2 + 4x - 3}{(x+1)(2x-3)} \equiv 1 + \dfrac{1}{x+1} + \dfrac{3}{2x-3}$.

Method 2   If you divide out first, you start with the form

$\dfrac{2x^2 + 4x - 3}{(x+1)(2x-3)} \equiv A + \dfrac{Px+Q}{(x+1)(2x-3)}$. Multiplying by $(x+1)(2x-3)$ and equating the coefficients of x^2 gives $A = 1$.

Thus $\dfrac{2x^2 + 4x - 3}{(x+1)(2x-3)} \equiv 1 + \dfrac{Px+Q}{(x+1)(2x-3)}$, so $\dfrac{2x^2 + 4x - 3}{(x+1)(2x-3)} - 1 \equiv \dfrac{Px+Q}{(x+1)(2x-3)}$.

Simplifying the left side,

$$\dfrac{2x^2 + 4x - 3 - (x+1)(2x-3)}{(x+1)(2x-3)} \equiv \dfrac{2x^2 + 4x - 3 - \left(2x^2 - x - 3\right)}{(x+1)(2x-3)} \equiv \dfrac{5x}{(x+1)(2x-3)}.$$

Any of the standard methods for partial fractions now shows that

$$\dfrac{5x}{(x+1)(2x-3)} \equiv \dfrac{1}{x+1} + \dfrac{3}{2x-3}, \quad \text{so} \quad \dfrac{2x^2 + 4x - 3}{(x+1)(2x-3)} \equiv 1 + \dfrac{1}{x+1} + \dfrac{3}{2x-3}.$$

Similar methods work when the denominator contains a repeated factor $(rx+s)^2$ or a quadratic factor $rx^2 + s$. An example is given of the first type.

Example 15.7.2

Split $f(x) \equiv \dfrac{4x^3 + 10x^2 + 8x - 1}{(2x+1)^2(x+2)}$ into partial fractions.

Dividing out first, and equating coefficients of x^3 in the identity

$$4x^3 + 10x^2 + 8x - 1 \equiv A(2x+1)^2(x+2) + \left(Px^2 + Qx + R\right)$$

gives $4 = 4A$, so $A = 1$. Then

$$Px^2 + Qx + R \equiv \left(4x^3 + 10x^2 + 8x - 1\right) - \left(4x^3 + 12x^2 + 9x + 2\right)$$
$$\equiv -2x^2 - x - 3.$$

So $\dfrac{-2x^2 - x - 3}{(2x+1)^2(x+2)}$ has to be put into the form $\dfrac{B}{(2x+1)^2} + \dfrac{C}{2x+1} + \dfrac{D}{x+2}$.

Multiplying by $x + 2$ gives $\dfrac{-2x^2 - x - 3}{(2x+1)^2} \equiv \dfrac{B(x+2)}{(2x+1)^2} + \dfrac{C(x+2)}{2x+1} + D$, and putting

$x = -2$ gives $\dfrac{-8 - (-2) - 3}{(-3)^2} = D$, so $D = -1$.

Multiplying instead by $(2x+1)^2$ gives $\dfrac{-2x^2 - x - 3}{x+2} \equiv B + C(2x+1) - \dfrac{(2x+1)^2}{x+2}$, and

putting $x = -\frac{1}{2}$ gives $\dfrac{-2\left(-\frac{1}{2}\right)^2 - \left(-\frac{1}{2}\right) - 3}{-\frac{1}{2} + 2} = B$, so $B = -2$.

So $\dfrac{-2x^2 - x - 3}{(2x+1)^2(x+2)} \equiv -\dfrac{2}{(2x+1)^2} + \dfrac{C}{2x+1} - \dfrac{1}{x+2}$.

Then $\dfrac{C}{2x+1} \equiv \dfrac{-2x^2 - x - 3}{(2x+1)^2(x+2)} + \dfrac{2}{(2x+1)^2} + \dfrac{1}{x+2}$

$\equiv \dfrac{-2x^2 - x - 3 + 2(x+2) + (2x+1)^2}{(2x+1)^2(x+2)}$

$\equiv \dfrac{-2x^2 - x - 3 + 2x + 4 + 4x^2 + 4x + 1}{(2x+1)^2(x+2)}$

$\equiv \dfrac{2x^2 + 5x + 2}{(2x+1)^2(x+2)} \equiv \dfrac{(2x+1)(x+2)}{(2x+1)^2(x+2)} \equiv \dfrac{1}{2x+1}$.

So $\mathrm{f}(x) \equiv \dfrac{4x^3 + 10x^2 + 8x - 1}{(2x+1)^2(x+2)} \equiv 1 - \dfrac{2}{(2x+1)^2} + \dfrac{1}{2x+1} - \dfrac{1}{x+2}$.

Exercise 15E

1 Express each of the following in divided out form. In parts (d), (e) and (f) go on to split them into partial fractions.

(a) $\dfrac{x+1}{x}$

(b) $\dfrac{x}{x+1}$

(c) $\dfrac{x+1}{x-1}$

(d) $\dfrac{x^2 + 1}{x^2 - 1}$

(e) $\dfrac{6x^2 - 22x + 18}{(2x-3)(x-2)}$

(f) $\dfrac{24x^2 + 67x + 11}{(2x+5)(3x+1)}$

2 Express each of the following in partial fractions.

(a) $\dfrac{x^3 - 1}{x^2(x+1)}$

(b) $\dfrac{x^3 + 2x + 1}{x(x^2 + 1)}$

(c) $\dfrac{x^3 + 3x^2 + x - 14}{(x+4)(x^2 + 1)}$

(d) $\dfrac{2x^3 + 6x^2 - 3x - 2}{(x-2)(x+2)^2}$

(e) $\dfrac{6x^3 + x + 10}{(x-2)(x+2)(2x-1)}$

(f) $\dfrac{-4x^3 + 16x^2 + 15x - 50}{x(4x^2 - 25)}$

(g) $\dfrac{x^3 + 2x + 1}{x(x+1)(x-1)}$

(h) $\dfrac{x^3 - 2x^2 + 3x + 6}{x^2(2+x)}$

(i) $\dfrac{12x^3 - 20x^2 + 31x - 49}{(4x^2 + 9)(x-1)}$

Miscellaneous exercise 15

1 Express $\dfrac{4}{(x-3)(x+1)}$ in partial fractions.   (OCR)

2 Express $\dfrac{2}{x(x-1)(x+1)}$ in partial fractions.   (OCR)

3 Express $\dfrac{2x^2+1}{x(x-1)^2}$ in partial fractions.   (OCR)

4 Express $\dfrac{x^2-11}{(x+2)^2(3x-1)}$ in partial fractions.   (OCR)

5 Find $\displaystyle\int \dfrac{1}{x(x+1)}\,dx$.   (OCR)

6 Express $\dfrac{3+2x}{x^2(3-x)}$ in partial fractions.

7 Find $\displaystyle\int \dfrac{x}{(x+1)(x+2)}\,dx$.   (OCR)

8 Find $\displaystyle\int \dfrac{3-x-3x^3}{x^2(1-x)}\,dx$.

9 Expand $\dfrac{6x^3+21x+7}{(2x+1)(x^2+4)}$ in ascending powers of x up to and including the term in x^2.

10 Express $\dfrac{1}{x^2(x-1)}$ in the form $\dfrac{A}{x}+\dfrac{B}{x^2}+\dfrac{C}{x-1}$, where A, B and C are constants. Hence

   find $\displaystyle\int \dfrac{1}{x^2(x-1)}\,dx$.   (OCR)

11 Simplify $\sqrt{\dfrac{3x^2+5x-2}{4x-3}\div\dfrac{4x^3+13x^2+4x-12}{2x(x+3)-(2-x)(1+x)}}$.

12 Simplify $\dfrac{x-21}{x^2-9}-\dfrac{3}{x+3}+\dfrac{4}{x-3}$.

13 Express $\dfrac{8x^3-12x^2-18x+15}{(4x^2-9)(2x-3)}$ in partial fractions.

14 Express in partial fractions

   (a) $\dfrac{x^2-x+3}{x(2x^2+3)}$,   (b) $\dfrac{7x^2-2x+5}{(x-1)(3x^2+2)}$.

15 Express $\dfrac{3}{(2x+1)(x-1)}$ in partial fractions. Hence find the exact value of

$\displaystyle\int_2^3 \dfrac{3}{(2x+1)(x-1)}\,dx$, giving your answer as a single logarithm.   (OCR)

16 Express $\dfrac{1}{(x+3)(4-x)}$ in partial fractions. Hence find the exact value of

$\displaystyle\int_0^2 \dfrac{1}{(x+3)(4-x)}\,dx$, giving your answer as a single logarithm.   (OCR)

17 Express $f(x) \equiv \dfrac{2}{2-3x+x^2}$ in partial fractions and hence, or otherwise, obtain $f(x)$ as a

series of ascending powers of x, giving the first four non-zero terms of this expansion.
State the set of values of x for which this expansion is valid.   (OCR)

18 Express $\dfrac{3-x}{(2+x)(1-2x)}$ in partial fractions and hence, or otherwise, obtain the first three

terms in the expansion of this expression in ascending powers of x. State the range of
values of x for which the expansion is valid.   (OCR)

19 Given that $\dfrac{18-4x-x^2}{(4-3x)(1+x)^2} \equiv \dfrac{A}{4-3x} + \dfrac{B}{1+x} + \dfrac{C}{(1+x)^2}$, show that $A = 2$, and obtain the

values of B and C. Hence show that $\displaystyle\int_0^1 \dfrac{18-4x-x^2}{(4-3x)(1+x)^2}\,dx = \tfrac{7}{3}\ln 2 + \tfrac{3}{2}$.   (OCR)

20 Let $y = \dfrac{4+7x}{(2-x)(1+x)^2}$. Express y in the form $\dfrac{A}{2-x} + \dfrac{B}{1+x} + \dfrac{C}{(1+x)^2}$, where the

numerical values of A, B and C are to be found. Hence, or otherwise, expand y in a
series of ascending powers of x up to and including the term in x^3, simplifying the
coefficients. Use your result to find the value of $\dfrac{dy}{dx}$ when $x = 0$.

21 Express $\dfrac{15-13x+4x^2}{(1-x)^2(4-x)}$ in partial fractions. Hence evaluate $\displaystyle\int_2^3 \dfrac{15-13x+4x^2}{(1-x)^2(4-x)}\,dx$ giving

the exact value in terms of logarithms.   (OCR)

22 Split $\dfrac{1}{x^4 - 13x^2 + 36}$ into partial fractions.

23* Let $f(x) = \dfrac{x^2+5x}{(1+x)(1-x)^2}$. Express $f(x)$ in the form $\dfrac{A}{1+x} + \dfrac{B}{1-x} + \dfrac{C}{(1-x)^2}$ where A, B

and C are constants. The expansion of $f(x)$, in ascending powers of x, is
$c_0 + c_1 x + c_2 x^2 + c_3 x^3 + \ldots + c_r x^r + \ldots$. Find c_0, c_1, c_2 and show that $c_3 = 11$. Express c_r
in terms of r.   (OCR)

24 It is given that $g(x) = (2x-1)(x+2)(x-3)$.

(a) Express $g(x)$ in the form $Ax^3 + Bx^2 + Cx + D$, giving the values of the constants A, B, C and D.

(b) Find the value of the constant a, given that $x+3$ is a factor of $g(x) + ax$.

(c) Express $\dfrac{x-3}{g(x)}$ in partial fractions.   (OCR)

25 The diagram shows part of the graph of $y = \dfrac{3}{\sqrt{x}(x-3)}$. The shaded region is bounded by the curve and the lines $y = 0$, $x = 4$ and $x = 6$. Find the volume of the solid formed when the shaded region is rotated through four right angles about the x-axis.

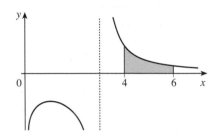

26 (a) Find the values of A, B and C for which $\dfrac{x^2 - 2}{(x-2)^2} \equiv A + \dfrac{B}{x-2} + \dfrac{C}{(x-2)^2}$.

(b) The region bounded by the curve with equation $y = \dfrac{x^2 - 2}{(x-2)^2}$, the x-axis and the lines $x = 3$ and $x = 4$ is denoted by R. Show that R has area $(2 + 4\ln 2)$ square units.   (OCR)

27 (a) Express $\dfrac{2}{(x-1)(x-3)}$ in partial fractions, and use the result to express $\dfrac{4}{(x-1)^2(x-3)^2}$ in partial fractions.

The finite region bounded by the curve with equation $y = \dfrac{2}{(x-1)(x-3)}$ and the lines $x = 4$ and $y = \frac{1}{4}$ is denoted by R.

(b) Show that the area of R is $\ln\frac{3}{2} - \frac{1}{4}$.

(c) Calculate the volume of the solid formed when R is rotated through 2π radians about the x-axis.   (OCR)

16 Complex numbers

In this chapter the concept of number is extended so that all numbers have square roots. When you have completed it, you should

- understand that new number systems can be created, provided that the definitions are algebraically consistent
- appreciate that complex number algebra excludes inequalities
- be able to do calculations with complex numbers
- know the meaning of conjugate complex numbers, and that non-real roots of equations with real coefficients occur in conjugate pairs
- know how to represent complex numbers as translations or as points
- know the meaning of modulus, and be able to use it algebraically
- be able to use complex numbers to prove geometrical results
- be able to solve simple equations with complex coefficients.

16.1 Extending the number system

Before negative numbers were developed, it was impossible to subtract a from b if $a > b$. If only positive numbers exist, then there is no number x such that $a + x = b$.

This was a serious drawback for mathematics and science, so a new kind of number was invented which could be either positive or negative. It was found that this results in a consistent system of numbers in which the ordinary rules of algebra apply, provided that you also make up the right rules for combining numbers, such as $b - a = -(a - b)$ and $(-a) \times (-b) = +(ab)$.

There is a similar problem with real numbers, that square roots only exist for positive numbers and zero. That is, if $a < 0$, there is no real number x such that $x^2 = a$.

The mathematical response to this is to invent a new kind of number, called a **complex number**. It turns out that this can be done very simply, by introducing just one new number, usually denoted by i, whose square is -1. If you also require that this number combines with the real numbers by the usual rules of algebra, this creates a whole new system of numbers.

Notice first that you don't need a separate symbol for the square root of -2, since the rules of algebra require that $\sqrt{-2} = \sqrt{2} \times \sqrt{-1}$, so that $\sqrt{-2}$ is just $\sqrt{2}\,i$.

Since you must be able to combine i with all the real numbers, the complex numbers must include all the products $b\,i$ where b is any real number. They must also include all the sums $a + b\,i$, where a is any real number.

> The **complex numbers** consist of numbers of the form $a + b\,i$, where a and b are real numbers and $i^2 = -1$.
>
> Complex numbers of the form $a + 0\,i$ are called **real numbers**; complex numbers of the form $0 + b\,i$ are called **imaginary numbers**.
>
> In a general complex number $a + b\,i$, a is called the **real part** and b the **imaginary part**. This is written $\operatorname{Re}(a + b\,i) = a$, $\operatorname{Im}(a + b\,i) = b$.

Some people prefer to use j rather than i for the square root of -1. Also, some books define the imaginary part of $a + b\,i$ as $b\,i$ rather than b.

Two questions need to be asked before going further: is algebra with complex numbers consistent, and are complex numbers useful? The answers are 'yes, but … ' and 'yes, very'. Complex numbers have an important place in modern physics and electronics.

The reason for the 'but' is that with complex numbers you cannot use the inequality symbols $>$ and $<$. One of the rules for inequalities is that, if $a > b$ and $c > 0$, then $ac > bc$. So if $a > 0$ and $a > 0$, then $aa > 0a$, that is $a^2 > 0$. What about the number i? Is $i > 0$ or $i < 0$?

Try following through the consequences of assumptions (a) and (b) in turn:

(a) if $i > 0$ then $i^2 > 0$, so $-1 > 0$;

(b) $i < 0 \;\Leftrightarrow\; i + (-i) < 0 + (-i) \;\Leftrightarrow\; 0 < -i \;\Leftrightarrow\; -i > 0$,
so if $i < 0$ then $-i > 0$ and $(-i)^2 > 0$, giving $-1 > 0$.

Either assumption, (a) or (b), leads to the conclusion that $-1 > 0$, which you know to be false. The way out of the dilemma is to make the rule:

> The relations $>$ and $<$ cannot be used to compare pairs of complex numbers.

16.2 Operations with complex numbers

It is remarkable, and not at all obvious, that when you add, subtract, multiply or divide two complex numbers $a + b\,i$ and $c + d\,i$, the result is another complex number.

Addition and subtraction   By the usual rules of algebra,

$$(a + b\,i) \pm (c + d\,i) = a + b\,i \pm c \pm d\,i = a \pm c + b\,i \pm d\,i = (a \pm c) + (b \pm d)\,i.$$

Since a, b, c, d are real numbers, so are $a \pm c$ and $b \pm d$. The expression at the end of the line therefore has the form $p + q\,i$ where p and q are real.

Uniqueness   If $a + b\,i = 0$, then $a = -b\,i$, so that $a^2 = (-b\,i)^2 = -b^2$. Now a and b are real, so that $a^2 \geqslant 0$ and $-b^2 \leqslant 0$. They can only be equal if $a^2 = 0$ and $b^2 = 0$. That is, $a = 0$ and $b = 0$. So if a complex number is zero, both its real and imaginary parts are zero.

Combining this with the rule for subtraction shows that

$$(a + b\,i) = (c + d\,i) \quad \Leftrightarrow \quad (a + b\,i) - (c + d\,i) = 0 \quad \Leftrightarrow \quad (a - c) + (b - d)\,i = 0$$
$$\Leftrightarrow \quad a - c = 0 \quad \text{and} \quad b - d = 0$$
$$\Leftrightarrow \quad a = c \quad \text{and} \quad b = d.$$

That is:

> If two complex numbers are equal, their real parts are equal and their imaginary parts are equal.

Multiplication   By the usual rules for multiplying out brackets,

$$(a + b\,i) \times (c + d\,i) = ac + a(d\,i) + (b\,i)c + (b\,i)(d\,i) = ac + ad\,i + bc\,i + bd\,i^2$$
$$= (ac - bd) + (ad + bc)\,i.$$

Since a, b, c, d are real numbers, so are $ac - bd$ and $ad + bc$. The product is therefore of the form the form $p + q\,i$ where p and q are real.

An important special case is

$$(a + b\,i) \times (a - b\,i) = (aa - b(-b)) + (a(-b) + ba)\,i = \left(a^2 + b^2\right) + 0\,i,$$

a real number. So with complex numbers the sum of two squares, $a^2 + b^2$, can be factorised as $(a + b\,i)(a - b\,i)$.

Division   First, there are two special cases to consider.

If $d = 0$, then

$$\frac{a + b\,i}{c + 0\,i} = \frac{a + b\,i}{c} = \frac{a}{c} + \frac{b}{c}\,i.$$

And if $c = 0$, you can simplify the expression by multiplying numerator and denominator by i:

$$\frac{a + b\,i}{0 + d\,i} = \frac{a + b\,i}{d\,i} = \frac{(a + b\,i)i}{(d\,i)i} = \frac{a\,i + b\,i^2}{d\,i^2}$$
$$= \frac{a\,i - b}{-d} = \frac{-b}{-d} + \left(\frac{a\,i}{-d}\right) = \frac{b}{d} - \frac{a}{d}\,i.$$

In the general case $\dfrac{a+b\,\mathrm{i}}{c+d\,\mathrm{i}}$ the trick is to multiply numerator and denominator by $c-d\,\mathrm{i}$, and to use the result just proved, that $(c+d\,\mathrm{i})(c-d\,\mathrm{i})=c^2+d^2$.

$$\frac{a+b\,\mathrm{i}}{c+d\,\mathrm{i}}=\frac{(a+b\,\mathrm{i})(c-d\,\mathrm{i})}{(c+d\,\mathrm{i})(c-d\,\mathrm{i})}=\frac{(ac+bd)+(-ad+bc)\,\mathrm{i}}{c^2+d^2}=\frac{ac+bd}{c^2+d^2}+\left(\frac{bc-ad}{c^2+d^2}\right)\mathrm{i}.$$

In every case the result has the form $p+q\,\mathrm{i}$ where p and q are real numbers. The only exception is when $c^2+d^2=0$, and since c and d are both real this can only occur if c and d are both 0, so that $c+d\,\mathrm{i}=0+0\,\mathrm{i}=0$. With complex numbers, as with real numbers, you cannot divide by zero.

Do not try to remember the formulae for $(a+b\,\mathrm{i})(c+d\,\mathrm{i})$ and $\dfrac{a+b\,\mathrm{i}}{c+d\,\mathrm{i}}$ in this section.

As long as you understand the method, it is simple to apply it when you need it.

Exercise 16A

1 If $p=2+3\,\mathrm{i}$ and $q=2-3\,\mathrm{i}$, express the following in the form $a+b\,\mathrm{i}$, where a and b are real numbers.

   (a) $p+q$ (b) $p-q$ (c) pq (d) $(p+q)(p-q)$

   (e) p^2-q^2 (f) p^2+q^2 (g) $(p+q)^2$ (h) $(p-q)^2$

2 If $r=3+\mathrm{i}$ and $s=1-2\,\mathrm{i}$, express the following in the form $a+b\,\mathrm{i}$, where a and b are real numbers.

   (a) $r+s$ (b) $r-s$ (c) $2r+s$ (d) $r+s\mathrm{i}$

   (e) rs (f) r^2 (g) $\dfrac{r}{s}$ (h) $\dfrac{s}{r}$

   (i) $\dfrac{r}{\mathrm{i}}$ (j) $(1+\mathrm{i})r$ (k) $\dfrac{s}{1+\mathrm{i}}$ (l) $\dfrac{1-\mathrm{i}}{s}$

3 If $(2+\mathrm{i})(x+y\,\mathrm{i})=1+3\,\mathrm{i}$, where x and y are real numbers, write two equations connecting x and y, and solve them.

   Compare your answer with that given by dividing $1+3\,\mathrm{i}$ by $2+\mathrm{i}$ using the method described in the text.

4 Evaluate the following.

   (a) $\mathrm{Re}(3+4\,\mathrm{i})$ (b) $\mathrm{Im}(4-3\,\mathrm{i})$ (c) $\mathrm{Re}(2+\mathrm{i})^2$

   (d) $\mathrm{Im}(3-\mathrm{i})^2$ (e) $\mathrm{Re}\dfrac{1}{1+\mathrm{i}}$ (f) $\mathrm{Im}\dfrac{1}{\mathrm{i}}$

5 If $s=a+b\,\mathrm{i}$ and $t=c+d\,\mathrm{i}$ are complex numbers, which of the following are always true?

   (a) $\mathrm{Re}\,s+\mathrm{Re}\,t=\mathrm{Re}(s+t)$ (b) $\mathrm{Re}\,3s=3\,\mathrm{Re}\,s$ (c) $\mathrm{Re}(\mathrm{i}\,s)=\mathrm{Im}\,s$

   (d) $\mathrm{Im}(\mathrm{i}\,s)=\mathrm{Re}\,s$ (e) $\mathrm{Re}\,s\times\mathrm{Re}\,t=\mathrm{Re}(st)$ (f) $\dfrac{\mathrm{Im}\,s}{\mathrm{Im}\,t}=\mathrm{Im}\dfrac{s}{t}$

16.3 Solving equations

Now that $a + b\,\mathrm{i}$ is recognised as a number in its own right, there is no need to go on writing it out in full. You can use a single letter, such as s (or any other letter you like, except i), to represent it. If you write $s = a + b\,\mathrm{i}$, it is understood that s is a complex number and that a and b are real numbers.

Just as x is often used to stand for a general real number, it is conventional to use z for a general complex number, and to write $z = x + y\,\mathrm{i}$, where x and y are real numbers. When you see z as the unknown in an equation, you know that there is a possibility that at least some of the roots may be complex numbers. (But if you see some other letter, don't assume that the solution is not complex.)

Example 16.3.1

Solve the quadratic equation $z^2 + 4z + 13 = 0$.

> **Method 1** In the usual notation $a = 1$, $b = 4$ and $c = 13$, so that
> $b^2 - 4ac = 16 - 52 = -36$. Previously you would have said that there are no roots, but you can now write $\sqrt{-36} = 6\,\mathrm{i}$. Using the quadratic formula gives
>
> $$z = \frac{-b \pm \sqrt{b^2 - 4ac}}{2a} = \frac{-4 \pm 6\,\mathrm{i}}{2} = -2 \pm 3\,\mathrm{i}.$$
>
> **Method 2** In completed square form, $z^2 + 4z + 13 = (z + 2)^2 + 9$. This is the sum of two squares, which you can now factorise as
>
> $$((z + 2) - 3\,\mathrm{i})((z + 2) + 3\,\mathrm{i}).$$
>
> So you can write the equation as
>
> $$(z + 2 - 3\,\mathrm{i})(z + 2 + 3\,\mathrm{i}) = 0, \quad \text{with roots} \quad z = -2 + 3\,\mathrm{i} \text{ and } -2 - 3\,\mathrm{i}.$$

You can use a similar method with any quadratic equation, $az^2 + bz + c = 0$, where the coefficients a, b and c are real. If $b^2 - 4ac > 0$ there are two roots, both real numbers. But if $b^2 - 4ac < 0$, you can write $b^2 - 4ac$ as $-\left(4ac - b^2\right)$, whose square root is

$\sqrt{4ac - b^2}\,\mathrm{i}$, so that the roots are $\dfrac{-b \pm \sqrt{4ac - b^2}\,\mathrm{i}}{2a}$, both complex numbers.

If the roots are complex numbers, then they have the form $x \pm y\,\mathrm{i}$ with the same real parts but opposite imaginary parts. Pairs of numbers like this are called **conjugate complex numbers**. If $x + y\,\mathrm{i}$ is written as z, then $x - y\,\mathrm{i}$ is denoted by z^* (which is read as 'z-star').

Complex numbers $z = x + y\,\mathrm{i}$, $z^* = x - y\,\mathrm{i}$ are conjugate complex numbers.

Their sum $z + z^* = 2x$ and product $zz^* = x^2 + y^2$ are real numbers, and their difference $z - z^* = 2y\,\mathrm{i}$ is an imaginary number.

If a quadratic equation with real coefficients has two complex roots, these roots are conjugate.

Conjugate complex numbers have some important properties. Suppose, for example, that $s = a + b\,\mathrm{i}$ and $t = c + d\,\mathrm{i}$ are two complex numbers, so that $s^* = a - b\,\mathrm{i}$ and $t^* = c - d\,\mathrm{i}$. Using the results in the last section and replacing b by $-b$ and d by $-d$, you get

$$s \pm t = (a \pm c) + (b \pm d)\,\mathrm{i} \qquad \text{and} \qquad s^* \pm t^* = (a \pm c) - (b \pm d)\,\mathrm{i};$$

$$st = (ac - bd) + (ad + bc)\,\mathrm{i} \qquad \text{and} \qquad s^* t^* = (ac - bd) - (ad + bc)\,\mathrm{i};$$

$$\frac{s}{t} = \frac{ac + bd + (bc - ad)\,\mathrm{i}}{c^2 + d^2} \qquad \text{and} \qquad \frac{s^*}{t^*} = \frac{ac + bd - (bc - ad)\,\mathrm{i}}{c^2 + d^2}.$$

You can see that the outcomes in each case are conjugate pairs. That is:

> If s and t are complex numbers, then
>
> $$(s \pm t)^* = s^* \pm t^*,$$
>
> $$(st)^* = s^* t^*,$$
>
> $$\left(\frac{s}{t}\right)^* = \frac{s^*}{t^*}.$$

If in the second of these rules you set both s and t equal to z, you get $(zz)^* = z^* z^*$, that is $\left(z^2\right)^* = (z^*)^2$. Then, setting $s = z^2$ and $t = z$, it follows that $\left(z^2 z\right)^* = \left(z^2\right)^* z^* = (z^*)^2 z^*$, that is $\left(z^3\right)^* = (z^*)^3$; and so on. The general result is:

> If z is a complex number, and n is a positive integer, then $\left(z^n\right)^* = (z^*)^n$.
>
> If a is a real number, then $\left(az^n\right)^* = a(z^*)^n$.

Now suppose that you have a polynomial of degree n,

$$\mathrm{p}(z) = a_n z^n + a_{n-1} z^{n-1} + \ldots + a_2 z^2 + a_1 z + a_0,$$

whose coefficients a_n, a_{n-1}, $\ldots$, a_2, a_1 and a_0 are all real numbers. Then $\mathrm{p}(z^*)$ is the sum of $n + 1$ terms of the form $a_r (z^*)^r$, which by the statement in the box you can write as $\left(a_r z^r\right)^*$. So each term is the conjugate of the corresponding term in $\mathrm{p}(z)$. Since the sum of conjugate numbers is the conjugate of the sum, it follows that $\mathrm{p}(z^*)$ is the conjugate of $\mathrm{p}(z)$.

It is only a short step from this last statement to an important result about equations of the form $\mathrm{p}(z) = 0$. Suppose that $z = s$ is a non-real root of this equation. Then $\mathrm{p}(s) = 0$, so $\mathrm{p}(s^*) = (\mathrm{p}(s))^* = 0^* = 0$, which means that $z = s^*$ is also a root of the equation.

You saw an example of this result in Example 16.3.1 for a quadratic equation. You can now see that this was a special case of a far more general result, for polynomial equations of any degree.

> If $p(z)$ is a polynomial with real coefficients, then $p(z*) = (p(z))*$.
>
> If s is a non-real root of the equation $p(z) = 0$, then $s*$ is also a root;
>
> that is, the non-real roots of the equation $p(z) = 0$ occur as conjugate pairs.

Example 16.3.2

Show that $(1+i)^4 = -4$. Hence find all the roots of the equation $z^4 + 4 = 0$.

$(1+i)^2 = 1 + 2i + i^2 = 1 + 2i - 1 = 2i$, so $(1+i)^4 = (2i)^2 = 4i^2 = -4$.

This shows that $1 + i$ is a root of the equation $z^4 + 4 = 0$, so another root must be $1 - i$. You can deduce that $z - 1 - i$ and $z - (1 - i) = z - 1 + i$ are both factors of $z^4 + 4$.

Now $(z - 1 - i)(z - 1 + i) \equiv (z - 1)^2 - i^2 \equiv z^2 - 2z + 2$. This means that $z^4 + 4$ must be the product of $z^2 - 2z + 2$ and another quadratic factor. By the usual method (see Section 14) you can find that

$$z^4 + 4 \equiv (z^2 - 2z + 2)(z^2 + 2z + 2).$$

The other two roots are therefore the roots of the quadratic equation

$z^2 + 2z + 2 = 0$, that is $z = \dfrac{-2 \pm \sqrt{4 - 4 \times 1 \times 2}}{2} = \dfrac{-2 \pm 2i}{2} = -1 \pm i$.

There are two conjugate pairs of roots: $1 + i$, $1 - i$, and $-1 + i$, $-1 - i$.

With hindsight, and a great deal of ingenuity, you might spot that if $z^4 + 4$ is written as $(z^4 + 4z^2 + 4) - 4z^2 = (z^2 + 2)^2 - (2z)^2$, then this is the difference of two squares, with factors $(z^2 + 2 - 2z)(z^2 + 2 + 2z)$.

Example 16.3.3

Solve the equation $z^5 - 6z^3 - 2z^2 + 17z - 10 = 0$.

Denote the left side by $p(z)$ and begin by trying to find some real factors, using the factor theorem.

Try $z = 1$: $p(1) = 0$, so $p(z) = (z - 1)q(z)$, where $q(z) = z^4 + z^3 - 5z^2 - 7z + 10$.

Try $z = 1$ again: $q(1) = 0$, so $q(z) = (z - 1)r(z)$, where $r(z) = z^3 + 2z^2 - 3z - 10$.

Try $z = 1$ again: $r(1) \neq 0$. So try $z = 2$: $r(2) = 0$, so $r(z) = (z - 2)(z^2 + 4z + 5)$.

Completing the square, $z^2 + 4z + 5 = (z + 2)^2 + 1 = (z + 2 - i)(z + 2 + i)$.

Thus $p(z) = (z - 1)^2(z - 2)(z + 2 - i)(z + 2 + i)$, and the roots are 1 (a repeated root, counted twice), 2 and the conjugate complex pair $-2 \pm i$.

Exercise 16B

1 If $p = 3 + 4i$, $q = 1 - i$ and $r = -2 + 3i$, solve the following equations for the complex number z.

    (a) $p + z = q$ (b) $2r + 3z = p$ (c) $qz = r$ (d) $pz + q = r$

2 Solve these pairs of simultaneous equations for the complex numbers z and w.

    (a) $(1 + i)z + (2 - i)w = 3 + 4i$ (b) $5z - (3 + i)w = 7 - i$

        $iz + (3 + i)w = -1 + 5i$ $(2 - i)z + 2iw = -1 + i$

3 Solve the following quadratic equations, giving answers in the form $a + bi$, where a and b are real numbers.

    (a) $z^2 + 9 = 0$ (b) $z^2 + 4z + 5 = 0$ (c) $z^2 - 6z + 25 = 0$ (d) $2z^2 + 2z + 13 = 0$

4 Write down the conjugates of

    (a) $1 + 7i$,   (b) $-2 + i$,   (c) 5,   (d) $3i$.

    For each of these complex numbers z find the values of

    (i) $z + z^*$,   (ii) $z - z^*$,   (iii) zz^*,   (iv) $\dfrac{z}{z^*}$.

5 Write the following polynomials as products of linear factors.

    (a) $z^2 + 25$ (b) $9z^2 - 6z + 5$ (c) $4z^2 + 12z + 13$ (d) $z^4 - 16$

    (e) $z^4 - 8z^2 - 9$ (f) $z^3 + z - 10$ (g) $z^3 - 3z^2 + z + 5$ (h) $z^4 - z^2 - 2z + 2$

6 Prove that $1 + i$ is a root of the equation $z^4 + 3z^2 - 6z + 10 = 0$. Find all the other roots.

7 Prove that $-2 + i$ is a root of the equation $z^4 + 24z + 55 = 0$. Find all the other roots.

8 Let $z = a + bi$, where a and b are real numbers. If $\dfrac{z}{z^*} = c + di$, where c and d are real, prove that $c^2 + d^2 = 1$.

9 Prove that, for any complex number z, $zz^* = (\operatorname{Re} z)^2 + (\operatorname{Im} z)^2$.

10 If $z = a + bi$, where a and b are real, use the binomial theorem to find the real and imaginary parts of z^5 and $(z^*)^5$.

16.4 Geometrical representations

There are two ways of thinking geometrically about positive and negative numbers: as translations of a line or as points on a line.

A business which loses $\$500$ in April and then gains $\$1200$ in May has over the two months a net gain of $\$700$. You could write this as $(-500) + (+1200) = (+700)$, and represent it by a diagram like Fig. 16.1. It makes no difference whether it started with its bank account in credit or with an overdraft; the diagram merely shows by how much the bank balance changes.

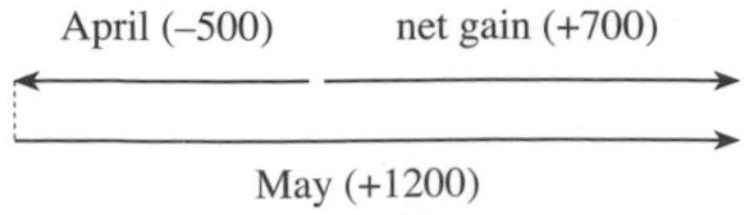

April (−500)   net gain (+700)

May (+1200)

Fig. 16.1

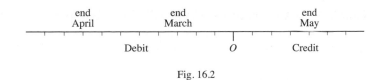

Fig. 16.2

If in fact the business had an overdraft of $300 at the beginning of April, there would be an overdraft of $800 at the end of April, and a credit balance of $400 at the end of May. You could represent the bank balance by a diagram like Fig. 16.2, in which each number is associated with a point on the line and overdrafts are treated as negative.

Similarly there are two ways of representing complex numbers, but now you need a plane rather than a line. The number $s = a + bi$ can be shown either as a translation of the plane, a units in the x-direction and b in the y-direction (see Fig. 16.3), or as the point S with coordinates (a, b) (Fig. 16.4).

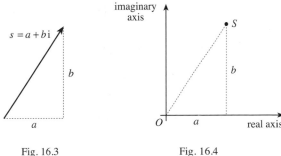

Fig. 16.3   Fig. 16.4

The second of these representations is called an **Argand diagram**, named after John-Robert Argand (1768–1822), a Parisian bookkeeper and mathematician. The axes are often called the **real axis** and the **imaginary axis**; these contain all the points representing real numbers and imaginary numbers respectively. Points representing conjugate pairs $a \pm bi$ are reflections of each other in the real axis.

Example 16.4.1
Show in an Argand diagram the roots of
(a) $z^4 + 4 = 0$, (b) $z^5 - 6z^3 - 2z^2 + 17z - 10 = 0$.

These are the equations in Examples 16.3.2 and 16.3.3, and the corresponding diagrams are Figs. 16.5 and 16.6. The symmetry of both diagrams about the real axis shows geometrically the property that the non-real roots occur in conjugate pairs.

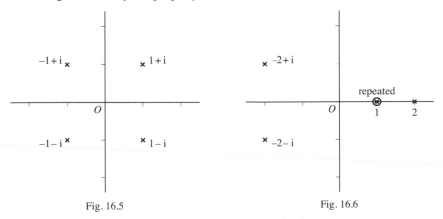

Fig. 16.5   Fig. 16.6

You have seen diagrams like Fig. 16.3 and Fig. 16.4 before. Complex numbers are represented just like vectors in two dimensions.

You can refresh your memory of adding vectors from P1 Sections 13.2 to 13.5. The equivalent operation with complex numbers is illustrated in Figs. 16.7 and 16.8.

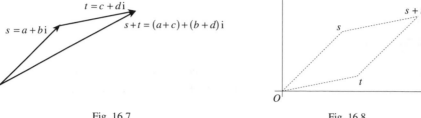

Fig. 16.7   Fig. 16.8

Complex numbers, as translations, are added by a triangle rule; in an Argand diagram you use a parallelogram rule.

For subtraction, it is enough to note that $z = t - s \iff s + z = t$. This is shown geometrically in Figs. 16.9 and 16.10.

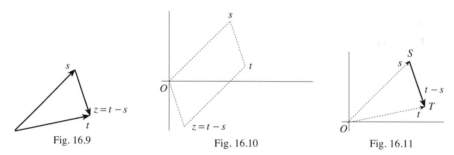

Fig. 16.9   Fig. 16.10   Fig. 16.11

You need to be able to switch between these two pictures of complex numbers. The link between them is the idea of a position vector. In an Argand diagram it doesn't matter whether you think of the complex number s as represented by the point S or by the translation $\overrightarrow{OS}$; and, as translations, $\overrightarrow{OS} + \overrightarrow{ST} = \overrightarrow{OT}$ (see Fig. 16.11). So the translation $\overrightarrow{ST}$ represents the complex number $t - s$, where S and T are the points representing the complex numbers s and t in an Argand diagram.

16.5 The modulus

In Fig. 16.3 the distance covered by the translation $s = a + b\mathrm{i}$ is $\sqrt{a^2 + b^2}$. In an Argand diagram (Fig. 16.4) this is the distance of the point S from O. This is called the **modulus** of s, and is denoted by $|s|$.

You have, of course, met this notation before for the modulus of a real number. But there is no danger of confusion; if s is the real number $a + 0\mathrm{i}$, then $|s|$ is $\sqrt{a^2 + 0^2} = \sqrt{a^2}$, which is $|a|$ as defined for the real number a (see Section 2.7). So the modulus of a complex number is just a generalisation of the modulus you have used previously.

But beware! If s is complex, then $|s|$ does *not* equal $\sqrt{s^2}$.

In fact, you have met the expression $a^2 + b^2$ already, as $(a + b\mathrm{i})(a - b\mathrm{i})$, or ss^*. So the correct generalisation of $|a| = \sqrt{a^2}$ is $|s| = \sqrt{ss^*}$. (Notice that, if a is real, then $a^* = a$, so that $a^2 = aa^*$. Thus the rule $|s| = \sqrt{ss^*}$ holds whether s is real or complex.)

> If s is a complex number,
>
> $$|s|^2 = ss^*.$$

You can use the modulus and an Argand diagram to link complex numbers with coordinate geometry. For example, in Fig. 16.11, if $s = a + b\,\mathrm{i}$ and $t = c + d\,\mathrm{i}$, then

$$\text{distance } ST = |t - s| = \sqrt{(t-s)(t-s)^*}$$
$$= \sqrt{((c-a) + (d-b)\,\mathrm{i})((c-a) - (d-b)\,\mathrm{i})}$$
$$= \sqrt{(c-a)^2 + (d-b)^2},$$

which is the familiar expression for the distance between the points (a,b) and (c,d).

Example 16.5.1

In an Argand diagram, points S and T represent 4 and $2\,\mathrm{i}$ respectively (see Fig. 16.12). Identify the points P such that $PS < PT$.

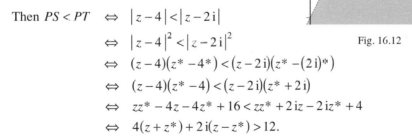

Fig. 16.12

Let P represent a complex number $z = x + y\,\mathrm{i}$. The lengths PS and PT are given by $PS = |z - 4|$ and $PT = |z - 2\,\mathrm{i}|$.

Then
$$PS < PT \iff |z - 4| < |z - 2\,\mathrm{i}|$$
$$\iff |z - 4|^2 < |z - 2\,\mathrm{i}|^2$$
$$\iff (z - 4)(z^* - 4^*) < (z - 2\,\mathrm{i})(z^* - (2\,\mathrm{i})^*)$$
$$\iff (z - 4)(z^* - 4) < (z - 2\,\mathrm{i})(z^* + 2\,\mathrm{i})$$
$$\iff zz^* - 4z - 4z^* + 16 < zz^* + 2\,\mathrm{i}z - 2\,\mathrm{i}z^* + 4$$
$$\iff 4(z + z^*) + 2\,\mathrm{i}(z - z^*) > 12.$$

You can put this into cartesian form by using the relations $z + z^* = 2x$ and $z - z^* = 2y\,\mathrm{i}$. Then

$$PS < PT \iff 8x + 2\,\mathrm{i}(2y\,\mathrm{i}) > 12$$
$$\iff 8x - 4y > 12 \iff 2x - y > 3.$$

The line $2x - y = 3$ is the perpendicular bisector of ST, which cuts the axes at $\left(\tfrac{3}{2}, 0\right)$ and $(0, -3)$; as complex numbers, these are the points $\tfrac{3}{2}$ and $-3\,\mathrm{i}$. You can check that these points are equidistant from the points 4 and $2\,\mathrm{i}$; that is, $\left|\tfrac{3}{2} - 4\right| = \left|\tfrac{3}{2} - 2\,\mathrm{i}\right|$ and $|-3\,\mathrm{i} - 4| = |-3\,\mathrm{i} - 2\,\mathrm{i}|$.

In examples like this, try to keep the algebra in complex number form as long as you can, before introducing x and y.

Example 16.5.2

In an Argand diagram the point S represents 3 (see Fig. 16.13). Show that the points P such that $OP = 2SP$ lie on a circle.

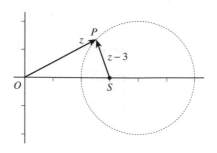

Fig. 16.13

If P represents z,

$$OP = 2SP \quad \Leftrightarrow \quad |z| = 2|z-3|$$
$$\Leftrightarrow \quad |z|^2 = 4|z-3|^2$$
$$\Leftrightarrow \quad zz^* = 4(z-3)(z^*-3^*)$$
$$\Leftrightarrow \quad zz^* = 4(z-3)(z^*-3)$$
$$\Leftrightarrow \quad zz^* = 4(zz^* - 3z - 3z^* + 9)$$
$$\Leftrightarrow \quad 3zz^* - 12z - 12z^* + 36 = 0$$
$$\Leftrightarrow \quad zz^* - 4z - 4z^* + 12 = 0.$$

You can now untangle this by a method similar to the completed square method for quadratics. Notice that

$$|z-4|^2 = (z-4)(z^*-4) = zz^* - 4z - 4z^* + 16,$$

so that

$$OP = 2SP \quad \Leftrightarrow \quad zz^* - 4z - 4z^* + 16 = 4$$
$$\Leftrightarrow \quad |z-4|^2 = 4 \quad \Leftrightarrow \quad |z-4| = 2.$$

The interpretation of this last equation is that the distance of P from 4 is equal to 2. That is, P lies on a circle with centre 4 and radius 2.

Other important properties of the modulus are:

$$|st| = |s||t|, \quad \left|\frac{s}{t}\right| = \frac{|s|}{|t|}, \quad |s+t| \leqslant |s| + |t|, \quad |s-t| \geqslant |s| - |t|.$$

The first two of these are easy to prove algebraically:

$$|st|^2 = (st)(st)^* = (st)(s^*t^*) = (ss^*)(tt^*) = |s|^2|t|^2,$$

$$\left|\frac{s}{t}\right|^2 = \left(\frac{s}{t}\right)\left(\frac{s}{t}\right)^* = \left(\frac{s}{t}\right)\left(\frac{s^*}{t^*}\right) = \frac{ss^*}{tt^*} = \frac{|s|^2}{|t|^2},$$

and the results follow by taking the square roots.

Figs. 16.7 and 16.9 show that the inequalities for the sum and difference are equivalent to the geometrical theorem that the sum of two sides of a triangle is greater than the third side. They are called **triangle inequalities**, but are not so easy to prove algebraically. (See Exercise 16C Questions 10 and 12.)

Example 16.5.3

S and T are the points representing -4 and 4 in an Argand diagram. A point P moves so that $SP + TP = 10$ (see Fig. 16.14). Prove that $3 \leqslant OP \leqslant 5$.

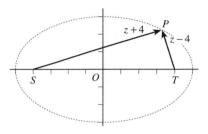

If P represents z, then
$SP + TP = |z + 4| + |z - 4|$, so that z satisfies
$|z + 4| + |z - 4| = 10$.

Then

Fig. 16.14

$$OP = |z| = \left| \tfrac{1}{2}\big((z + 4) + (z - 4)\big) \right|$$
$$\leqslant \tfrac{1}{2}\big(|z + 4| + |z - 4|\big) \qquad (\text{using } |s + t| \leqslant |s| + |t|)$$
$$= \tfrac{1}{2} \times 10 = 5.$$

Also, $\big(|z + 4| + |z - 4|\big)^2 = 100$ and $\big(|z + 4| - |z - 4|\big)^2 \geqslant 0$; adding these gives

$$\big(|z + 4| + |z - 4|\big)^2 + \big(|z + 4| - |z - 4|\big)^2 \geqslant 100,$$
$$2|z + 4|^2 + 2|z - 4|^2 \geqslant 100,$$
$$(z + 4)(z^* + 4^*) + (z - 4)(z^* - 4^*) \geqslant 50,$$
$$2zz^* + 32 \geqslant 50, \quad \text{which is} \quad |z|^2 \geqslant 9; \quad \text{that is,} \quad OP \geqslant 3.$$

You probably know that the property $SP + TP = 10$ defines an ellipse. OP takes its greatest value of 5 when S, T and P are in a straight line, at the ends of the major axis; and it takes its smallest value of 3 when $|z + 4| = |z - 4|$, that is when $SP = TP$, at the ends of the minor axis.

Exercise 16C

1 Draw   (i) vector diagrams,   (ii) Argand diagrams to represent the following relationships.

(a) $(3 + i) + (-1 + 2i) = (2 + 3i)$ (b) $(1 + 4i) - (3i) = (1 + i)$

2 Draw Argand diagrams to illustrate the following properties of complex numbers.

(a) $z + z^* = 2\operatorname{Re}z$ (b) $z - z^* = 2i\operatorname{Im}z$ (c) $(s + t)^* = s^* + t^*$

(d) $\operatorname{Re}z \leqslant |z| \leqslant \operatorname{Re}z + \operatorname{Im}z$ (e) $(kz)^* = kz^*$, where k is a real number

3 Draw Argand diagrams showing the roots of the following equations.

(a) $z^4 - 1 = 0$ (b) $z^3 + 1 = 0$ (c) $z^3 + 6z + 20 = 0$

(d) $z^4 + 4z^3 + 4z^2 - 9 = 0$ (e) $z^4 + z^3 + 5z^2 + 4z + 4 = 0$

4 Represent the roots of the equation $z^4 - z^3 + z - 1 = 0$ in an Argand diagram, and show that they all have the same modulus.

5 Identify in an Argand diagram the points corresponding to the following equations.

(a) $|z| = 5$ (b) $\operatorname{Re} z = 3$ (c) $z + z^* = 6$

(d) $z - z^* = 2i$ (e) $|z - 2| = 2$ (f) $|z - 4| = |z|$

(g) $|z + 2i| = |z + 4|$ (h) $|z + 4| = 3|z|$ (i) $1 + \operatorname{Re} z = |z - 1|$

6 Identify in an Argand diagram the points corresponding to the following inequalities.

(a) $|z| > 2$ (b) $|z - 3i| \leqslant 1$ (c) $|z + 1| \leqslant |z - i|$ (d) $|z - 3| > 2|z|$

7 P is a point in an Argand diagram corresponding to a complex number z, and $|z - 5| + |z + 5| = 26$. Prove that $12 \leqslant OP \leqslant 13$, and illustrate this result.

8 P is a point in an Argand diagram corresponding to a complex number z which satisfies the equation $|4 + z| - |4 - z| = 6$. Prove that $|4 + z|^2 - |4 - z|^2 \geqslant 48$, and deduce that $\operatorname{Re} z \geqslant 3$. Draw a diagram to illustrate this result.

9 Show that, if z is a complex number and n is a positive integer, $|z^n| = |z|^n$.

Hence show that all the roots of the equation $z^n + a^n = 0$, where a is a positive real number, have modulus a.

10 Assuming that the inequality $|s + t| \leqslant |s| + |t|$ holds for any two complex numbers s and t, prove that $|z - w| \geqslant |z| - |w|$ holds for any two complex numbers z and w.

11 If s and t are two complex numbers, prove the following.

(a) $|s + t|^2 = |s|^2 + |t|^2 + (st^* + s^* t)$ (b) $|s - t|^2 = |s|^2 + |t|^2 - (st^* + s^* t)$

(c) $|s + t|^2 + |s - t|^2 = 2|s|^2 + 2|t|^2$

Interpret the result of part (c) as a geometrical theorem.

12 If s and t are two complex numbers, prove the following.

(a) $(s^* t)^* = s t^*$.

(b) $st^* + s^* t$ is a real number, and $st^* - s^* t$ is an imaginary number.

(c) $(st^* + s^* t)^2 - (st^* - s^* t)^2 = (2|st|)^2$.

(d) $(|s| + |t|)^2 - |s + t|^2 = 2|st| - (st^* + s^* t)$.

Use these results to deduce that $|s + t| \leqslant |s| + |t|$.

16.6 Equations with complex coefficients

Complex numbers were introduced so that all real numbers should have square roots. But do complex numbers themselves have square roots?

Example 16.6.1

Find the square roots of   (a) $8i$,   (b) $3 - 4i$.

The problem is to find real numbers a and b such that $(a + bi)^2$ is equal to (a) $8i$ and (b) $3 - 4i$. To do this, note that $(a + bi)^2 = (a^2 - b^2) + 2abi$ and remember that, for two complex numbers to be equal, their real and imaginary parts must be equal.

(a) If $(a+b\,\mathrm{i})^2 = 8\,\mathrm{i}$, then $a^2 - b^2 = 0$ and $2ab = 8$. Therefore $b = \dfrac{4}{a}$, and so $a^2 - \left(\dfrac{4}{a}\right)^2 = 0$, or $a^4 = 16$. Since a is real, this implies that $a^2 = 4$, $a = \pm 2$.

If $a = 2$, $b = \dfrac{4}{a} = 2$; if $a = -2$, $b = -2$. So $8\,\mathrm{i}$ has two square roots, $2 + 2\,\mathrm{i}$ and $-2 - 2\,\mathrm{i}$.

(b) If $(a+b\,\mathrm{i})^2 = 3 - 4\,\mathrm{i}$, then $a^2 - b^2 = 3$ and $2ab = -4$. Therefore $b = -\dfrac{2}{a}$, and so $a^2 - \left(-\dfrac{2}{a}\right)^2 = 3$, or $a^4 - 3a^2 - 4 = 0$.

This is a quadratic equation in a^2, which can be solved by factorising the left side to give $\left(a^2 - 4\right)\left(a^2 + 1\right) = 0$. Since a is real, a^2 cannot equal -1; but for $a^2 = 4$, $a = \pm 2$.

If $a = 2$, $b = -\dfrac{2}{a} = -1$; if $a = -2$, $b = 1$. So $3 - 4\,\mathrm{i}$ has two square roots, $2 - \mathrm{i}$ and $-2 + \mathrm{i}$.

You can use this method to find the square root of any complex number. The equation for a^2 will always have two roots, one of which is positive, leading to two values of a, and hence two square roots.

You will see also that, if one root is $a + b\,\mathrm{i}$ the other is $-a - b\,\mathrm{i}$, so that the roots can be written as $\pm(a + b\,\mathrm{i})$. But of course you cannot say that one of the roots is 'positive' and the other 'negative', because these words have no meaning for complex numbers.

In Section 16.3 you saw that complex numbers make it possible to solve any quadratic equation with real coefficients. Now that you can find square roots of complex numbers, you can solve any quadratic equation even if the coefficients are complex numbers.

Example 16.6.2
Solve the quadratic equation $(2 - \mathrm{i})z^2 + (4 + 3\,\mathrm{i})z + (-1 + 3\,\mathrm{i}) = 0$.

Method 1 Using the quadratic formula with $a = 2 - \mathrm{i}$, $b = 4 + 3\,\mathrm{i}$ and $c = -1 + 3\,\mathrm{i}$ gives

$$z = \frac{-(4 + 3\,\mathrm{i}) \pm \sqrt{(4 + 3\,\mathrm{i})^2 - 4 \times (2 - \mathrm{i}) \times (-1 + 3\,\mathrm{i})}}{2(2 - \mathrm{i})}$$

$$= \frac{-(4 + 3\,\mathrm{i}) \pm \sqrt{7 + 24\,\mathrm{i} - 4 \times (1 + 7\,\mathrm{i})}}{2(2 - \mathrm{i})}$$

$$= \frac{-(4 + 3\,\mathrm{i}) \pm \sqrt{3 - 4\,\mathrm{i}}}{2(2 - \mathrm{i})}$$

$$= \frac{-(4 + 3\,\mathrm{i}) \pm (2 - \mathrm{i})}{2(2 - \mathrm{i})} \qquad \text{(using Example 16.6.1(b))}$$

$$= \frac{-2-4\,\mathrm{i}}{2(2-\mathrm{i})} \quad \text{or} \quad \frac{-6-2\,\mathrm{i}}{2(2-\mathrm{i})}$$

$$= \frac{-1-2\,\mathrm{i}}{2-\mathrm{i}} \quad \text{or} \quad \frac{-3-\mathrm{i}}{2-\mathrm{i}}$$

$$= \frac{(-1-2\,\mathrm{i})(2+\mathrm{i})}{(2-\mathrm{i})(2+\mathrm{i})} \quad \text{or} \quad \frac{(-3-\mathrm{i})(2+\mathrm{i})}{(2-\mathrm{i})(2+\mathrm{i})}$$

$$= \frac{-5\,\mathrm{i}}{5} \quad \text{or} \quad \frac{-5-5\,\mathrm{i}}{5}$$

$$= -\mathrm{i} \quad \text{or} \quad -1-\mathrm{i}.$$

Method 2   You can often reduce the work by first making the coefficient of z^2 real. Multiplying through by $(2-\mathrm{i})* = 2+\mathrm{i}$, the equation becomes

$$(2+\mathrm{i})(2-\mathrm{i})z^2 + (2+\mathrm{i})(4+3\,\mathrm{i})z + (2+\mathrm{i})(-1+3\,\mathrm{i}) = 0,$$
$$5z^2 + (5+10\,\mathrm{i})z + (-5+5\,\mathrm{i}) = 0,$$
$$z^2 + (1+2\,\mathrm{i})z + (-1+\mathrm{i}) = 0.$$

The quadratic formula with $a = 1$, $b = 1+2\,\mathrm{i}$ and $c = -1+\mathrm{i}$ then gives

$$z = \frac{-(1+2\,\mathrm{i}) \pm \sqrt{(1+2\,\mathrm{i})^2 - 4 \times 1 \times (-1+\mathrm{i})}}{2}$$

$$= \frac{-(1+2\,\mathrm{i}) \pm \sqrt{-3+4\,\mathrm{i} - 4(-1+\mathrm{i})}}{2}$$

$$= \frac{-(1+2\,\mathrm{i}) \pm \sqrt{1}}{2} = \frac{-(1+2\,\mathrm{i}) \pm 1}{2}$$

$$= \frac{-2\,\mathrm{i}}{2} \quad \text{or} \quad \frac{-2-2\,\mathrm{i}}{2}$$

$$= -\mathrm{i} \quad \text{or} \quad -1-\mathrm{i}.$$

It is important to notice that although the quadratic equation has two roots, these are not now conjugate complex numbers. The property in Section 16.3 that the roots occur in conjugate pairs holds only for equations whose coefficients are real.

The fact that, with complex numbers, every quadratic equation has two roots is a particular case of a more general result:

> Every polynomial equation of degree n has n roots.

You need to understand that for this to be true, repeated roots have to count more than once. If the polynomial $\mathrm{p}(z)$ has a factor $(z-s)^k$ with $k > 1$, then in the equation $\mathrm{p}(z) = 0$ the root $z = s$ has to count as k roots. For example, the equation of degree 5 in Example 16.3.3 has only 4 different roots (1, 2, $-2+\mathrm{i}$ and $-2-\mathrm{i}$) but the repeated root 1 counts twice because $(z-1)^2$ is a factor of $\mathrm{p}(z)$.

This remarkable result is one of the main reasons that complex numbers are important. Unfortunately the proof is too difficult to give here.

Exercise 16D

1 Find the square roots of

    (a) $-2i$,   (b) $-3+4i$,   (c) $5+12i$,   (d) $8-6i$.

2 Solve the following quadratic equations.

    (a) $z^2+z+(1-i)=0$ (b) $z^2+(1-i)z+(-6+2i)=0$

    (c) $z^2+4z+(4+2i)=0$ (d) $(1+i)z^2+2iz+4i=0$

    (e) $(2-i)z^2+(3+i)z-5=0$

3 Find the fourth roots of

    (a) -64,   (b) $7+24i$.

    Show your answers on an Argand diagram.

4 If $(x+yi)^3=8i$, where x and y are real numbers, prove that either $x=0$ or $x=\pm\sqrt{3}\,y$. Hence find all the cube roots of $8i$. Show your answers on an Argand diagram.

5 If $(x+yi)^3=2-2i$, where x and y are real numbers, prove that

$$x\left(x^2-3y^2\right)=y\left(y^2-3x^2\right)=2.$$

    Show that these equations have one solution in which $x=y$, and hence find one cube root of $2-2i$.

    Find the quadratic equation satisfied by the other cube roots of $2-2i$, and solve it.

    Show all the roots on an Argand diagram.

Miscellaneous exercise 16

1 Given that z is a complex number such that $z+3z^*=12+8i$, find z.   (OCR)

2 Given that $3i$ is a root of the equation $3z^3-5z^2+27z-45=0$, find the other two roots.   (OCR)

3 Two of the roots of a cubic equation, in which all the coefficients are real, are 2 and $1+3i$. State the third root and find the cubic equation.   (OCR)

4 It is given that $3-i$ is a root of the quadratic equation $z^2-(a+bi)z+4(1+3i)=0$, where a and b are real. In either order,

    (a) find the values of a and b,

    (b) find the other root of the quadratic equation, given that it is of the form ki, where k is real.   (OCR)

5 Find the roots of the equation $z^2=21-20i$.   (OCR)

6 Verify that $(3-2\mathrm{i})^2 = 5-12\mathrm{i}$. Find the two roots of the equation $(z-\mathrm{i})^2 = 5-12\mathrm{i}$.

(OCR)

7 You are given the complex number $w = 1-\mathrm{i}$. Express w^2, w^3 and w^4 in the form $a+b\mathrm{i}$.

(a) Given that $w^4 + 3w^3 + pw^2 + qw + 8 = 0$, where p and q are real numbers, find the values of p and q.

(b) Write down two roots of the equation $z^4 + 3z^3 + pz^2 + qz + 8 = 0$, where p and q are the real numbers found in part (a), and hence solve the equation completely.

(MEI, adapted)

8 Two complex numbers, z and w, satisfy the inequalities $|z-3-2\mathrm{i}| \leqslant 2$ and $|w-7-5\mathrm{i}| \leqslant 1$. By drawing an Argand diagram, find the least possible value of $|z-w|$.

(OCR)

9 A sequence of complex numbers z_1, z_2, z_3, ... is defined by $z_1 = 1-2\mathrm{i}$ and

$$z_{n+1} = \frac{z_n^2 + 2z_n + 5n^2}{2n}$$ for $n \geqslant 1$. Show that $z_2 = 2z_1$ and find similar expressions for z_3

and z_4 in terms of z_1. Suggest a conjecture for the value of z_n, and test whether your conjecture is correct when $n = 5$.

(MEI, adapted)

10 (a) The complex number z satisfies the equation $\left(z + \dfrac{2\mathrm{i}}{k}\right)\left(\dfrac{1}{z} - \dfrac{2\mathrm{i}}{k}\right) = 1$, where k is a

positive real number. Obtain a quadratic equation for z, and show that its solution can be expressed in the form $\mathrm{i}kz = a \pm \sqrt{bk^2 + ck + d}$ for suitable real numbers a, b, c, d. Show that z is purely imaginary when $k \leqslant 1$.

(b) A second complex number α is defined in terms of z by $\alpha = 1 + \dfrac{2\mathrm{i}}{kz}$. What can be

said about α when $k \leqslant 1$? Show that $|\alpha| = 1$ when $k \geqslant 1$.

(c) A third complex number β is defined by $\dfrac{1}{\beta} = 1 - \dfrac{\mathrm{i}}{k}$. By finding the real and imaginary

parts of $\beta - \frac{1}{2}k\mathrm{i}$, show that β lies on a circle with centre $\frac{1}{2}k\mathrm{i}$ and radius $\frac{1}{2}k$. (OCR)

11 Solve Example 15.6.1 on page 212 by writing the denominator as $(x+2)(x+2\mathrm{i})(x-2\mathrm{i})$, using the substitution method for partial fractions with simple denominators, and finally combining the two complex fractions.

17 Complex numbers in polar form

New insights about complex numbers come by expressing them in polar coordinate form. When you have completed this chapter, you should

- know the meaning of the argument of a complex number
- be able to multiply and divide complex numbers in modulus-argument form
- know how to represent multiplication and division geometrically
- be able to solve geometrical problems involving angles using complex numbers
- be able to write square roots in modulus-argument form
- know that complex numbers can be written as exponentials.

17.1 Modulus and argument

If s and t are complex numbers, the sum $s+t$ can be shown geometrically by the triangle rule for adding translations, or by the parallelogram rule in an Argand diagram. But you do not yet have a way of representing multiplication.

To do this, the clue is to describe points in the Argand diagram in a new way. Instead of locating a point P by its cartesian coordinates x and y, you can fix its position by giving the distance OP and the angle which the vector $\overrightarrow{OP}$ makes with the x-axis (measured in radians). These quantities are denoted by r and θ, as shown in Fig. 17.1. They are known as the **polar coordinates** of the point P.

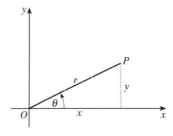

Fig. 17.1

If P represents the complex number z in the Argand diagram, then r is the same as $|z|$. The angle θ is called the **argument** of z; it is denoted by $\arg z$. To make the definition precise, the argument is chosen so that $-\pi < \theta \leqslant \pi$. (The argument is not defined if $z = 0$.)

You can see from Fig. 17.1 that $x = r\cos\theta$ and $y = r\sin\theta$. These equations still hold if θ is obtuse or negative, because of the way in which the definitions of cosine and sine were extended in P1 Chapter 10. So you can write

$$z = x + y\,i = r\cos\theta + (r\sin\theta)\,i = r(\cos\theta + i\sin\theta).$$

> The complex number $z\,(\neq 0)$ can be written in **modulus-argument form**
> as $z = r(\cos\theta + i\sin\theta)$, where $r = |z| > 0$ is the modulus and $\theta = \arg z$,
> with $-\pi < \theta \leqslant \pi$, is the argument.

Example 17.1.1

Write in modulus-argument form

(a) i,   (b) -2,   (c) $-2+i$,   (d) $-1-i$.

The points are plotted in the Argand
diagram in Fig. 17.2. Comparing this
with Fig. 17.1, you can see that the
values of r and θ for the four points are:

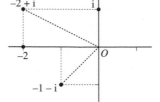

(a) $r = 1$, $\theta = \frac{1}{2}\pi$;

(b) $r = 2$, $\theta = \pi$;

(c) $r = \sqrt{5}$, $\theta = \pi - \tan^{-1}\frac{1}{2} = 2.677\ldots$;

Fig. 17.2

(d) $r = \sqrt{2}$, $\theta = -\frac{3}{4}\pi$.

So, in modulus-argument form, the complex numbers are:

(a) $1\left(\cos\frac{1}{2}\pi + i\sin\frac{1}{2}\pi\right)$;

(b) $2\left(\cos\pi + i\sin\pi\right)$;

(c) $\sqrt{5}\left(\cos 2.677\ldots + i\sin 2.677\ldots\right)$;

(d) $\sqrt{2}\left(\cos\left(-\frac{3}{4}\pi\right) + i\sin\left(-\frac{3}{4}\pi\right)\right)$.

Example 17.1.2

If $\arg z = \frac{1}{4}\pi$ and $\arg(z-3) = \frac{1}{2}\pi$, find $\arg(z-6i)$.

This is best done with an Argand diagram (Fig. 17.3). Since
$\arg z = \frac{1}{4}\pi$, the point z lies on the half-line u starting at O
at an angle $\frac{1}{4}\pi$ to the real axis. (Points on the other half of
the line, in the third quadrant, have $\arg z = -\frac{3}{4}\pi$.)

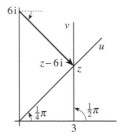

As $\arg(z-3) = \frac{1}{2}\pi$, the translation from 3 to z makes an
angle $\frac{1}{2}\pi$ with the real axis, so the point z lies on the half-
line v in the direction of the imaginary axis.

Fig. 17.3

These two half-lines meet at $z = 3+3i$, so $\arg(z-6i) = \arg(3-3i)$. The
translation $3-3i$ is at an angle $\frac{1}{4}\pi$ with the real axis in the clockwise sense, so
that $\arg(z-6i) = -\frac{1}{4}\pi$.

Exercise 17A

1   Show these numbers on an Argand diagram, and write them in the form $a+bi$. Where
appropriate leave surds in your answers, or give answers correct to 2 decimal places.

(a) $2\left(\cos\frac{1}{3}\pi + i\sin\frac{1}{3}\pi\right)$ (b) $10\left(\cos\frac{3}{4}\pi + i\sin\frac{3}{4}\pi\right)$ (c) $5\left(\cos\left(-\frac{1}{2}\pi\right) + i\sin\left(-\frac{1}{2}\pi\right)\right)$

(d) $3(\cos\pi + i\sin\pi)$ (e) $10(\cos 2 + i\sin 2)$ (f) $\cos(-3) + i\sin(-3)$

2 Write these complex numbers in modulus-argument form. Where appropriate express the argument as a rational multiple of π, otherwise give the modulus and argument correct to 2 decimal places.

(a) $1+2i$ (b) $3-4i$ (c) $-5+6i$ (d) $-7-8i$

(e) 1 (f) $2i$ (g) -3 (h) $-4i$

(i) $\sqrt{2}-\sqrt{2}i$ (j) $-1+\sqrt{3}i$

3 Show in an Argand diagram the sets of points satisfying the following equations.

(a) $\arg z = \frac{1}{5}\pi$ (b) $\arg z = -\frac{2}{3}\pi$ (c) $\arg z = \pi$

(d) $\arg(z-2) = \frac{1}{2}\pi$ (e) $\arg(2z-1) = 0$ (f) $\arg(z+i) = \pi$

(g) $\arg(z-1-2i) = \frac{3}{4}\pi$ (h) $\arg(z+1-i) = -\frac{2}{5}\pi$

4 Show in an Argand diagram the sets of points satisfying the following inequalities. Use a solid line to show boundary points which are included, and a dotted line for boundary points which are not included.

(a) $0 < \arg z \leqslant \frac{1}{6}\pi$ (b) $\frac{1}{2}\pi \leqslant \arg z \leqslant \pi$

(c) $\frac{1}{3}\pi < \arg(z-1) \leqslant \frac{2}{3}\pi$ (d) $-\frac{1}{4}\pi < \arg(z+1-1) < \frac{1}{4}\pi$

5 Use an Argand diagram to find, in the form $a+bi$, the complex numbers which satisfy the following pairs of equations.

(a) $\arg(z+2) = \frac{1}{2}\pi, \arg z = \frac{2}{3}\pi$ (b) $\arg(z+1) = \frac{1}{4}\pi, \arg(z-3) = \frac{3}{4}\pi$

(c) $\arg(z-3) = -\frac{3}{4}\pi, \arg(z+3) = -\frac{1}{2}\pi$ (d) $\arg(z+2i) = \frac{1}{6}\pi, \arg(z-2i) = -\frac{1}{3}\pi$

(e) $\arg(z-2-3i) = -\frac{5}{6}\pi, \arg(z-2+i) = \frac{5}{6}\pi$ (f) $\arg z = \frac{7}{12}\pi, \arg(z-2-2i) = \frac{11}{12}\pi$

6 Use an Argand diagram to find, in the form $a+bi$, the complex number(s) satisfying the following pairs of equations.

(a) $\arg z = \frac{1}{6}\pi, |z| = 2$ (b) $\arg(z-3) = \frac{1}{2}\pi, |z| = 5$

(c) $\arg(z-4i) = \pi, |z+6| = 5$ (d) $\arg(z-2) = \frac{3}{4}\pi, |z+2| = 3$

7 If $\arg\left(z-\frac{1}{2}\right) = \frac{1}{5}\pi$, what is $\arg(2z-1)$?

8 If $\arg\left(\frac{1}{3}-z\right) = \frac{1}{6}\pi$, what is $\arg(3z-1)$?

9 If $\arg(z-1) = \frac{1}{3}\pi$ and $\arg(z-i) = \frac{1}{6}\pi$, what is $\arg z$?

10 If $\arg(z+1) = \frac{1}{6}\pi$ and $\arg(z-1) = \frac{2}{3}\pi$, what is $\arg z$?

11 If $\arg(z+i) = 0$ and $\arg(z-i) = -\frac{1}{4}\pi$, what is $|z|$?

12 If $\arg(z-2) = \frac{2}{3}\pi$ and $|z| = 2$, what is $\arg z$?

17.2 Multiplication and division

Suppose that s has modulus p and argument α, and t has modulus q and argument β. Then

$$st = p(\cos\alpha + i\sin\alpha) \times q(\cos\beta + i\sin\beta)$$
$$= pq(\cos\alpha\cos\beta + i\sin\alpha\cos\beta + i\cos\alpha\sin\beta + i^2\sin\alpha\sin\beta)$$
$$= pq((\cos\alpha\cos\beta - \sin\alpha\sin\beta) + i(\sin\alpha\cos\beta + \cos\alpha\sin\beta)).$$

The expressions inside the brackets are $\cos(\alpha+\beta)$ and $\sin(\alpha+\beta)$ (see Section 5.3). Therefore

$$st = pq(\cos(\alpha+\beta) + i\sin(\alpha+\beta)).$$

Therefore pq (which is positive, since $p>0$ and $q>0$) is the modulus of st. It may also be true that $\alpha+\beta$ is the argument of st; but if addition takes $\alpha+\beta$ outside the interval $-\pi < \theta \leqslant \pi$, then you must adjust $\alpha+\beta$ by 2π to bring it inside the interval.

Example 17.2.1

Show $s = -\sqrt{3} + i$ and $t = \sqrt{2} + \sqrt{2}\,i$ as points in an Argand diagram. Find st and $\dfrac{s}{t}$ in modulus-argument form, and put them into the diagram.

You will recognise in Fig. 17.4 triangles with angles of $\frac{1}{6}\pi$ and $\frac{1}{4}\pi$, so that

s has modulus $p = 2$ and argument $\alpha = \frac{5}{6}\pi$,

t has modulus $q = 2$ and argument $\beta = \frac{1}{4}\pi$.

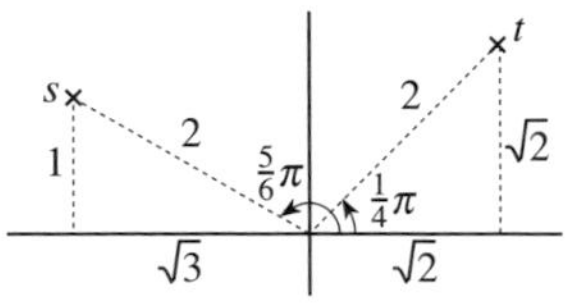

It follows that st has modulus $pq = 4$.

Fig. 17.4

Since $\alpha + \beta = \frac{5}{6}\pi + \frac{1}{4}\pi = \frac{13}{12}\pi > \pi$, the argument of st must be adjusted to $\frac{13}{12}\pi - 2\pi = -\frac{11}{12}\pi$.

Let $w = \dfrac{s}{t}$ have modulus m and argument γ. Since $tw = s$, equating moduli gives $qm = p$, so that $m = \dfrac{p}{q} = \dfrac{2}{2} = 1$. The argument of tw is $\beta + \gamma = \frac{1}{4}\pi + \gamma$, which (adjusted by 2π if necessary) must equal $\frac{5}{6}\pi$. In this case no adjustment is needed, since $\frac{5}{6}\pi - \frac{1}{4}\pi = \frac{7}{12}\pi$ lies inside the interval $-\pi < \theta \leqslant \pi$.

So $\quad st = 4\left(\cos\left(-\frac{11}{12}\pi\right) + i\sin\left(-\frac{11}{12}\pi\right)\right)$ and $\dfrac{s}{t} = \cos\frac{7}{12}\pi + i\sin\frac{7}{12}\pi$.

The corresponding points are shown on an Argand diagram in Fig. 17.5.

You will notice that, since complex numbers are multiplied by multiplying the moduli and adding the arguments, they are divided by dividing the moduli and subtracting the arguments.

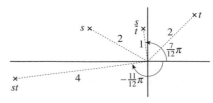

Fig. 17.5

The rules for multiplication and division in modulus-argument form are

$$|st| = |s||t|, \quad \arg(st) = \arg s + \arg t + k(2\pi),$$

$$\left|\frac{s}{t}\right| = \frac{|s|}{|t|}, \quad \arg\left(\frac{s}{t}\right) = \arg s - \arg t + k(2\pi),$$

where in each case the number k $(= -1, 0$ or $1)$ is chosen to ensure that the argument lies in the interval $-\pi < \theta \leqslant \pi$.

It would be a good idea to get used to these rules by working through some of the questions at the beginning of Exercise 17B before going on to the next example.

Example 17.2.2

If z has modulus 1 and argument θ, where $0 < \theta < \pi$, find the modulus and argument

of   (a) $z + 1$,   (b) $z - 1$,   (c) $\dfrac{z-1}{z+1}$.

Since $z = \cos\theta + i\sin\theta$,

$$z + 1 = \cos\theta + i\sin\theta + 1 = (1 + \cos\theta) + i\sin\theta$$

and $z - 1 = \cos\theta + i\sin\theta - 1 = -(1 - \cos\theta) + i\sin\theta$.

These expressions can be simplified by using the identities $1 + \cos 2A \equiv 2\cos^2 A$ and $1 - \cos 2A \equiv 2\sin^2 A$ (see Section 5.5) with θ in place of $2A$, that is $A = \frac{1}{2}\theta$. You will then also need to express $\sin\theta$ in terms of $\frac{1}{2}\theta$, using $\sin 2A \equiv 2\sin A \cos A$.

(a) $z + 1 = (1 + \cos\theta) + i\sin\theta = 2\cos^2\frac{1}{2}\theta + 2i\sin\frac{1}{2}\theta\cos\frac{1}{2}\theta$

$$= 2\cos\frac{1}{2}\theta\left(\cos\frac{1}{2}\theta + i\sin\frac{1}{2}\theta\right).$$

This suggests that $|z + 1| = 2\cos\frac{1}{2}\theta$ and $\arg(z + 1) = \frac{1}{2}\theta$.

To be sure, you need to check that $2\cos\frac{1}{2}\theta > 0$ and that $-\pi < \frac{1}{2}\theta \leqslant \pi$. Since you are given that $0 < \theta < \pi$, so that $0 < \frac{1}{2}\theta < \frac{1}{2}\pi$, these inequalities are satisfied.

(b) $z - 1 = -(1 - \cos\theta) + i\sin\theta = -2\sin^2\frac{1}{2}\theta + 2i\sin\frac{1}{2}\theta\cos\frac{1}{2}\theta$

$$= 2\sin\frac{1}{2}\theta\left(-\sin\frac{1}{2}\theta + i\cos\frac{1}{2}\theta\right).$$

To get this into the standard modulus-argument form, notice that $-\sin\frac{1}{2}\theta + i\cos\frac{1}{2}\theta$ can be written as $\left(\cos\frac{1}{2}\theta + i\sin\frac{1}{2}\theta\right) \times i$.

Since i has modulus 1 and argument $\frac{1}{2}\pi$,

$$\left(\cos\tfrac{1}{2}\theta + i\sin\tfrac{1}{2}\theta\right)\times i = \left(\cos\tfrac{1}{2}\theta + i\sin\tfrac{1}{2}\theta\right)\!\left(\cos\tfrac{1}{2}\pi + i\sin\tfrac{1}{2}\pi\right)$$

$$= \left(\cos\!\left(\tfrac{1}{2}\theta + \tfrac{1}{2}\pi\right) + i\sin\!\left(\tfrac{1}{2}\theta + \tfrac{1}{2}\pi\right)\right),$$

so

$$z - 1 = 2\sin\tfrac{1}{2}\theta\!\left(\cos\!\left(\tfrac{1}{2}\theta + \tfrac{1}{2}\pi\right) + i\sin\!\left(\tfrac{1}{2}\theta + \tfrac{1}{2}\pi\right)\right).$$

You again need to check that $2\sin\tfrac{1}{2}\theta > 0$ and that $-\pi < \tfrac{1}{2}\theta + \tfrac{1}{2}\pi \leqslant \pi$. Since $0 < \theta < \pi$, these inequalities are certainly satisfied. It follows that

$$|z - 1| = 2\sin\tfrac{1}{2}\theta \quad \text{and} \quad \arg(z - 1) = \tfrac{1}{2}\pi + \tfrac{1}{2}\theta.$$

(c) The rules for division give

$$\left|\frac{z-1}{z+1}\right| = \frac{|z-1|}{|z+1|} = \frac{2\sin\tfrac{1}{2}\theta}{2\cos\tfrac{1}{2}\theta} = \tan\tfrac{1}{2}\theta$$

and $\quad \arg\!\left(\dfrac{z-1}{z+1}\right) = \arg(z-1) - \arg(z+1) = \left(\tfrac{1}{2}\theta + \tfrac{1}{2}\pi\right) - \tfrac{1}{2}\theta = \tfrac{1}{2}\pi.$

This is shown in an Argand diagram in Fig. 17.6. Since the modulus of z is 1 and $0 < \theta < \pi$, the point z lies on the upper half of the unit circle, with centre O and radius 1.

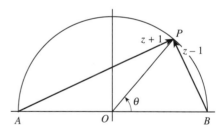

Fig. 17.6

The points A, B and P represent the complex numbers -1, 1 and z respectively.

The translation $\overrightarrow{BP}$ then represents $z - 1$; and, writing $z + 1$ as $z - (-1)$, the translation $\overrightarrow{AP}$ represents $z + 1$. Since the triangles OAP and OBP are isosceles, it is easy to see that the arguments of $z + 1$ and $z - 1$ are $\tfrac{1}{2}\theta$ and

$\tfrac{1}{2}\pi + \tfrac{1}{2}\theta$ respectively. The argument of $\dfrac{z-1}{z+1}$ is the angle you have to turn through to get from $\overrightarrow{AP}$ to $\overrightarrow{BP}$, that is $\tfrac{1}{2}\pi$ anticlockwise, or $+\tfrac{1}{2}\pi$.

Also the moduli of $z + 1$ and $z - 1$ are the lengths AP and BP. From the triangle ABP, in which $AB = 2$ and angle $BAP = \tfrac{1}{2}\theta$,

$$AP = 2\cos\tfrac{1}{2}\theta \quad \text{and} \quad BP = 2\sin\tfrac{1}{2}\theta.$$

Exercise 17B

1 If $s = 2\left(\cos\frac{1}{3}\pi + i\sin\frac{1}{3}\pi\right)$, $t = \cos\frac{1}{4}\pi + i\sin\frac{1}{4}\pi$ and $u = 4\left(\cos\left(-\frac{5}{6}\pi\right) + i\sin\left(-\frac{5}{6}\pi\right)\right)$, write the following in modulus-argument form.

(a) st (b) $\dfrac{s}{t}$ (c) $\dfrac{t}{s}$ (d) su

(e) $\dfrac{u}{s}$ (f) $\dfrac{t}{u}$ (g) s^2 (h) u^2

(i) st^2u (j) $\dfrac{2s}{tu}$ (k) s^* (l) st^*

(m) ts^* (n) $\dfrac{u^*}{t^2}$ (o) $\dfrac{1}{t}$ (p) $\dfrac{2}{s}$

(q) $\dfrac{4i}{u^*}$ (r) $\dfrac{s^3}{u}$

2 If $s = 3\left(\cos\frac{1}{5}\pi + i\sin\frac{1}{5}\pi\right)$ and if $s^2t = 18\left(\cos\left(-\frac{4}{5}\pi\right) + i\sin\left(-\frac{4}{5}\pi\right)\right)$, express t in modulus-argument form.

3 If $s = \cos\frac{2}{3}\pi + i\sin\frac{2}{3}\pi$ and $t = \cos\frac{1}{4}\pi + i\sin\frac{1}{4}\pi$, show in an Argand diagram

(a) s,   (b) st,   (c) st^2,

(d) st^3,   (e) $\dfrac{s}{t}$,   (f) $\dfrac{s}{t^2}$.

4 Repeat Question 3 with $s = 4\left(\cos\frac{2}{3}\pi + i\sin\frac{2}{3}\pi\right)$ and $t = 2\left(\cos\frac{1}{4}\pi + i\sin\frac{1}{4}\pi\right)$.

5 Give the answers to the following questions in modulus-argument form.

(a) If $s = \cos\theta + i\sin\theta$, express s^* in terms of θ.

(b) If $s = \cos\theta + i\sin\theta$, express $\dfrac{1}{s}$ in terms of θ.

(c) If $t = r(\cos\theta + i\sin\theta)$, express t^* in terms of r and θ.

(d) If $t = r(\cos\theta + i\sin\theta)$, express $\dfrac{1}{t}$ in terms of r and θ.

6 Write $1 + \sqrt{3}\,i$ and $1 - i$ in modulus-argument form. Hence express $\dfrac{\left(1 + \sqrt{3}\,i\right)^4}{(1-i)^6}$ in the form $a + bi$.

7 By converting into and out of modulus-argument form, evaluate the following with the aid of a calculator. Use the binomial theorem to check your answers.

(a) $(1 + 2i)^7$ (b) $(3i - 2)^5$ (c) $(3 - i)^{-8}$

8 Show in an Argand diagram the points representing the complex numbers i, $-i$ and $\sqrt{3}$. Hence write down the values of

(a) $\arg\left(\sqrt{3} - i\right)$,   (b) $\arg\left(\sqrt{3} + i\right)$,   (c) $\arg\dfrac{\sqrt{3} + i}{\sqrt{3} - i}$,   (d) $\arg\dfrac{2i}{\sqrt{3} + i}$.

9 A and B are points in an Argand diagram representing the complex numbers 1 and i. P is a point on the circle having AB as a diameter. If P represents the complex number z, find the value of $\arg\dfrac{z-1}{z-\mathrm{i}}$ if P is in

(a) the first quadrant,   (b) the second quadrant,   (c) the fourth quadrant.

10 Identify the set of points in an Argand diagram for which $\arg\dfrac{z-3}{z-4\mathrm{i}}=\tfrac{1}{2}\pi$.

11 If A and B represent complex numbers a and b in an Argand diagram, identify the set of points for which $\arg\dfrac{z-a}{z-b}=\pi$.

12 Identify the set of points in an Argand diagram for which $\arg\dfrac{z-\mathrm{i}}{z+\mathrm{i}}=\tfrac{1}{4}\pi$.

13 Find the modulus and argument of $1+\mathrm{i}\tan\theta$ in the cases

(a) $0<\theta<\tfrac{1}{2}\pi$,   (b) $\tfrac{1}{2}\pi<\theta<\pi$,   (c) $\pi<\theta<\tfrac{3}{2}\pi$,   (d) $\tfrac{3}{2}\pi<\theta<2\pi$.

14 If $z=\cos 2\theta+\mathrm{i}\sin 2\theta$, find the modulus and argument of $1-z$ in the cases

(a) $0<\theta<\pi$,   (b) $-\pi<\theta<0$.

Illustrate your answer using an Argand diagram.

17.3 Spiral enlargement

Suppose that a number $t=q(\cos\beta+\mathrm{i}\sin\beta)$ is represented by a translation $\overrightarrow{AT}$ (see Fig. 17.7). How would you represent the number $w=st$, where $s=p(\cos\alpha+\mathrm{i}\sin\alpha)$?

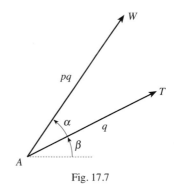

Fig. 17.7

Since multiplication by s multiplies the modulus by p, and increases the angle with the real axis by α, w could be represented by a translation $\overrightarrow{AW}$ whose length is p times the length of $\overrightarrow{AT}$ and whose direction makes an angle α with the direction of $\overrightarrow{AT}$ in the anticlockwise sense.

A transformation of a plane which multiplies lengths of vectors by a scale factor of p and rotates them through an angle α is called a **spiral enlargement**.

> If complex numbers are represented by translations of a plane, multiplication by a complex number s has the effect of a spiral enlargement of scale factor $|s|$ and angle $\arg s$.

Example 17.3.1

Fig. 17.8 shows an Argand diagram in which A and T are the points $1+i$ and $4-i$. W is a point such that $AW = 2AT$ and angle $TAW = \frac{1}{3}\pi$. Find the number represented by W.

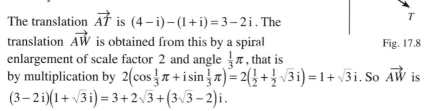

Fig. 17.8

The translation $\overrightarrow{AT}$ is $(4-i)-(1+i)=3-2i$. The translation $\overrightarrow{AW}$ is obtained from this by a spiral enlargement of scale factor 2 and angle $\frac{1}{3}\pi$, that is by multiplication by $2\left(\cos\frac{1}{3}\pi + i\sin\frac{1}{3}\pi\right) = 2\left(\frac{1}{2}+\frac{1}{2}\sqrt{3}\,i\right) = 1+\sqrt{3}\,i$. So $\overrightarrow{AW}$ is $(3-2i)\left(1+\sqrt{3}\,i\right) = 3+2\sqrt{3}+\left(3\sqrt{3}-2\right)i$.

The point W is therefore $(1+i)+\left(3+2\sqrt{3}+\left(3\sqrt{3}-2\right)i\right) = 4+2\sqrt{3}+\left(3\sqrt{3}-1\right)i$.

Example 17.3.2*

ABC is a triangle. Fig. 17.9 shows three similar triangles BUC, CVA and AWB drawn external to ABC. Prove that the centroids of the triangles ABC and UVW coincide.

The centroid of a triangle, the point where the medians intersect, was defined in P1 Example 13.5.2. There it was given as the point with position vector $\frac{1}{3}(\mathbf{a}+\mathbf{b}+\mathbf{c})$, where $\mathbf{a}$, $\mathbf{b}$ and $\mathbf{c}$ are the position vectors of the vertices of the triangle. But since complex numbers under addition behave exactly like two-dimensional vectors, the result carries straight over to an Argand diagram. If A, B and C correspond to complex numbers a, b and c, then the centroid of the triangle corresponds to $\frac{1}{3}(a+b+c)$. Try writing out the proof for yourself using complex numbers instead of position vectors.

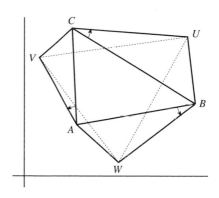

Fig. 17.9

What is new in the Argand diagram is that you now have a way of dealing with the similar triangles. Suppose that multiplication by a number s gives the spiral enlargement that transforms $\overrightarrow{CB}$ into $\overrightarrow{CU}$. Then the same spiral enlargement transforms $\overrightarrow{AC}$ into $\overrightarrow{AV}$, and $\overrightarrow{BA}$ into $\overrightarrow{BW}$. So, if U, V and W correspond to complex numbers u, v and w, you can write

$$u-c = s(b-c), \quad v-a = s(c-a) \quad \text{and} \quad w-b = s(a-b).$$

If you add these three equations, you get

$$(u-c)+(v-a)+(w-b) = s((b-c)+(c-a)+(a-b)) = 0,$$

so $u+v+w = a+b+c$.

Therefore $\frac{1}{3}(u+v+w) = \frac{1}{3}(a+b+c)$. That is, ABC and UVW have the same centroid.

At the beginning of this section spiral enlargements were described in terms of general translations, and this is how they are used in these examples. But if you take the point A as the origin, you can also use them in an Argand diagram. Fig. 17.10 shows s, t and $w = st$ represented by points S, T and W. Also shown is the point U representing the number 1. Then multiplication by s gives a spiral enlargement which transforms $\overrightarrow{OT}$ to $\overrightarrow{OW}$; it also transforms $\overrightarrow{OU}$ to $\overrightarrow{OS}$, since $s = s \times 1$. Therefore the triangles OTW and OUS are similar.

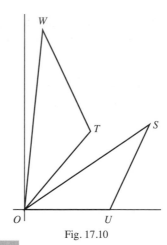

Fig. 17.10

> In an Argand diagram, the triangles formed by the points representing 0, t, st and 0, 1, s are similar.

This gives you a geometrical method of drawing the point st on an Argand diagram when you know the points s and t.

Exercise 17C

1   A is the point in an Argand diagram representing $1 + 3i$. Find the complex numbers represented by the two points B such that $OB = \sqrt{2} \times OA$ and angle $AOB = \frac{1}{4}\pi$.

2   A is the point in an Argand diagram representing $3 - 2i$. Find the complex numbers represented by the two points B such that $OB = 2OA$ and angle $AOB = \frac{1}{3}\pi$.

3   Points A and B represent $1 + i$ and $2 - i$ in an Argand diagram. C is a point such that $AC = 2AB$ and angle $BAC = \frac{2}{3}\pi$. Find two possibilities for the complex number represented by C.

4   Points A, B and C represent i, $3 - i$ and $4 + 2i$ in an Argand diagram. D is the reflection of C in the line AB. Find the complex number which is represented by D.

5   A point S represents the complex number s in an Argand diagram. Draw diagrams to show how to construct the points which represent

     (a) s^2,   (b) s^3,   (c) $\dfrac{1}{s}$.

6   Points S and T represent the complex numbers s and t in an Argand diagram. Draw a diagram to show how to construct the point which represents $\dfrac{s}{t}$.

7   A snail starts at the origin of an Argand diagram and walks along the real axis for an hour, covering a distance of 8 metres. At the end of each hour it changes its direction by $\frac{1}{2}\pi$ anticlockwise; and in each hour it walks half as far as it did in the previous hour. Find where it is

     (a) after 4 hours,   (b) after 8 hours,   (c) eventually.

8 Repeat Question 7 if the change of direction is through $\frac{1}{4}\pi$ and it walks $\dfrac{1}{\sqrt{2}}$ times as far as in the previous hour.

9* A and B are two points on a computer screen. A program produces a trace on the screen to execute the following algorithm.

Step 1 Start at any point P on the screen.

Step 2 From the current position rotate through a quarter circle about A.

Step 3 From the current position rotate through a quarter circle about B.

Step 4 Repeat Step 2.

Step 5 Repeat Step 3, and stop.

Show that the trace ends where it began.

10* A, B, C and D are four points on a computer screen. A program selects a point P on the screen at random and then produces a trace by rotating successively through a right angle about A, B, C and D. Show that, if the trace ends where it began, the line segments AC and BD are equal in length and perpendicular to each other.

11* ABC is a triangle such that the order A, B, C takes you anticlockwise round the triangle. Squares $ACPQ$, $BCRS$ are drawn outside the triangle ABC. If A, B and C represent complex numbers a, b and c in an Argand diagram, find the complex numbers represented by Q, S and the mid-point M of QS. Show that the position of M doesn't depend on the position of C, and find how it is related to the points A and B.

12* Points in an Argand diagram representing the complex numbers $-2i$, 4, $2+4i$ and $2i$ form a convex quadrilateral. Squares are drawn outside the quadrilateral on each of the four sides. Find the numbers represented by the centres P, Q, R and S of the four squares. Hence prove that

(a) $PR = QS$,   (b) PR is perpendicular to QS.

Show that the same conclusion follows for any four points forming a convex quadrilateral.

13* In an Argand diagram points A, B, C, U, V and W represent complex numbers a, b, c, u, v and w. Prove that, if the triangles ABC and UVW are directly similar, then $aw + bu + cv = av + bw + cu$. ('Directly similar' means that, if you go round the triangles in the order A, B, C and U, V, W, then you go round both triangles in the same sense.) Prove that the converse result is also true. Hence show that a triangle is equilateral if and only if $a^2 + b^2 + c^2 = bc + ca + ab$.

17.4 Square roots of complex numbers

Section 16.6 gave a method of finding square roots of complex numbers in the form $a + bi$. You can also use a method based on modulus-argument form.

A special case of the rule for multiplying two complex numbers is that, if $s = p(\cos\alpha + i\sin\alpha)$, then $s^2 = p^2(\cos 2\alpha + i\sin 2\alpha)$. That is, to square a complex number, you square the modulus and double the argument (adjusting by 2π if necessary).

It follows that, to find a square root, you take the
square root of the modulus and halve the argument.
That is,

$$\sqrt{s} = \sqrt{p}\left(\cos\tfrac{1}{2}\alpha + i\sin\tfrac{1}{2}\alpha\right).$$

This is illustrated on an Argand diagram in Fig. 17.11.

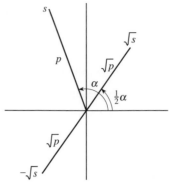

This gives only one of the two square roots. Since the
two square roots of s are of the form $\pm\sqrt{s}$, the two
square roots are symmetrically placed around the
origin in the Argand diagram. So the other root also
has modulus $\sqrt{p}$, and has argument $\tfrac{1}{2}\alpha \pm \pi$, where
the $+$ or $-$ sign is chosen so that the argument is
between $-\pi$ and π.

Fig. 17.11

> The square roots of a complex number s have
> modulus $\sqrt{|s|}$ and arguments $\tfrac{1}{2}\arg s$ and $\tfrac{1}{2}\arg s \pm \pi$
> where the sign is $+$ if $\arg s < 0$ and $-$ if $\arg s \geqslant 0$.

Example 17.4.1
Find the square roots of   (a) $8i$,   (b) $3 - 4i$ (see Example 16.6.1).

(a) $8i$ has modulus 8 and argument $\tfrac{1}{2}\pi$. So its square roots have modulus
$\sqrt{8} = 2\sqrt{2}$ and arguments $\tfrac{1}{4}\pi$ and $\tfrac{1}{4}\pi - \pi = -\tfrac{3}{4}\pi$. In cartesian form, these are

$$2\sqrt{2}\left(\frac{1}{\sqrt{2}} + \frac{1}{\sqrt{2}}i\right) \quad \text{and} \quad 2\sqrt{2}\left(-\frac{1}{\sqrt{2}} - \frac{1}{\sqrt{2}}i\right),$$

that is $2 + 2i$ and $-2 - 2i$.

(b) $3 - 4i$ has modulus $\sqrt{3^2 + (-4)^2} = 5$ and argument $-\tan^{-1}\tfrac{4}{3} = -0.927\,295\,2\ldots$.
The square roots therefore have modulus $\sqrt{5}$ and arguments $-0.463\,647\,6\ldots$ and
$-0.463\,647\,6\ldots + \pi$. In cartesian form, these are $2 - i$ and $-2 + i$.

17.5 The exponential form
The rule that, when you multiply two numbers, you add the arguments may have reminded
you of a similar rule for logarithms, that $\log mn = \log m + \log n$. There is a reason for this;
the arguments of complex numbers and logarithms can both be thought of as indices.

To show this, let $z = \cos\theta + i\sin\theta$. You can differentiate this with respect to θ to get

$$\frac{dz}{d\theta} = -\sin\theta + i\cos\theta.$$

It is easy to see that the right side of this is $i(\cos\theta + i\sin\theta)$, which is iz. So you can describe the relation between z and θ by the equation

$$\frac{dz}{d\theta} = iz.$$

Can you think of a function which, when differentiated, gives you i times what you started with?

If the multiplier were a real number a, rather than i, then the answer is simple. You know that this is the property of the exponential function: if $y = e^{ax}$, then $\dfrac{dy}{dx} = ae^{ax} = ay$. But this is not the complete answer. It is proved in Section 19.4 that the most general function such that $\dfrac{dy}{dx} = ay$ has the form $y = Ke^{ax}$, where K is an arbitrary constant.

Coming back to the complex number problem, this suggests that $z = \cos\theta + i\sin\theta$ has the form $Ke^{i\theta}$ for some value of K.

To find K, consider the special value $\theta = 0$. Then $z = \cos 0 + i\sin 0 = 1 + 0i = 1$, and $Ke^{i0} = Ke^0 = K \times 1 = K$. So K must equal 1. It follows that:

> $\cos\theta + i\sin\theta$ can be written as $e^{\theta i}$.

Is this a 'proof'? Not strictly, since e^x has so far only been defined when x is real. But notice that, if e^{yi} is written as $\cos y + i\sin y$, then

$$e^{\alpha i} \times e^{\beta i} = (\cos\alpha + i\sin\alpha)(\cos\beta + i\sin\beta)$$
$$= \cos(\alpha + \beta) + i\sin(\alpha + \beta)$$
$$= e^{(\alpha+\beta)i} = e^{\alpha i + \beta i},$$

which is the usual multiplication rule for indices. This suggests that it would be a good idea to *define* e^{yi} as $\cos y + i\sin y$. Then many of the properties of the exponential function that you know already, such as the multiplication rule and the rule for differentiation, would still hold.

Most people who use complex numbers prefer the compact notation e^{yi} to the rather clumsy form $\cos y + i\sin y$ that you have used so far in this chapter. For instance, in Example 17.2.2 you could write z as $e^{\theta i}$ and $z+1$ as $\left(2\cos\frac{1}{2}\theta\right)e^{\frac{1}{2}\theta i}$.

Before stating this as a definition, it is worthwhile going one stage further, and defining $e^{x+y\mathrm{i}}$ as $e^x \times e^{y\mathrm{i}}$. You then have:

> If $z = x + y\mathrm{i}$, the **exponential function** e^z is defined by
>
> $$e^z = e^x(\cos y + \mathrm{i} \sin y).$$
>
> Thus $\left| e^z \right| = e^x$, and $\arg e^z = y + k(2\pi)$,
> where k is chosen so that $-\pi < \arg e^z \leqslant \pi$.

A special case of this, with $x = 0$ and $y = \pi$, leads to a famous equation which connects five of the most important numbers in mathematics:

> $$e^{\pi\mathrm{i}} + 1 = 0.$$

Exercise 17D

1 Use the modulus-argument method to find the square roots of the following complex numbers.

(a) $4\left(\cos \frac{2}{5}\pi + \mathrm{i}\sin \frac{2}{5}\pi\right)$ (b) $9\left(\cos \frac{4}{7}\pi - \mathrm{i}\sin \frac{4}{7}\pi\right)$ (c) $-2\mathrm{i}$

(d) $20\mathrm{i} - 21$ (e) $1 + \mathrm{i}$ (f) $5 - 12\mathrm{i}$

2 Find

(a) the square roots of $8\sqrt{3}\,\mathrm{i} - 8$,   (b) the fourth roots of $8\sqrt{3}\,\mathrm{i} - 8$.

3 Prove that, if z is a complex number, then $e^z + e^{z^*}$ and $e^z \times e^{z^*}$ are both real. What can you say about $e^z - e^{z^*}$ and $e^z \div e^{z^*}$?

4 In an Argand diagram, plot the complex numbers

(a) $e^{\pi\mathrm{i}}$,   (b) $e^{\frac{1}{2}\pi\mathrm{i}}$,   (c) $2e^{-\mathrm{i}}$,   (d) $e^{4\mathrm{i}}$,

(e) $e^{1+\mathrm{i}}$,   (f) $e^{-1+\mathrm{i}}$,   (g) $e^{1-\mathrm{i}}$.

5 Find the square roots of

(a) $e^{\frac{2}{3}\pi\mathrm{i}}$,   (b) $e^{1+2\mathrm{i}}$.

6 If y is real, simplify $\dfrac{e^{y\mathrm{i}} - e^{-y\mathrm{i}}}{e^{y\mathrm{i}} + e^{-y\mathrm{i}}}$.

7 Prove that, if $s = a + b\mathrm{i}$, then

(a) $\left(e^s\right)^2 = e^{2s}$,   (b) $\left(e^s\right)^3 = e^{3s}$,   (c) $\left(e^s\right)^{-1} = e^{-s}$.

8 Use the definition of e^z to prove that, if $s = a + b\mathrm{i}$ and $t = c + d\mathrm{i}$, then $e^s \times e^t = e^{s+t}$.

9 Use the definition of e^z to prove that, if $s = a + bi$, then $\dfrac{d}{dx} e^{sx} = se^{sx}$.

10 If $z = \cos\theta + i\sin\theta$, find the modulus and argument of e^z in terms of θ.

11 If $z = e^{\theta i}$, show that $z + 1 = e^{\frac{1}{2}\theta i}\left(e^{\frac{1}{2}\theta i} - e^{-\frac{1}{2}\theta i}\right)$. Deduce the result of Example 17.2.2(a).

Use a similar method to prove the result of Example 17.2.2(b).

Miscellaneous exercise 17

1 Given that $z = \tan\alpha + i$, where $0 < \alpha < \frac{1}{2}\pi$, and $w = 4\left(\cos\frac{1}{10}\pi + i\sin\frac{1}{10}\pi\right)$, find in their simplest forms

   (a) $|z|$,   (b) $|zw|$,   (c) $\arg z$,   (d) $\arg\dfrac{z}{w}$.   (OCR)

2 Given that $(5 + 12i)z = 63 + 16i$, find $|z|$ and $\arg z$, giving this answer in radians correct to 3 significant figures. Given also that $w = 3\left(\cos\frac{1}{3}\pi + i\sin\frac{1}{3}\pi\right)$, find

   (a) $\left|\dfrac{z}{w}\right|$,   (b) $\arg(zw)$.   (OCR)

3 In an Argand diagram, the point P represents the complex number z. On a single diagram, illustrate the set of possible positions of P for each of the cases

   (a) $|z - 3i| \leqslant 3$,   (b) $\arg(z + 3 - 3i) = \frac{1}{4}\pi$.

Given that z satisfies both (a) and (b), find the greatest possible value of $|z|$.   (OCR)

4 A complex number z satisfies $|z - 3 - 4i| = 2$. Describe in geometrical terms, with the aid of a sketch, the locus of the point which represents z in an Argand diagram. Find

   (a) the greatest value of $|z|$,

   (b) the difference between the greatest and least values of $\arg z$.   (OCR)

5 Given that $|z - 5| = |z - 2 - 3i|$, show on an Argand diagram the locus of the point which represents z. Using your diagram, show that there is no value of z satisfying both $|z - 5| = |z - 2 - 3i|$ and $\arg z = \frac{1}{4}\pi$.   (OCR)

6 A complex number z satisfies the inequality $|z + 2 - 2\sqrt{3}i| \leqslant 2$. Describe, in geometrical terms, with the aid of a sketch, the corresponding region in an Argand diagram. Find

   (a) the least possible value of $|z|$,

   (b) the greatest possible value of $\arg z$.   (OCR)

7 The quadratic equation $z^2 + 6z + 34 = 0$ has complex roots α and β.

   (a) Find the roots, in the form $a + bi$.

   (b) Find the modulus and argument of each root, and illustrate the two roots on an Argand diagram.

   (c) Find $|\alpha - \beta|$.   (MEI)

8 The complex numbers α and β are given by $\dfrac{\alpha+4}{\alpha}=2-\mathrm{i}$ and $\beta=-\sqrt{6}+\sqrt{2}\,\mathrm{i}$.

(a) Show that $\alpha=2+2\,\mathrm{i}$ and that $|\alpha|=|\beta|$. Find $\arg\alpha$ and $\arg\beta$.

(b) Find the modulus and argument of $\alpha\beta$. Illustrate the complex numbers α, β and $\alpha\beta$ on an Argand diagram.

(c) Describe the locus of points in the Argand diagram representing complex numbers z for which $|z-\alpha|=|z-\beta|$. Draw this locus on your diagram.

(d) Show that $z=\alpha+\beta$ satisfies $|z-\alpha|=|z-\beta|$. Mark the point representing $\alpha+\beta$ on your diagram, and find the exact value of $\arg(\alpha+\beta)$.   (MEI)

9 (a) Given that $\alpha=-1+2\,\mathrm{i}$, express α^2 and α^3 in the form $a+b\,\mathrm{i}$. Hence show that α is a root of the cubic equation $z^3+7z^2+15z+25=0$. Find the other two roots.

(b) Illustrate the three roots of the cubic equation on an Argand diagram, and find the modulus and argument of each root.

(c) L is the locus of points in the Argand diagram representing complex numbers z for which $\left|z+\tfrac{5}{2}\right|=\tfrac{5}{2}$. Show that all three roots of the cubic equation lie on L and draw the locus L on your diagram.   (MEI)

10 Complex numbers α and β are given by $\alpha=2\left(\cos\tfrac{1}{8}\pi+\mathrm{i}\sin\tfrac{1}{8}\pi\right)$ and $\beta=4\sqrt{2}\left(\cos\tfrac{5}{8}\pi+\mathrm{i}\sin\tfrac{5}{8}\pi\right)$.

(a) Write down the modulus and argument of each of the complex numbers α, β, $\alpha\beta$ and $\dfrac{\alpha}{\beta}$. Illustrate these four complex numbers on an Argand diagram.

(b) Express $\alpha\beta$ in the form $a+b\,\mathrm{i}$, giving a and b in their simplest forms.

(c) Indicate a length on your diagram which is equal to $|\beta-\alpha|$, and show that $|\beta-\alpha|=6$.

(d) On your diagram, draw

(i) the locus L of points representing complex numbers z such that $|z-\alpha|=6$,

(ii) the locus M of points representing complex numbers z such that $\arg(z-\alpha)=\tfrac{5}{8}\pi$.   (MEI)

11 You are given that $\alpha=1+\sqrt{3}\,\mathrm{i}$ is a root of the cubic equation $3z^3-4z^2+8z+8=0$.

(a) Write down another complex root β, and hence solve the cubic equation.

(b) Find the modulus and argument of each of the complex numbers α, β, $\alpha\beta$ and $\dfrac{\alpha}{\beta}$. Illustrate these four complex numbers on an Argand diagram.

(c) Describe the locus of points in the Argand diagram representing the complex numbers z for which $|z-\alpha|=\sqrt{3}$. Sketch this locus on your diagram.

(d) Express $\dfrac{6+\alpha}{2\alpha-\mathrm{i}}$ in the form $a+b\,\mathrm{i}$, where a and b are real numbers.   (MEI)

12* The fixed points A and B represent the complex numbers a and b in an Argand diagram with origin O.

(a) The variable point P represents the complex number z, and λ is a real variable. Describe the locus of P in relation to A and B in the following cases, illustrating your loci in separate diagrams.

    (i) $z - a = \lambda b$ (ii) $z - a = \lambda(z - b)$ (iii) $z - a = i\lambda(z - b)$

(b) By writing $a = |a|e^{i\alpha}$ and $b = |b|e^{i\beta}$, show that $|\operatorname{Im}(ab)| = 2\Delta$, where Δ is the area of triangle OAB.

13* A function f has the set of complex numbers for its domain and range. It has the property that, for any two complex numbers z and w, $|f(z) - f(w)| = |z - w|$. Explain why $f(1) - f(0)$ must be non-zero.

The function g is defined by $g(z) = \dfrac{f(z) - f(0)}{f(1) - f(0)}$. Show that g has the same property as f, plus the additional properties $g(0) = 0$ and $g(1) = 1$. Prove, by making two suitable choices for w, that $|g(z)| = |z|$ and $|g(z) - 1| = |z - 1|$. By writing $z = x + yi$ and $g(z) = u + vi$, show that, for each z, $g(z)$ must equal either z or z^*.

If z and w have non-zero imaginary parts, why is it impossible for $g(z)$ to equal z and $g(w)$ to equal w^*? Use your answer to write down the most general form for the function g. What is the most general form for the function f? Interpret your answer geometrically.

(OCR)

18 Integration

This chapter is about two methods of integration, one derived from the chain rule for differentiation, and the other derived from the rule for differentiating a product. When you have completed it, you should

- understand and be able to find integrals using both direct and reverse substitution
- be able to find new limits of integration when a definite integral is evaluated by substitution
- recognise the form $\int \dfrac{f'(x)}{f(x)}\,dx$, and be able to write down the integral at sight
- know and be able to apply the method of integration by parts.

18.1 Direct substitution

None of the methods of integration described so far could be used to find

$$\int \frac{1}{x + \sqrt{x}}\,dx.$$

However, by expressing x in terms of a new variable, integrals like this can be put into a form which you know how to integrate.

Denote the integral by I, so that

$$\frac{dI}{dx} = \frac{1}{x + \sqrt{x}}.$$

The difficulty in solving this equation for I lies in the square root, so write $x = u^2$. Then, by the chain rule,

$$\frac{dI}{du} = \frac{dI}{dx} \times \frac{dx}{du} = \frac{1}{u^2 + u} \times 2u = \frac{2}{u + 1}.$$

From this you can integrate to find I in terms of u, as

$$I = 2\ln|u + 1| + k.$$

The solution to the original equation is then found by replacing u by $\sqrt{x}$, so that

$$I = 2\ln\left(\sqrt{x} + 1\right) + k.$$

(You do not need the modulus sign, since $\sqrt{x} + 1$ is always positive.)

You can easily check by differentiation that this integral is correct.

This method is called **integration by substitution**. It is the equivalent for integrals of the chain rule for differentiation.

In general, to find $I = \displaystyle\int f(x)\,dx$, the equation $\dfrac{dI}{dx} = f(x)$ is changed by writing x as some

function $s(u)$. Then $\dfrac{dI}{du} = f(x) \times \dfrac{dx}{du} = g(u) \times \dfrac{dx}{du}$, where $g(u) = f(s(u))$. If you can find

$\displaystyle\int g(u) \times \dfrac{dx}{du}\,du$, then you can find the original integral by replacing u by $s^{-1}(x)$.

> If $x = s(u)$ and $g(u) = f(s(u))$, then $\displaystyle\int f(x)\,dx$ is equal to $\displaystyle\int g(u) \times \dfrac{dx}{du}\,du$, with u
>
> replaced by $s^{-1}(x)$.

Do not try to memorise this as a formal statement; what is important is to learn how to use the method. Notice how the notation helps; although the dx and du in the integrals have no meaning in themselves, the replacement of dx in the first integral by $\dfrac{dx}{du}\,du$ in the second makes the method easy to apply.

Example 18.1.1

Find $\displaystyle\int \frac{1}{x}\ln x\,dx$ using the substitution $x = e^u$.

The difficulty lies in the logarithm factor, which is removed by using the substitution $x = e^u$ (that is, $\ln x = u$). Then $\dfrac{dx}{du} = e^u$, and the integral becomes

$$\int \frac{1}{e^u} \ln(e^u) \times e^u\,du = \int u\,du = \tfrac{1}{2}u^2 + k.$$

Replacing u in this expression by $\ln x$, the original integral is

$$\int \frac{1}{x} \ln x\,dx = \tfrac{1}{2}(\ln x)^2 + k.$$

Example 18.1.2

Find $\displaystyle\int \frac{6x}{\sqrt{2x+1}}\,dx$ using the substitution $2x+1 = u^2$.

The awkward bit of the integral is the expression $\sqrt{2x+1}$. If $2x+1$ is written as u^2, then $\sqrt{2x+1}$ is equal to u. The equation $2x+1 = u^2$ is equivalent to $x = \tfrac{1}{2}u^2 - \tfrac{1}{2}$. This gives $\dfrac{dx}{du} = \tfrac{1}{2}(2u) = u$. So $\displaystyle\int \frac{6x}{\sqrt{2x+1}}\,dx$ becomes

$$\int \frac{6\left(\tfrac{1}{2}u^2 - \tfrac{1}{2}\right)}{u} \times u\,du = \int \left(3u^2 - 3\right)du = u^3 - 3u + k.$$

You want this in terms of x, so substituting $\sqrt{2x+1}$ for u gives

$$\int \frac{6x}{\sqrt{2x+1}}\,dx = \left(\sqrt{2x+1}\right)^3 - 3\sqrt{2x+1} + k.$$

It is quite acceptable to leave the answer in this form, but it would be neater to note that $\left(\sqrt{2x+1}\right)^3 = (2x+1)\sqrt{2x+1}$, so

$$\int \frac{6x}{\sqrt{2x+1}}\,dx = (2x+1)\sqrt{2x+1} - 3\sqrt{2x+1} + k$$

$$= (2x+1-3)\sqrt{2x+1} + k = 2(x-1)\sqrt{2x+1} + k.$$

Since this is quite a complicated piece of algebra, it is worth checking it by using the product rule to differentiate $2(x-1)\sqrt{2x+1}$, and showing that the result is $\dfrac{6x}{\sqrt{2x+1}}$.

The method used in this example is sometimes described as 'substituting $u = \sqrt{2x+1}$' and sometimes as 'substituting $x = \frac{1}{2}\left(u^2-1\right)$'. In the course of the calculation you use the relation both ways round, so either description is equally appropriate.

The next example requires you to find a suitable substitution for yourself.

Example 18.1.3

Find $\displaystyle\int \sqrt{4-x^2}\,dx$.

You need a substitution for x such that $4-x^2$ simplifies to an exact square. A function with this property is $2\sin u$, since if $x = 2\sin u$, then

$$4 - x^2 = 4 - 4\sin^2 u = 4\left(1 - \sin^2 u\right) = 4\cos^2 u.$$

Therefore $\sqrt{4-x^2} = 2\cos u$. Also $\dfrac{dx}{du} = 2\cos u$. The integral then becomes

$$\int 2\cos u \times 2\cos u\,du = \int 4\cos^2 u\,du = \int 2(1 + \cos 2u)\,du$$

$$= 2u + \sin 2u + k.$$

To get the original integral, note that $\sin u = \frac{1}{2}x$, so that $2u = 2\sin^{-1}\left(\frac{1}{2}x\right)$. But rather than using this form in the second term, it is simpler to expand $\sin 2u$ as $2\sin u \cos u$, which is $x \times \frac{1}{2}\sqrt{4-x^2}$. Therefore

$$\int \sqrt{4-x^2}\,dx = 2\sin^{-1}\left(\frac{1}{2}x\right) + \frac{1}{2}x\sqrt{4-x^2} + k.$$

Notice one further detail. The reference in the general statement (near the top of page 259) to the inverse function s^{-1} should alert you to the need for the substitution function s to be one–one. This is arranged in the usual way, by restricting the domain of s.

In the introductory example on page 258, for $x = u^2$ to have an inverse you can restrict u to be non-negative. This justifies writing $\sqrt{x}$ as u (since by definition $\sqrt{x} \geqslant 0$) when the variable was changed from x to u, and then replacing u by $\sqrt{x}$ (rather than $-\sqrt{x}$) at the final stage.

In Example 18.1.3, the domain of u is restricted to the interval $-\frac{1}{2}\pi \leqslant u \leqslant \frac{1}{2}\pi$, so that the substitution function $x = 2\sin u$ has inverse $u = \sin^{-1}\left(\frac{1}{2}x\right)$, with $\sin^{-1}$ defined as in P1 Section 18.4. Over this interval $\cos u \geqslant 0$, which justifies taking $2\cos u$ to be the positive square root of $4 - x^2$.

Exercise 18A

In Questions 3 and 5 you are required to find the substitution for yourself.

1 Use the given substitutions to find the following integrals.

(a) $\displaystyle\int \frac{1}{x - 2\sqrt{x}}\,dx$ $\qquad x = u^2$

(b) $\displaystyle\int \frac{1}{(3x+4)^2}\,dx$ $\qquad 3x + 4 = u$

(c) $\displaystyle\int \sin\left(\frac{1}{3}\pi - \frac{1}{2}x\right)dx$ $\qquad \frac{1}{3}\pi - \frac{1}{2}x = u$

(d) $\displaystyle\int x(x-1)^5\,dx$ $\qquad x = 1 + u$

(e) $\displaystyle\int \frac{e^x}{1 + e^x}\,dx$ $\qquad x = \ln u$

(f) $\displaystyle\int \frac{1}{3\sqrt{x} + 4x}\,dx$ $\qquad x = u^2$

(g) $\displaystyle\int 3x\sqrt{x+2}\,dx$ $\qquad x = u^2 - 2$

(h) $\displaystyle\int \frac{x}{\sqrt{x-3}}\,dx$ $\qquad x = 3 + u^2$

(i) $\displaystyle\int \frac{1}{x \ln x}\,dx$ $\qquad x = e^u$

(j) $\displaystyle\int \frac{1}{\sqrt{4 - x^2}}\,dx$ $\qquad x = 2\sin u$

2 Use a substitution of the form $ax + b = u$ to find the following integrals.

(a) $\displaystyle\int x(2x+1)^3\,dx$

(b) $\displaystyle\int (x+2)(2x-3)^5\,dx$

(c) $\displaystyle\int x\sqrt{2x-1}\,dx$

(d) $\displaystyle\int \frac{x-2}{\sqrt{x-4}}\,dx$

(e) $\displaystyle\int \frac{x}{(x+1)^2}\,dx$

(f) $\displaystyle\int \frac{x}{2x+3}\,dx$

3* Use a suitable substitution to find the following integrals.

(a) $\displaystyle\int \frac{1}{\sqrt{1 - 9x^2}}\,dx$

(b) $\displaystyle\int \sqrt{16 - 9x^2}\,dx$

(c) $\displaystyle\int \frac{1}{2 + e^{-x}}\,dx$

(d) $\displaystyle\int \frac{x}{\sqrt[3]{1+x}}\,dx$

(e) $\displaystyle\int \left(1 - x^2\right)^{-\frac{3}{2}}\,dx$

(f) $\displaystyle\int \frac{1}{2 - \sqrt{x}}\,dx$

4 (a) Use the substitution $x = \tan u$ to show that $\displaystyle\int \frac{1}{1 + x^2}\,dx = \tan^{-1}x + k$.

(b) Use the substitution $x = \ln u$ to find $\displaystyle\int \frac{e^x}{1 + e^{2x}}\,dx$.

5* For the following integrals, use a substitution to produce an integrand which is a rational function of u, then use partial fractions to complete the integration.

(a) $\displaystyle\int \frac{1}{e^x - e^{-x}}\,dx$

(b) $\displaystyle\int \frac{1}{x\left(\sqrt{x} + 1\right)}\,dx$

18.2 Definite integrals

The most difficult part in Example 18.1.3 was not the integration, but getting the result back from an expression in u to an expression in x. If you have a definite integral to find, this last step is not necessary. Instead you can use the substitution equation to change the interval of integration from values of x to values of u.

$$\text{If } x = s(u), \text{ then } \int_a^b f(x)\,dx = \int_p^q g(u) \times \frac{dx}{du}\,du,$$

where $g(u) = f\big(s(u)\big)$, and $p = s^{-1}(a)$, $q = s^{-1}(b)$.

Once again, it is more important to be able to use the result than to remember it in this form.

Example 18.2.1

Find $\displaystyle\int_0^1 \sqrt{4 - x^2}\ dx,$

(a) using the substitution $x = 2\sin u$,
(b) by relating it to an area.

(a) Follow Example 18.1.3 as far as the form of the integral in terms of u, and note that the new limits of integration are $\sin^{-1}\big(\tfrac{1}{2} \times 0\big) = 0$ and $\sin^{-1}\big(\tfrac{1}{2} \times 1\big) = \tfrac{1}{6}\pi$.

Therefore

$$\int_0^1 \sqrt{4 - x^2}\ dx = \int_0^{\frac{1}{6}\pi} 4\cos^2 u\,du = [2u + \sin 2u]_0^{\frac{1}{6}\pi}$$

$$= \tfrac{1}{3}\pi + \sin\tfrac{1}{3}\pi = \tfrac{1}{3}\pi + \tfrac{1}{2}\sqrt{3}.$$

(b) If $y = \sqrt{4 - x^2}$, then $x^2 + y^2 = 4$, which is the equation of a circle with centre the origin and radius 2. The integral therefore represents the area of the region under the upper semicircle from $x = 0$ to $x = 1$, shown shaded in Fig. 18.1. This region consists of a sector with angle $\tfrac{1}{6}\pi$ and a triangle with base 1 and height $\sqrt{3}$. The value of the integral is therefore

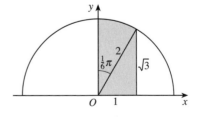

Fig. 18.1

$$\tfrac{1}{2} \times 2^2 \times \big(\tfrac{1}{6}\pi\big) + \tfrac{1}{2} \times 1 \times \sqrt{3} = \tfrac{1}{3}\pi + \tfrac{1}{2}\sqrt{3}.$$

Example 18.2.2

Find $\displaystyle\int_0^1 \frac{1}{1+x^2}\,dx$ using the substitution $x = \tan u$.

The substitution $x = \tan u$ makes $1 + x^2 = 1 + \tan^2 u = \sec^2 u$ (see Section 5.2). Also $\dfrac{dx}{du} = \sec^2 u$ (see Section 7.2), $\tan^{-1} 0 = 0$ and $\tan^{-1} 1 = \frac{1}{4}\pi$. Therefore

$$\int_0^1 \frac{1}{1+x^2}\,dx = \int_0^{\frac{1}{4}\pi} \frac{1}{\sec^2 u} \times \sec^2 u\,du = \int_0^{\frac{1}{4}\pi} 1\,du = [u]_0^{\frac{1}{4}\pi} = \frac{1}{4}\pi.$$

Exercise 18B

In Questions 3 and 4 you are required to find the substitution for yourself.

1 Use the given substitutions to find the following integrals.

(a) $\displaystyle\int_0^1 \frac{e^x}{1+e^x}\,dx$ $\qquad x = \ln u$ $\qquad$ (b) $\displaystyle\int_9^{16} \frac{1}{x - 2\sqrt{x}}\,dx$ $\qquad x = u^2$

(c) $\displaystyle\int_1^2 x(x-1)^5\,dx$ $\qquad x = 1 + u$ $\qquad$ (d) $\displaystyle\int_1^2 x\sqrt{x-1}\,dx$ $\qquad x = 1 + u$

(e) $\displaystyle\int_0^1 \frac{1}{\sqrt{4-x^2}}\,dx$ $\qquad x = 2\sin u$ $\qquad$ (f) $\displaystyle\int_6^9 \frac{x^2}{\sqrt{x-5}}\,dx$ $\qquad x = 5 + u$

(g) $\displaystyle\int_{-4}^4 \sqrt{16 - x^2}\,dx$ $\qquad x = 4\sin u$ $\qquad$ (h) $\displaystyle\int_1^6 \frac{1}{4+x^2}\,dx$ $\qquad x = 2\tan u$

(i) $\displaystyle\int_e^{e^2} \frac{1}{x(\ln x)^2}\,dx$ $\qquad x - e^u$ $\qquad$ (j) $\displaystyle\int_0^{\frac{1}{2}} \frac{1}{\left(1 - x^2\right)^{\frac{3}{2}}}\,dx$ $\qquad x = \sin u$

(k) $\displaystyle\int_1^8 \frac{1}{x\left(1 + \sqrt[3]{x}\right)}\,dx$ $\qquad x = u^3$

2 Use the substitution $x = \sin^2 u$ to calculate $\displaystyle\int_0^{\frac{1}{2}} \sqrt{\frac{x}{1-x}}\,dx$.

3* Use trigonometric substitutions to evaluate the following infinite and improper integrals.

(a) $\displaystyle\int_0^\infty \frac{1}{x^2+4}\,dx$ $\qquad$ (b) $\displaystyle\int_0^3 \frac{1}{\sqrt{9-x^2}}\,dx$ $\qquad$ (c) $\displaystyle\int_{-\infty}^\infty \frac{1}{9x^2+4}\,dx$

(d) $\displaystyle\int_0^1 \frac{1}{\sqrt{x(1-x)}}\,dx$ $\qquad$ (e) $\displaystyle\int_1^\infty \frac{1}{\left(1+x^2\right)^{\frac{3}{2}}}\,dx$ $\qquad$ (f) $\displaystyle\int_1^\infty \frac{1}{x\sqrt{x^2-1}}\,dx$

4* Evaluate the following infinite integrals by using suitable substitutions.

(a) $\displaystyle\int_2^\infty \frac{1}{x(\ln x)^3}\,dx$ (b) $\displaystyle\int_{\ln 2}^\infty \frac{1}{e^x - 1}\,dx$ (c) $\displaystyle\int_1^\infty \frac{1}{x(x + \sqrt{x})}\,dx$ (d) $\displaystyle\int_0^\infty x e^{-\frac{1}{2}x^2}\,dx$

5* Use a substitution, followed by a change of letter in the integrand, to show that, if $\sigma > 0$,

$$\int_{-\infty}^{\infty} e^{-\frac{x^2}{2\sigma^2}}\,dx = \sigma \int_{-\infty}^{\infty} e^{-\frac{1}{2}x^2}\,dx.$$

6* Use the substitution $x = \dfrac{a^2}{u}$, where $a > 0$, to show that

$$\int_0^a \frac{1}{a^2 + x^2}\,dx = \int_a^{\infty} \frac{1}{a^2 + x^2}\,dx.$$

7* Use a substitution to prove that $\displaystyle\int_0^{\pi} x \sin x\,dx = \int_0^{\pi} (\pi - x)\sin x\,dx$. Hence show that

$$\int_0^{\pi} x \sin x\,dx = \tfrac{1}{2}\pi \int_0^{\pi} \sin x\,dx, \text{ and evaluate this.}$$

18.3 Reverse substitution

If $y = \sqrt{1 + \ln x}$, then the chain rule gives $\dfrac{dy}{dx} = \dfrac{1}{2\sqrt{1 + \ln x}} \times \dfrac{1}{x}$. So, turning this into integral form,

$$\int \frac{1}{\sqrt{1 + \ln x}} \times \frac{1}{x}\,dx = 2\sqrt{1 + \ln x} + k.$$

But how could you find this integral if you didn't know the answer to start with? You can see that the integrand is the product of two factors. The first of these has the form of a composite function of x; you could write it as $\dfrac{1}{\sqrt{u}}$, where $u = 1 + \ln x$. The lucky break is that the second factor, $\dfrac{1}{x}$, is the derivative $\dfrac{du}{dx}$. So the integral can be written as

$$\int \frac{1}{\sqrt{u}} \times \frac{du}{dx}\,dx,$$

which can be worked out as

$$\int \frac{1}{\sqrt{u}}\,du = 2\sqrt{u} + k = 2\sqrt{1 + \ln x} + k.$$

This seems to be a different form of integration by substitution, in which you can already see the derivative $\dfrac{du}{dx}$ as part of the integrand.

To describe it in general terms, write $\dfrac{1}{\sqrt{1 + \ln x}}$ as $\mathrm{f}(x)$, $u = 1 + \ln x$ as $\mathrm{r}(x)$ and $\dfrac{1}{\sqrt{u}}$ as $\mathrm{g}(u)$, so $\mathrm{f}(x) = \mathrm{g}(\mathrm{r}(x))$ and $\dfrac{du}{dx} = \dfrac{1}{x}$. You then get:

If $u = \mathrm{r}(x)$, and if $\mathrm{g}(\mathrm{r}(x)) = \mathrm{f}(x)$, then $\displaystyle\int \mathrm{f}(x) \times \frac{du}{dx}\,dx$ is equal to $\displaystyle\int \mathrm{g}(u)\,du$, with u replaced by $\mathrm{r}(x)$.

You can check that this is in effect the same as the statement in Section 18.1, with f and g, x and u interchanged, and r written in place of s. But the method of applying it is different, because you need to begin by identifying the derivative $\dfrac{du}{dx}$ as a factor in the integrand.

As before, do not memorise the general statement, but learn to use the method by studying some examples.

Example 18.3.1

Find $\displaystyle\int x^2 \sqrt{1 + x^3}\,dx$.

Begin by noticing that the derivative of $1 + x^3$ is $3x^2$, so that if the integral is written as

$$\int \tfrac{1}{3}\sqrt{1 + x^3} \times 3x^2\,dx,$$

then it can be changed into the form

$$\int \tfrac{1}{3}\left(1 + x^3\right)^{\frac{1}{2}} \times \frac{du}{dx}\,dx$$

with $u = 1 + x^3$. This is equal to

$$\int \tfrac{1}{3}u^{\frac{1}{2}}\,du = \tfrac{2}{9}u^{\frac{3}{2}} + k,$$

with u replaced by $1 + x^3$. That is,

$$\int x^2 \sqrt{1 + x^3}\,dx = \tfrac{2}{9}\left(1 + x^3\right)^{\frac{3}{2}} + k.$$

Example 18.3.2

Find $\displaystyle\int_0^{\frac{1}{2}\pi} \cos^4 x \sin x\,dx$.

If the integrand is written as $-\cos^4 x \times (-\sin x)$, then the second factor is $\dfrac{du}{dx}$ with $u = \cos x$.

$$\int_0^{\frac{1}{2}\pi} \cos^4 x \sin x\,dx = \int_0^{\frac{1}{2}\pi} -\cos^4 x \times \frac{du}{dx}\,dx$$

$$= \int_1^0 -u^4\,du = -\left[\tfrac{1}{5}u^5\right]_1^0 = \tfrac{1}{5}.$$

Notice that the limits of integration change from 0, $\frac{1}{2}\pi$ to 1, 0 at the step where the integral changes from $\int_0^{\frac{1}{2}\pi} -\cos^4 x \times \dfrac{du}{dx}\,dx$ to $\int_1^0 -u^4\,du$. Since $\cos x$ is decreasing over the interval $0 \leqslant x \leqslant \frac{1}{2}\pi$, the limits for u appear in reversed order.

Example 18.3.3

Find $\displaystyle\int \dfrac{\cos x - \sin x}{\sin x + \cos x}\,dx$.

Write this as $\displaystyle\int \dfrac{1}{\sin x + \cos x} \times (\cos x - \sin x)\,dx$. If $u = \sin x + \cos x$, this is

$\displaystyle\int \dfrac{1}{\sin x + \cos x} \times \dfrac{du}{dx}\,dx$, which is $\displaystyle\int \dfrac{1}{u}\,du = \ln|u| + k$.

So $\displaystyle\int \dfrac{\cos x - \sin x}{\sin x + \cos x}\,dx = \ln|\sin x + \cos x| + k$.

In this last example the integral has the form $\displaystyle\int \dfrac{f'(x)}{f(x)}\,dx$, where $f(x) = \sin x + \cos x$. This type of integral often arises, and the result is important:

$$\int \dfrac{f'(x)}{f(x)}\,dx = \ln|f(x)| + k.$$

Example 18.3.4

Find $\displaystyle\int_0^1 \dfrac{e^x - e^{-x}}{e^x + e^{-x}}\,dx$.

The integrand is $\dfrac{f'(x)}{f(x)}$ with $f(x) = e^x + e^{-x}$, so the value of the integral is

$$\left[\ln|e^x + e^{-x}|\right]_0^1 = \ln\left(e + \dfrac{1}{e}\right) - \ln 2 = \ln\left(\dfrac{e^2 + 1}{2e}\right).$$

Exercise 18C

In Questions 2, 5 and 6 you are required to find the substitution for yourself.

1 Use the given substitutions to find the following integrals.

(a) $\displaystyle\int 2x(x^2 + 1)^3\,dx$ $u = x^2 + 1$ (b) $\displaystyle\int x\sqrt{4 + x^2}\,dx$ $u = 4 + x^2$

(c) $\displaystyle\int \sin^5 x \cos x\,dx$ $u = \sin x$ (d) $\displaystyle\int \tan^3 x \sec^2 x\,dx$ $u = \tan x$

(e) $\displaystyle\int \dfrac{2x^3}{\sqrt{1 - x^4}}\,dx$ $u = 1 - x^4$ (f) $\displaystyle\int \cos^3 2x \sin 2x\,dx$ $u = \cos 2x$

2* Find the following integrals by using suitable substitutions.

(a) $\displaystyle\int x\left(1-x^2\right)^5 dx$

(b) $\displaystyle\int x\sqrt{3-2x^2}\,dx$

(c) $\displaystyle\int x^2\left(5-3x^3\right)^6 dx$

(d) $\displaystyle\int \frac{x^2}{\sqrt{1+x^3}}\,dx$

(e) $\displaystyle\int \sec^4 x\tan x\,dx$

(f) $\displaystyle\int \sin^3 4x\cos 4x\,dx$

3 Without carrying out a substitution, write down the following indefinite integrals.

(a) $\displaystyle\int \frac{\cos x}{1+\sin x}\,dx$

(b) $\displaystyle\int \frac{x^2}{1+x^3}\,dx$

(c) $\displaystyle\int \cot x\,dx$

(d) $\displaystyle\int \frac{e^x}{4+e^x}\,dx$

(e) $\displaystyle\int \frac{2e^{3x}}{5-e^{3x}}\,dx$

(f) $\displaystyle\int \tan 3x\,dx$

4 Evaluate each of the following integrals, giving your answer in an exact form.

(a) $\displaystyle\int_1^2 \frac{e^x}{e^x-1}\,dx$

(b) $\displaystyle\int_4^5 \frac{x-2}{x^2-4x+5}\,dx$

(c) $\displaystyle\int_0^{\frac{1}{6}\pi} \frac{\sin 2x}{1+\cos 2x}\,dx$

5* Evaluate each of the following integrals, giving your answer in an exact form.

(a) $\displaystyle\int_0^{\frac{1}{2}\pi} \frac{\cos x}{\sqrt{1+3\sin x}}\,dx$

(b) $\displaystyle\int_0^2 x\left(x^2+1\right)^3 dx$

(c) $\displaystyle\int_0^{\frac{1}{4}\pi} \sin x\cos^2 x\,dx$

(d) $\displaystyle\int_1^8 (1+2x)\sqrt{x+x^2}\,dx$

(e) $\displaystyle\int_0^{\frac{1}{3}\pi} \frac{\sin x}{(1+\cos x)^2}\,dx$

(f) $\displaystyle\int_0^3 2x\sqrt{1+x^2}\,dx$

(g) $\displaystyle\int_0^{\frac{1}{4}\pi} \sec^2 x\tan^2 x\,dx$

(h) $\displaystyle\int_1^e \frac{(\ln x)^n}{x}\,dx$

(i) $\displaystyle\int_0^{\frac{1}{3}\pi} \sec^3 x\tan x\,dx$

6* Find an expression, in terms of n and a, for $\displaystyle\int_0^a \frac{x}{\left(1+x^2\right)^n}\,dx$. For what values of n does

$\displaystyle\int_0^\infty \frac{x}{\left(1+x^2\right)^n}\,dx$ exist? State its value in terms of n.

18.4 Integration by parts

Another method of integration depends on the product rule for differentiation. For example, from

$$\frac{d}{dx}(x\sin x)=\sin x+x\cos x$$

you can deduce that

$$x\sin x=\int \sin x\,dx+\int x\cos x\,dx=-\cos x+k+\int x\cos x\,dx.$$

You can rearrange this to give the new result

$$\int x\cos x\,\mathrm{d}x = x\sin x + \cos x - k.$$

But if you were asked to find $\int x\cos x\,\mathrm{d}x,$ you would not immediately guess that the answer comes from differentiating $x\sin x$. You can overcome this by applying the same argument to the general product rule.

From $\dfrac{\mathrm{d}}{\mathrm{d}x}(uv) = \dfrac{\mathrm{d}u}{\mathrm{d}x}v + u\dfrac{\mathrm{d}v}{\mathrm{d}x}$ you can deduce that

$$uv = \int \frac{\mathrm{d}u}{\mathrm{d}x}v\,\mathrm{d}x + \int u\frac{\mathrm{d}v}{\mathrm{d}x}\,\mathrm{d}x.$$

If you can find one of the integrals on the right, this equation tells you the other. It can be rearranged to give the rule:

Integration by parts

$$\int u\frac{\mathrm{d}v}{\mathrm{d}x}\,\mathrm{d}x = uv - \int \frac{\mathrm{d}u}{\mathrm{d}x}v\,\mathrm{d}x.$$

For example, if you want to integrate $x\cos x$, you write $u = x$ and find a function v such that $\dfrac{\mathrm{d}v}{\mathrm{d}x} = \cos x$. The simplest function is $v = \sin x$. The rule gives

$$\int x\cos x\,\mathrm{d}x = x\sin x - \int 1\times\sin x\,\mathrm{d}x$$
$$= x\sin x + \cos x + k.$$

Notice that the result at the top of the page has a constant $-k$, and the same integral here has a constant $+k$. It is not difficult to see that the two forms are equivalent.

Example 18.4.1

Find $\int xe^{3x}\,\mathrm{d}x.$

Take $u = x$ and find v such that $\dfrac{\mathrm{d}v}{\mathrm{d}x} = e^{3x}$. The simplest function for v is $\frac{1}{3}e^{3x}$. The rule gives

$$\int xe^{3x}\,\mathrm{d}x = x\times\frac{1}{3}e^{3x} - \int 1\times\frac{1}{3}e^{3x}\,\mathrm{d}x$$
$$= \frac{1}{3}xe^{3x} - \frac{1}{9}e^{3x} + k$$
$$= \frac{1}{9}(3x-1)e^{3x} + k.$$

The next example applies the method to a definite integral. The rule then takes the form:

$$\int_a^b u \frac{dv}{dx}\, dx = \left[uv \right]_a^b - \int_a^b \frac{du}{dx} v\, dx.$$

Example 18.4.2

Find $\displaystyle\int_2^8 x \ln x\, dx$.

If you write $u = x$, you need v to satisfy $\dfrac{dv}{dx} = \ln x$. But although Section 4.3 gave the derivative of $\ln x$, its integral is not yet known. (See Example 18.4.3.)

When this occurs, try writing the product the other way round. Take $u = \ln x$, and find a v such that $\dfrac{dv}{dx} = x$, which is $v = \frac{1}{2} x^2$. The rule then gives

$$\int_2^8 x \ln x\, dx = \left[\ln x \times \tfrac{1}{2} x^2 \right]_2^8 - \int_2^8 \frac{1}{x} \times \tfrac{1}{2} x^2\, dx$$

$$= 32 \ln 8 - 2 \ln 2 - \int_2^8 \tfrac{1}{2} x\, dx$$

$$= 32 \ln 2^3 - 2 \ln 2 - \left[\tfrac{1}{4} x^2 \right]_2^8$$

$$= 32(3 \ln 2) - 2 \ln 2 - (16 - 1)$$

$$= 94 \ln 2 - 15.$$

It is usually best to leave the answer in a simple exact form like this. If you need a numerical value, it is easy enough to calculate one.

Example 18.4.3

Find $\displaystyle\int \ln x\, dx$.

You wouldn't at first expect to use integration by parts for this, since it doesn't appear to be a product. But taking u as $u = \ln x$ and $\dfrac{dv}{dx} = 1$, so that $v = x$, the rule gives

$$\int \ln x\, dx = \ln x \times x - \int \frac{1}{x} \times x\, dx = x \ln x - \int 1\, dx$$

$$= x \ln x - x + k.$$

The integral of $\ln x$ is an important result. You need not remember the answer, but you should remember how to get it.

The next example concerns two integrals which are used in probability.

Example 18.4.4

Find $\displaystyle\int_0^\infty x e^{-ax}\,dx$ and $\displaystyle\int_0^\infty x^2 e^{-ax}\,dx$, where a is positive.

Begin by finding the integrals from 0 to s, and then consider their limits as $s \to \infty$.

For both integrals take $\dfrac{dv}{dx} = e^{-ax}$, so $v = -\dfrac{1}{a} e^{-ax}$.

$$\int_0^s x e^{-ax}\,dx = \left[x \times \left(-\frac{1}{a}\right) e^{-ax} \right]_0^s - \int_0^s 1 \times \left(-\frac{1}{a}\right) e^{-ax}\,dx$$

$$= -\frac{1}{a} s e^{-as} - \left[\frac{1}{a^2} e^{-ax} \right]_0^s$$

$$= -\frac{1}{a} s e^{-as} - \frac{1}{a^2} e^{-as} + \frac{1}{a^2}.$$

$$\int_0^s x^2 e^{-ax}\,dx = \left[x^2 \times \left(-\frac{1}{a}\right) e^{-ax} \right]_0^s - \int_0^s 2x \times \left(-\frac{1}{a}\right) e^{-ax}\,dx$$

$$= -\frac{1}{a} s^2 e^{-as} + \frac{2}{a} \int_0^s x e^{-ax}\,dx.$$

The integral in the last line here is the integral that has just been found, so you could now use the first answer to complete the evaluation of $\displaystyle\int_0^s x^2 e^{-ax}\,dx$.

For the infinite integral, you need to know the limits of e^{-as}, se^{-as} and $s^2 e^{-as}$ as $s \to \infty$. You know that $e^{-as} \to 0$, and Miscellaneous exercise 7 Question 11 shows that also $s e^{-as} \to 0$ and $s^2 e^{-as} \to 0$.

It follows that $\displaystyle\int_0^\infty x e^{-ax}\,dx = \frac{1}{a^2}$,

and that $\displaystyle\int_0^\infty x^2 e^{-ax}\,dx = \frac{2}{a} \int_0^\infty x e^{-ax}\,dx = \frac{2}{a^3}$.

Exercise 18D

1   Use integration by parts to integrate the following functions with respect to x.

    (a) $x \sin x$ (b) $3x e^x$ (c) $(x+4) e^x$

2   Use integration by parts to integrate the following functions with respect to x.

    (a) $x e^{2x}$ (b) $x \cos 4x$ (c) $x \ln 2x$

3   Find

    (a) $\displaystyle\int x^5 \ln 3x\,dx$,   (b) $\displaystyle\int x e^{2x+1}\,dx$,   (c) $\displaystyle\int \ln 2x\,dx$.

4 Find the exact values of

(a) $\displaystyle\int_1^e x\ln x\,dx,$

(b) $\displaystyle\int_0^{\frac{1}{2}\pi} x\sin\tfrac{1}{2}x\,dx,$

(c) $\displaystyle\int_1^e x^n\ln x\,dx \quad (n>0).$

5 Prove that $\displaystyle\int x^2\sin x\,dx = -x^2\cos x + 2\int x\cos x\,dx.$ Hence, by using integration by parts a second time, find $\displaystyle\int x^2\sin x\,dx.$ Use a similar method to integrate the following functions with respect to x.

(a) $x^2 e^{2x}$

(b) $x^2\cos\tfrac{1}{2}x$

6 Find the area bounded by the curve $y = xe^{-x}$, the x-axis and the lines $x=0$ and $x=2$. Find also the volume of the solid of revolution obtained by rotating this region about the x-axis.

7 Find the area between the x-axis and the curve $y = x\sin 3x$ for $0 \le x \le \tfrac{1}{3}\pi$. Leave your answer in terms of π. Find also the volume of the solid of revolution obtained by rotating this region about the x-axis.

8 Find

(a) $\displaystyle\int_0^{\pi} e^x\cos x\,dx,$

(b) $\displaystyle\int_{-\pi}^{\pi} e^{-4x}\sin 2x\,dx,$

(c) $\displaystyle\int_0^{2\pi} e^{-ax}\cos bx\,dx.$

9 Find

(a) $\displaystyle\int_0^{\infty} e^{-x}\sin x\,dx,$

(b) $\displaystyle\int_0^1 \frac{x}{\sqrt{1-x}}\,dx.$

Draw diagrams to illustrate the areas measured by these definite integrals.

Miscellaneous exercise 18

In Questions 3 and 4 you are required to find the substitution for yourself.

1 By using the substitution $u = 2x-1$, or otherwise, find $\displaystyle\int \frac{2x}{(2x-1)^2}\,dx.$ (OCR)

2 Use integration by parts to find the value of $\displaystyle\int_1^2 x\ln x\,dx.$ (OCR)

3* Find $\displaystyle\int \frac{1}{4x^2+9}\,dx.$ (OCR)

4* By using a suitable substitution, or otherwise, evaluate $\displaystyle\int_0^1 x(1-x)^9\,dx.$ (OCR)

5 Use integration by parts to determine $\displaystyle\int_0^{\frac{1}{3}} xe^{2x}\,dx.$ (OCR)

6 Use the given substitution and then use integration by parts to complete the integration.

(a) $\displaystyle\int \cos^{-1} x \, dx \quad x = \cos u$ (b) $\displaystyle\int \tan^{-1} x \, dx \quad x = \tan u$ (c) $\displaystyle\int (\ln x)^2 \, dx \quad x = e^u$

7 Use the substitution $x = \sin u$ to find $\displaystyle\int_0^1 \sin^{-1} x \, dx$.

8 Find $\displaystyle\int \frac{1}{e^x + 4e^{-x}} \, dx$, by means of the substitution $u = e^x$, followed by another substitution, or otherwise. (OCR, adapted)

9 Find $\displaystyle\int \frac{6x}{1+3x^2} \, dx$. (OCR)

10 Calculate the exact value of $\displaystyle\int_0^3 \frac{x}{1+x^2} \, dx$. (OCR)

11 By using the substitution $u = \sin x$, or otherwise, find $\displaystyle\int \sin^3 x \sin 2x \, dx$, giving your answer in terms of x. (OCR)

12 By means of the substitution $u = 1 + \sqrt{x}$, or otherwise, find $\displaystyle\int \frac{1}{1+\sqrt{x}} \, dx$, giving your answer in terms of x. (OCR)

13 Integrate with respect to x, (i) by using a substitution of the form $ax + b = u$, and (ii) by parts, and show that your answers are equivalent.

(a) $x\sqrt{4x-1}$, (b) $x\sqrt{2-x}$, (c) $x\sqrt{2x+3}$.

14 Use the substitution $u = \ln x$ to show that $\displaystyle\int_e^{e^2} \frac{1}{x\sqrt{\ln x}} \, dx = 2\sqrt{2} - 2$. (OCR)

15 Use the substitution $u = 4 + x^2$ to show that $\displaystyle\int_0^1 \frac{x^3}{\sqrt{4+x^2}} \, dx = \tfrac{1}{3}\left(16 - 7\sqrt{5}\right)$. (OCR)

16 Use the substitution $u = 3x - 1$ to express $\displaystyle\int x(3x-1)^4 \, dx$ as an integral in terms of u.

Hence, or otherwise, find $\displaystyle\int x(3x-1)^4 \, dx$, giving your answer in terms of x. (OCR)

17 Show, by means of the substitution $x = \tan\theta$, that $\displaystyle\int_0^1 \frac{1}{(x^2+1)^2} \, dx = \int_0^{\frac{1}{4}\pi} \cos^2\theta \, d\theta$.

Hence find the exact value of $\displaystyle\int_0^1 \frac{1}{(x^2+1)^2} \, dx$.

18 Find

(a) $\displaystyle\int x(1+x)^6 \, dx$, (b) $\displaystyle\int x(3x-1)^4 \, dx$, (c) $\displaystyle\int x(ax+b)^{12} \, dx$.

19 Evaluate $\displaystyle\int_0^1 xe^{-x} \, dx$, showing all your working. (OCR)

20 Showing your working clearly, use integration by parts to evaluate $\int_0^{\pi} 4x\sin\frac{1}{2}x\,dx$. (OCR)

21 By using the substitution $u = 3x + 1$, or otherwise, show that

$$\int_0^1 \frac{x}{(3x+1)^2}\,dx = \frac{2}{9}\ln 2 - \frac{1}{12}.$$

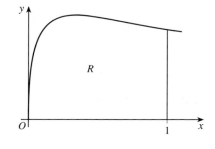

The diagram shows the finite region R in the first quadrant which is bounded by the curve $y = \dfrac{6\sqrt{x}}{3x+1}$, the x-axis and the line

$x = 1$. Find the volume of the solid formed when R is rotated completely about the x-axis, giving your answer in terms of π and $\ln 2$. (OCR)

22 Use integration by parts to determine $\int 3x\sqrt{x-1}\,dx$. (OCR)

23 Use the trapezium rule with subdivisions at $x = 3$ and $x = 5$ to obtain an approximation to $\int_1^7 \dfrac{x^3}{1+x^4}\,dx$, giving your answer correct to three places of decimals.

By evaluating the integral exactly, show that the error in the approximation is about 4.1%. (OCR)

24 The diagram (not to scale) shows the region R bounded by the axes, the curve $y = \left(x^2 + 1\right)^{-\frac{3}{2}}$ and the line $x = 1$. The integral

$$\int_0^1 \left(x^2 + 1\right)^{-\frac{3}{2}}\,dx$$

is denoted by I.

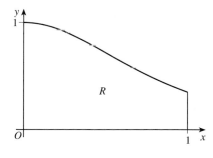

(a) Use the trapezium rule, with ordinates at $x = 0$, $x = \frac{1}{2}$ and $x = 1$, to estimate the value of I, giving your answer correct to 2 significant figures.

(b) Use the substitution $x = \tan\theta$ to show that $I = \frac{1}{2}\sqrt{2}$.

(c) By using the trapezium rule, with the same ordinates as in part (a), or otherwise, estimate the volume of the solid formed when R is rotated completely about the x-axis, giving your answer correct to 3 significant figures.

(d) Find the exact value of the volume in part (c), and compare your answers to parts (c) and (d). (OCR, adapted)

25 Use integration by parts to determine the exact value of $\int_0^{\frac{1}{2}\pi} 3x\sin 2x\,dx$. (OCR)

26* The figure shows part of a *cycloid*, given by the parametric equations

$$x = a(t - \sin t), \ y = a(1 - \cos t)$$

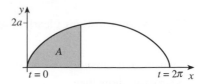

for $0 \leqslant t \leqslant 2\pi$. Show that, if A denotes the region between the cycloid and the x-axis as far as the value of x with parameter t, then

$\dfrac{dA}{dt} = y\dfrac{dx}{dt}$. Deduce that the total area of the region enclosed between this arch of the

cycloid and the x-axis is $\displaystyle\int_0^{2\pi} y\dfrac{dx}{dt}\,dt$. Calculate this area in terms of a.

Use a similar method to find the volume of the solid of revolution formed when this region is rotated about the x-axis.

27* The figure shows part of a *tractrix*, given by parametric equations

$$x = c\ln(\sec t + \tan t) - c\sin t, \ y = c\cos t$$

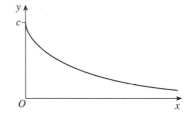

for $0 \leqslant t < \tfrac{1}{2}\pi$. Find the area of the region between the tractrix and the positive x-axis, and the volume of the solid of revolution formed when this region is rotated about the x-axis.

28* Show that the area enclosed by the astroid in Example 10.4.2 is given by the integral

$\displaystyle\int_0^{\frac{1}{2}\pi} 12a^2 \sin^4 t \cos^2 t\,dt$. Use the substitution $t = \tfrac{1}{2}\pi - u$ to show that the area could also

be calculated as $\displaystyle\int_0^{\frac{1}{2}\pi} 12a^2 \cos^4 t \sin^2 t\,dt$.

Prove that $\sin^4 t \cos^2 t + \cos^4 t \sin^2 t = \tfrac{1}{8}(1 - \cos 4t)$, and deduce that the area enclosed by

the astroid is equal to $\displaystyle a^2 \int_0^{\frac{1}{2}\pi} \tfrac{3}{4}(1 - \cos 4t)\,dt$. Evaluate this area.

29* (a) Given that

$$\frac{2}{(x-1)^2(x^2+1)} \equiv \frac{A}{x-1} + \frac{B}{(x-1)^2} + \frac{Cx}{x^2+1},$$

find the values of the constants A, B and C.

    (b) Show that $\displaystyle\int_2^3 \frac{2}{(x-1)^2(x^2+1)}\,dx = a + b\ln 2$, where a and b are constants whose

values you should find.   (OCR)

30 Calculate $\displaystyle\int_1^2 \frac{2x^3 + 3x^2 + 28}{(x+2)(x^2+4)}\,dx$ giving your answer in exact form.

19 Differential equations

This chapter is about differential equations of the form $\dfrac{dy}{dx} = f(x)$, $\dfrac{dy}{dx} = f(y)$ or $\dfrac{dy}{dx} = \dfrac{f(x)}{g(y)}$. When you have completed it, you should

- be able to find general solutions of these equations, or particular solutions satisfying given initial conditions
- know the relation connecting the derivatives $\dfrac{dy}{dx}$ and $\dfrac{dx}{dy}$, and understand its significance
- be able to formulate differential equations as models, and interpret the solutions.

19.1 Forming and solving equations

Many applications of mathematics involve two variables, and you want to find a relation between them. Often this relation is expressed in terms of the rate of change of one variable with respect to the other. This then leads to a **differential equation**. Its **solution** will be an equation connecting the two variables.

Example 19.1.1

At each point P of a curve for which $x > 0$ the tangent cuts the y-axis at T, and N is the foot of the perpendicular from P to the y-axis (see Fig. 19.1). If T is always 1 unit below N, find the equation of the curve.

Since $NP = x$, the gradient of the tangent is $\dfrac{1}{x}$, so that

$$\frac{dy}{dx} = \frac{1}{x}.$$

This can be integrated directly to give

$$y = \ln x + k.$$

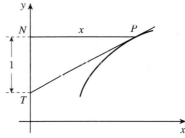

Fig. 19.1

The modulus sign is not needed, since $x > 0$.

The equation $y = \ln x + k$ is called the **general solution** of the differential equation $\dfrac{dy}{dx} = \dfrac{1}{x}$ for $x > 0$. It can be represented by a family of graphs, or **solution curves**, one for each value of k. Fig. 19.2 shows just a few typical graphs, but there are in fact infinitely many graphs with the property described.

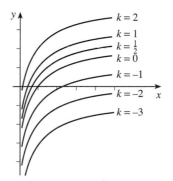

Fig. 19.2

A differential equation often originates from a scientific law or hypothesis. The equation is then called a **mathematical model** of the real-world situation.

Example 19.1.2

A rodent has mass 30 grams at birth. It reaches maturity in 3 months. The rate of growth is modelled by the differential equation $\dfrac{dm}{dt} = 120(t-3)^2$, where m grams is the mass of the rodent t months after birth. Find the mass of the rodent when fully grown.

The differential equation has general solution

$$m = 40(t-3)^3 + k.$$

However, only one equation from this family of solutions is right for this problem. It is given that, when $t = 0$ (at birth), $m = 30$. So k must satisfy the equation

$$30 = 40(0-3)^3 + k, \text{ giving } 30 = -1080 + k; \text{ so } k = 1110.$$

The mass of the rodent after t months is therefore $m = 40(t-3)^3 + 1110$.

The mass when fully grown is found by putting $t = 3$ in this formula, giving $m = 0 + 1110 = 1110$. So the mass of the rodent at maturity is 1110 grams.

In this example the variables have to satisfy an **initial condition**, or **boundary condition**, that $m = 30$ when $t = 0$. The equation $m = 40(t-3)^3 + 1110$ is the **particular solution** of the differential equation which satisfies the initial condition.

Example 19.1.3

A botanist working in the UK makes a hypothesis that the rate of growth of hot-house plants is proportional to the amount of daylight they receive. If t is the time in years after the shortest day of the year, the length of effective daylight is given by the formula $12 - 4\cos 2\pi t$ hours. On the shortest day in the December of one year the height of one plant is measured to be 123.0 cm; 55 days later the height is 128.0 cm. What will its height be on the longest day of the year in the following June?

If h is the height in centimetres, the rate of growth is given by

$$\frac{dh}{dt} = c(12 - 4\cos 2\pi t),$$

but c, the constant of proportionality, is not known. Nor can it be found directly from the data. However, integration gives

$$h = c\left(12t - \frac{2}{\pi}\sin 2\pi t\right) + k.$$

The initial condition is that $h = 123.0$ when $t = 0$, so

$$123.0 = c(0-0) + k, \text{ giving } k = 123.0. \text{ Therefore}$$

$$h = c\left(12t - \frac{2}{\pi}\sin 2\pi t\right) + 123.0.$$

After 55 days the value of t is $\frac{55}{365} = 0.150\ldots$, and it is given that at this time $h = 128.0$. So

$$128.0 = c\left(12 \times 0.150\ldots - \frac{2}{\pi}\sin(2\pi \times 0.150\ldots)\right) + 123.0$$

$$= c(1.80\ldots - 0.51\ldots) + 123.0, \text{which gives } c = 3.87\ldots.$$

The longest day occurs when $t = \frac{1}{2}$, and then

$$h = 3.87\ldots \times (6-0) + 123.0 = 146, \text{correct to 3 significant figures.}$$

According to the botanist's hypothesis, the height of the plant on the longest day will be 146 cm.

This last example is typical of many applications of differential equations. Often the form of a hypothetical law is known, but the values of the numerical constants are not. But once the differential equation has been solved, experimental data can be used to find values for the constants.

Exercise 19A

When you have found a solution to a differential equation, it is often helpful to sketch its graph and to look at its features in relation to the original differential equation.

1 Find general solutions of the following differential equations.

(a) $\dfrac{dy}{dx} = (3x - 1)(x - 3)$

(b) $\dfrac{dx}{dt} = \sin^2 3t$

(c) $\dfrac{dP}{dt} = 50e^{0.1t}$

(d) $e^{2t}\dfrac{du}{dt} = 100$

(e) $\sqrt{x}\,\dfrac{dy}{dx} = x + 1,$ for $x > 0$

(f) $\sin t\,\dfrac{dx}{dt} = \cos t + \sin 2t,$ for $0 < t < \pi$

2 Solve the following differential equations with the given initial conditions.

(a) $\dfrac{dx}{dt} = 2e^{0.4t},\ x = 1$ when $t = 0$

(b) $\dfrac{dv}{dt} = 6(\sin 2t - \cos 3t),\ v = 0$ when $t = 0$

(c) $\left(1 - t^2\right)\dfrac{dy}{dt} = 2t,\ y = 0$ when $t = 0,$ for $-1 < t < 1$

3 Find the solution curves of the following differential equations which pass through the given points.

(a) $\dfrac{dy}{dx} = \dfrac{x-1}{x^2},$ through $(1,0),$ for $x > 0$

(b) $\dfrac{dy}{dx} = \dfrac{1}{\sqrt{x}},$ through $(4,0),$ for $x > 0$

(c) $(x+1)\dfrac{dy}{dx} = x - 1,$ through the origin, for $x > -1$

4 In starting from rest, the driver of an electric car depresses the throttle gradually. If the speed of the car after t seconds is v m s^{-1}, the acceleration $\dfrac{dv}{dt}$ (in metre-second units) is given by $0.2t$. How long does it take for the car to reach a speed of 20 m s^{-1}?

5 The solution curve for a differential equation of the form $\dfrac{dy}{dx} = x - \dfrac{a}{x^2}$ for $x > 0$, passes through the points $(1,0)$ and $(2,0)$. Find the value of y when $x = 3$.

6 A point moves on the x-axis so that its coordinate at time t satisfies the differential equation $\dfrac{dx}{dt} = 5 + a\cos 2t$ for some value of a. It is observed that $x = 3$ when $t = 0$, and $x = 0$ when $t = \frac{1}{4}\pi$. Find the value of a, and the value of x when $t = \frac{1}{3}\pi$.

7 The normal to a curve at a point P cuts the y-axis at T, and N is the foot of the perpendicular from P to the y-axis. If, for all P, T is always 1 unit below N, find the equation of the curve.

8 Water is leaking slowly out of a tank. The depth of the water after t hours is h metres, and these variables are related by a differential equation of the form $\dfrac{dh}{dt} = -ae^{-0.1t}$. Initially the depth of water is 6 metres, and after 2 hours it has fallen to 5 metres. At what depth will the level eventually settle down?

Find an expression for $\dfrac{dh}{dt}$ in terms of h.

9 Four theories are proposed about the growth of an organism:

   (a) It grows at a constant rate of k units per year.

   (b) It only grows when there is enough daylight, so that its rate of growth at time t years is $k\left(1 - \frac{1}{2}\cos 2\pi t\right)$ units per year.

   (c) Its growth is controlled by the 10-year sunspot cycle, so that its rate of growth at time t years is $k\left(1 + \frac{1}{4}\cos\frac{1}{5}\pi t\right)$ units per year.

   (d) Both (b) and (c) are true, so that its rate of growth is $k\left(1 - \frac{1}{2}\cos 2\pi t\right)\left(1 + \frac{1}{4}\cos\frac{1}{5}\pi t\right)$ units per year.

   The size of the organism at time $t = 0$ is A units. For each model, find an expression for the size of the organism at time t years. Do they all give the same value for the size of the organism after 10 years?

19.2 Independent and dependent variables

In many applications there is little doubt which of two variables to regard as the independent variable (often denoted by x), and which as the dependent variable (y). But when a function is one–one, so that an inverse function exists, there are occasions when you can choose to treat either variable as the independent variable.

For example, you could record the progress of a journey either by noting the distance you have gone at certain fixed times, or by noting the time when you pass certain fixed landmarks. If x denotes the distance from the start and t the time, then the rate of change would be either $\dfrac{dx}{dt}$ (the speed) or $\dfrac{dt}{dx}$ (which would be measured in a unit such as minutes per kilometre).

What is the connection between $\dfrac{dy}{dx}$ and $\dfrac{dx}{dy}$? The notation suggests that

$$\frac{dx}{dy} = 1 \bigg/ \frac{dy}{dx},$$

and this is in fact correct.

Fig. 19.3 shows the graph of a relation connecting variables x and y, and triangles showing the increases δx and δy when you move from P to Q. If you are thinking of y as a function of x, then you would draw the triangle PNQ, and the gradient of the chord PQ would be $\dfrac{\delta y}{\delta x}$. For x as a function of y, you would draw the triangle PMQ, and the gradient of PQ would be $\dfrac{\delta x}{\delta y}$.

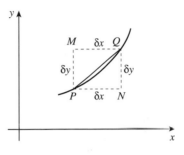

Fig. 19.3

The product of the gradients is $\dfrac{\delta y}{\delta x} \times \dfrac{\delta x}{\delta y}$, which clearly equals 1.

Now let Q tend to P, so that both δx and δy tend to 0. Then $\dfrac{\delta y}{\delta x}$ tends to $\dfrac{dy}{dx}$ and $\dfrac{\delta x}{\delta y}$ tends to $\dfrac{dx}{dy}$. Assuming (as in P1 Section 12.5) that the limit of a product is the product of the limits,

$$\frac{dy}{dx} \times \frac{dx}{dy} = \lim\left(\frac{\delta y}{\delta x}\right) \times \lim\left(\frac{\delta x}{\delta y}\right) = \lim\left(\frac{\delta y}{\delta x} \times \frac{\delta x}{\delta y}\right) = \lim 1 = 1.$$

Example 19.2.1

Verify that $\dfrac{dy}{dx} \times \dfrac{dx}{dy} = 1$ when $y = x^3$.

If $y = x^3$, then $\dfrac{dy}{dx} = 3x^2$. You can also write the relation $y = x^3$ as $x = y^{\frac{1}{3}}$, so

$$\frac{dx}{dy} = \tfrac{1}{3} y^{-\frac{2}{3}} = \tfrac{1}{3}\left(x^3\right)^{-\frac{2}{3}} = \tfrac{1}{3} x^{-2} = \frac{1}{3x^2}.$$

Therefore $\dfrac{dy}{dx} \times \dfrac{dx}{dy} = 3x^2 \times \dfrac{1}{3x^2} = 1$.

19.3 Switching variables in differential equations

In all the differential equations in Section 19.1 the derivative was given as a formula involving the independent variable, which was x in Example 19.1.1 and t in Examples 19.1.2 and 19.1.3.

Often, however, the derivative is known in terms of the dependent variable. When this occurs, you can use the relation $\dfrac{dy}{dx} \times \dfrac{dx}{dy} = 1$ to turn the differential equation into a form which you know how to solve. That is,

$$\frac{dy}{dx} = f(y) \quad \Leftrightarrow \quad \frac{dx}{dy} = \frac{1}{f(y)} \quad \Leftrightarrow \quad x = \int \frac{1}{f(y)}\,dy.$$

Example 19.3.1

A hot-air balloon can reach a maximum height of 1.25 km, and the rate at which it gains height decreases as it climbs, according to the formula

$$\frac{dh}{dt} = 20 - 16h,$$

where h is the height in km and t is the time in hours after lift-off. How long does the balloon take to reach a height of 1 km?

You can invert the differential equation to give

$$\frac{dt}{dh} = 1 \Big/ \frac{dh}{dt} = \frac{1}{20 - 16h},$$

so that $t = \displaystyle\int \frac{1}{20 - 16h}\,dh$.

The solution can be completed in either of two ways.

Method 1   The indefinite integral is

$$t = \int \frac{1}{20 - 16h}\,dh = -\frac{1}{16}\ln(20 - 16h) + k.$$

(Notice that $20 - 16h$ is always positive when $0 \leqslant h \leqslant 1$.) Since t is measured from the instant of lift-off, $h = 0$ when $t = 0$. The particular solution with this initial condition must therefore satisfy

$$0 = -\tfrac{1}{16}\ln 20 + k, \quad \text{so} \quad k = \tfrac{1}{16}\ln 20.$$

The equation connecting the variables h and t is therefore

$$t = -\tfrac{1}{16}\ln(20 - 16h) + \tfrac{1}{16}\ln 20 = \tfrac{1}{16}\ln\left(\frac{20}{20 - 16h}\right) = \tfrac{1}{16}\ln\left(\frac{5}{5 - 4h}\right).$$

When $h = 1$, $t = \tfrac{1}{16}\ln\left(\dfrac{5}{5 - 4}\right) = \tfrac{1}{16}\ln 5 \approx 0.10$.

Method 2   Since only the time at $h = 1$ is required, you need not find the general equation connecting h and t. Instead, you can find the time as a definite integral, from $h = 0$ to $h = 1$:

$$\int_0^1 \frac{1}{20 - 16h}\, dh = \left[-\tfrac{1}{16} \ln(20 - 16h) \right]_0^1 = -\tfrac{1}{16}(\ln 4 - \ln 20)$$

$$= -\tfrac{1}{16} \ln\left(\tfrac{4}{20}\right) = -\tfrac{1}{16} \ln\left(\tfrac{1}{5}\right) = \tfrac{1}{16} \ln 5 \approx 0.10.$$

The balloon takes 0.1 hours, or 6 minutes, to reach a height of 1 km.

Example 19.3.2

When a ball is dropped from the roof of a tall building, the greatest speed that it can reach (called the terminal speed) is u. One model for its speed v when it has fallen a distance x is given by the differential equation

$$\frac{dv}{dx} = c \, \frac{u^2 - v^2}{v}, \text{ where } c \text{ is a positive constant.}$$

Find an expression for v in terms of x.

No units are given, but the constants u and c will depend on the units in which v and x are measured.

Since $\dfrac{dv}{dx}$ is given in terms of v rather than x, invert the equation to give

$$\frac{dx}{dv} = \frac{1}{c} \times \frac{v}{u^2 - v^2}, \qquad \text{so that} \qquad x = \int \frac{1}{c} \times \frac{v}{u^2 - v^2}\, dv.$$

The integral can be found by writing the integrand as $-\dfrac{1}{2c} \times \dfrac{-2v}{u^2 - v^2}$.

Note that v must be less than u, so $u^2 - v^2 > 0$. The second factor has the form $\dfrac{f'(v)}{f(v)}$, where $f(v) = u^2 - v^2$. It can therefore be integrated using the result in Section 18.3, as

$$x = -\frac{1}{2c} \ln\left(u^2 - v^2\right) + k.$$

The ball is not moving at the instant when it is dropped, so $v = 0$ when $x = 0$. This initial condition gives an equation for k:

$$0 = -\frac{1}{2c} \ln\left(u^2\right) + k, \qquad \text{so} \qquad k = \frac{\ln u^2}{2c}.$$

The equation connecting v and x is therefore

$$x = \frac{1}{2c}\left(\ln u^2 - \ln\left(u^2 - v^2\right)\right) = \frac{1}{2c} \ln\left(\frac{u^2}{u^2 - v^2}\right).$$

You must now turn this equation round to get v in terms of x:

$$2cx = \ln\left(\frac{u^2}{u^2 - v^2}\right),$$

$$\frac{u^2}{u^2 - v^2} = e^{2cx},$$

$$u^2 = \left(u^2 - v^2\right)e^{2cx},$$

$$v^2 e^{2cx} = u^2\left(e^{2cx} - 1\right),$$

$$v^2 = u^2\left(1 - e^{-2cx}\right).$$

Therefore, since $v > 0$, the required expression is $v = u\sqrt{1 - e^{-2cx}}$.

Example 19.3.3

A steel ball is heated to a temperature of 700 degrees Celsius and dropped into a drum of powdered ice. The temperature of the ball falls to 500 degrees in 30 seconds. Two models are suggested for the temperature, T degrees, after t seconds:

(a) the rate of cooling is proportional to T,
(b) the rate of cooling is proportional to $T^{1.2}$.

It is found that it takes a further 3 minutes for the temperature to fall from 500 to 100 degrees. Which model fits this information better?

The rate of cooling is measured by $\dfrac{dT}{dt}$, and this is negative.

(a) This model is described by the differential equation

$$\frac{dT}{dt} = -aT, \qquad \text{where } a \text{ is a positive constant.}$$

Inverting, $\dfrac{dt}{dT} = -\dfrac{1}{aT}$, which has solution $t = -\dfrac{1}{a}\ln T + k$.

Since $T = 700$ when $t = 0$, $0 = -\dfrac{1}{a}\ln 700 + k$, so $k = \dfrac{1}{a}\ln 700$.

The equation connecting T and t is therefore

$$t = \frac{1}{a}\left(\ln 700 - \ln T\right) = \frac{1}{a}\ln\frac{700}{T}.$$

The value of a can be found from the fact that $T = 500$ when $t = 30$:

$$30 = \frac{1}{a}\ln\frac{700}{500}, \qquad \text{which gives } a = \frac{\ln 1.4}{30} \approx 0.0112.$$

(b) For this model

$$\frac{dT}{dt} = -bT^{1.2}, \text{ so } \frac{dt}{dT} = -\frac{1}{b}T^{-1.2}, \text{ and } t = \frac{1}{0.2b}T^{-0.2} + k.$$

From the initial condition, that $T = 700$ when $t = 0$,

$$0 = \frac{5}{b} 700^{-0.2} + k, \quad \text{so} \quad t = \frac{5}{b}\left(T^{-0.2} - 700^{-0.2}\right).$$

From the other boundary condition, that $T = 500$ when $t = 30$,

$$30 = \frac{5}{b}\left(500^{-0.2} - 700^{-0.2}\right), \quad \text{so} \quad b = \frac{500^{-0.2} - 700^{-0.2}}{6} \approx 0.00313.$$

Fig. 19.4 shows the two models to be compared, whose equations are

(a) $t = \dfrac{1}{0.0112} \ln \dfrac{700}{T}$, and

(b) $t = \dfrac{5}{0.00313}\left(T^{-0.2} - 700^{-0.2}\right)$.

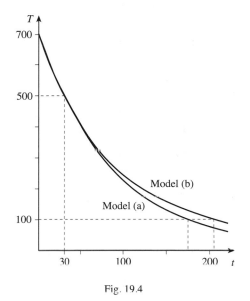

That is,

(a) $T = 700e^{-0.0112t}$, and

(b) $T = \dfrac{1}{\left(0.000\,626t + 0.270\right)^5}$.

To choose between the models, use the other piece of data, that $T = 100$ after a further 3 minutes, which is when $t = 30 + 180 = 210$. Try putting $T = 100$ in the two equations: model (a) gives $t = 174$, and model (b) gives $t = 205$.

Fig. 19.4

This suggests that (b) is the better model.

19.4 The equation for exponential growth

The key feature of exponential growth or decay is that the rate of increase or decrease of a quantity is proportional to its current value. Denoting the quantity by Q, this is expressed mathematically by the differential equation $\dfrac{dQ}{dt} = aQ$, where t stands for the time and a is a constant. The sign of a is positive for exponential growth, and negative for exponential decay.

You can solve this equation by writing $\dfrac{dt}{dQ} = \dfrac{1}{aQ}$ and integrating:

$$t = \int \frac{1}{aQ}\,dQ = \frac{1}{a}\ln|Q| + k.$$

Suppose that Q has the value Q_0 when $t = 0$. Then $0 = \dfrac{1}{a}\ln|Q_0| + k$, so that

$$t = \frac{1}{a}\ln|Q| - \frac{1}{a}\ln|Q_0|, \quad \text{so} \quad \left|\frac{Q}{Q_0}\right| = e^{at}.$$

Now Q must have the same sign as Q_0. In the equation $\dfrac{dt}{dQ} = \dfrac{1}{aQ}$ the value $Q = 0$ has to be excluded, since $\dfrac{1}{a \times 0}$ has no meaning. So if a solution begins at a value $Q_0 > 0$, Q remains positive; and if a solution begins at a value $Q_0 < 0$, Q remains negative. So $\dfrac{Q}{Q_0}$ is always positive, and you can replace $\left|\dfrac{Q}{Q_0}\right|$ in the above equation by $\dfrac{Q}{Q_0}$.

It follows that:

> If $\dfrac{dQ}{dt} = aQ$, where a is a non-zero constant,
>
> and $Q = Q_0$ when $t = 0$, then $Q = Q_0 e^{at}$.

You will meet this differential equation so often that it is worthwhile learning this result. In a particular application, you can then write down the solution without going through the theory each time.

If e^a is written as b, then $b > 0$ whether a is positive or negative. The equation $Q = Q_0 e^{at}$ can then be written as $Q = Q_0\left(e^a\right)^t = Q_0 b^t$, which has the form of the definition of exponential growth given in Chapter 3.

Exercise 19B

When you have found a solution to a differential equation, it is often helpful to sketch its graph and to look at its features in relation to the original differential equation.

1   Find general solutions of the following differential equations, expressing the dependent variable as a function of the independent variable.

   (a) $\dfrac{dy}{dx} = y^2$ (b) $\dfrac{dy}{dx} = \tan y$, for $-\frac{1}{2}\pi < y < \frac{1}{2}\pi$ (c) $\dfrac{dx}{dt} = 4x$

   (d) $\dfrac{dz}{dt} = \dfrac{1}{z}$, for $z > 0$ (e) $\dfrac{dx}{dt} = \operatorname{cosec} x$, for $0 < x < \pi$ (f) $u^2 \dfrac{du}{dx} = a$, for $u > 0$

2   Solve the following differential equations with the given initial conditions.

   (a) $\dfrac{dx}{dt} = -2x$, $x = 3$ when $t = 0$ (b) $\dfrac{du}{dt} = u^3$, $u = 1$ when $t = 0$

3 Find the solution curves of the following differential equations which pass through the given points. Suggest any restrictions which should be placed on the values of x.

(a) $\dfrac{dy}{dx} = y + y^2$, through $(0,1)$ (b) $\dfrac{dy}{dx} = e^y$, through $(2,0)$

4 A girl lives 500 metres from school. She sets out walking at 2 m s^{-1}, but when she has walked a distance of x metres her speed has dropped to $\left(2 - \frac{1}{400}x\right)$ m s^{-1}. How long does she take to get to school?

5 A boy is eating a 250 gram burger. When he has eaten a mass m grams, his rate of consumption is $100 - \frac{1}{900}m^2$ grams per minute. How long does he take to finish his meal?

6 A sculler is rowing a 2 kilometre course. She starts rowing at 5 m s^{-1}, but gradually tires, so that when she has rowed x metres her speed has dropped to $5e^{-0.0001x}$ m s^{-1}. How long will she take to complete the course?

7 A tree is planted as a seedling of negligible height. The rate of increase in its height, in metres per year, is given by the formula $0.2\sqrt{25-h}$, where h is the height of the tree, in metres, t years after it is planted.

(a) Explain why the height of the tree can never exceed 25 metres.

(b) Write down a differential equation connecting h and t, and solve it to find an expression for t as a function of h.

(c) How long does it take for the tree to put on

    (i) its first metre of growth,   (ii) its last metre of growth?

(d) Find an expression for the height of the tree after t years. Over what interval of values of t is this model valid?

8 Astronomers observe a luminous cloud of stellar gas which appears to be expanding. When it is observed a month later, its radius is estimated to be 5 times the original radius. After a further 3 months, the radius appears to be 5 times as large again.

It is thought that the expansion is described by a differential equation of the form $\dfrac{dr}{dt} = cr^m$

where c and m are constants. There is, however, a difference of opinion about the appropriate value to take for m. Two hypotheses are proposed, that $m = \frac{1}{3}$ and $m = \frac{1}{2}$.

Investigate which of these models fits the observed data better.

19.5 Differential equations with separable variables

All the differential equations you have met so far in this chapter have had $\dfrac{dy}{dx}$ expressed as functions of either x or y, but not of both. Another common type of equation has the form

$$\frac{dy}{dx} = \frac{f(x)}{g(y)}.$$

This can be solved by reversing the process described in Section 11.3. Such an equation is said to have **separable variables**, because it can be rearranged to get just y on the left side and just x on the right. This process is called 'separating the variables'.

Multiplying by $g(y)$ gives

$$g(y)\frac{dy}{dx} = f(x),$$

and this is the kind of equation you get when you differentiate an implicit equation. If you can find functions $G(y)$ and $F(x)$ such that $G'(y) = g(y)$ and $F'(x) = f(x)$, then the equation can be written as

$$G'(y)\frac{dy}{dx} = F'(x).$$

The term on the left is $\dfrac{d}{dx}G(y)$, so you can integrate with respect to x to obtain the implicit equation

$$G(y) = F(x) + k.$$

This last step is based on

$$\int G'(y)\frac{dy}{dx}\,dx = \int G'(y)\,dy,$$

which you use when doing integration by substitution.

Example 19.5.1
The gradient of the tangent at each point P of a curve is equal to the square of the gradient of OP. Find the equation of the curve.

If (x,y) is a point on the curve, the gradient of OP is $\dfrac{y}{x}$, so the gradient of the tangent at P will be $\left(\dfrac{y}{x}\right)^2$. Therefore y and x satisfy the differential equation

$$\frac{dy}{dx} = \frac{y^2}{x^2}.$$

The variables can be separated by dividing by y^2, which gives $\dfrac{1}{y^2}\dfrac{dy}{dx} = \dfrac{1}{x^2}$.

Integrating with respect to x gives the general solution $-\dfrac{1}{y} = -\dfrac{1}{x} + k.$

This can be written as

$$-\frac{1}{y} = -\frac{1-kx}{x}, \quad \text{so} \quad y = \frac{x}{1-kx}.$$

It is interesting to see what happens when you take different values of k. All the solution curves in Fig. 19.5 have the property described above. But you have to exclude the origin, since the property has no meaning if P coincides with O.

Notice that all the curves have positive gradients at all points. This is expected, since at each point P the gradient is the square of the gradient of OP.

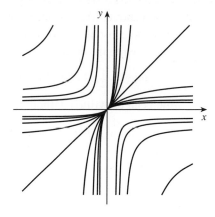

Fig. 19.5

Notice also that if you draw a line $y = mx$ through the origin ($m \neq 0$ or 1), it will cut a lot of the curves. At every point P of intersection the gradient of OP is m, so the gradient of the curve will be m^2. So the tangents to the curves at the points where the line cuts them are all parallel.

Example 19.5.2

For the differential equation $\dfrac{dy}{dx} = \dfrac{xy}{x^2 + 1}$, find

(a) the equation of the solution curve which passes through $(1,2)$,

(b) the general solution.

You can separate the variables by dividing by y, giving $\dfrac{1}{y}\dfrac{dy}{dx} = \dfrac{x}{x^2 + 1}$. Integrating

with respect to x gives $\displaystyle\int \frac{1}{y}\frac{dy}{dx}\,dx = \int \frac{x}{x^2 + 1}\,dx.$

The left side can be expressed as $\displaystyle\int \frac{1}{y}\,dy$, and the right side has the form

$\dfrac{1}{2}\displaystyle\int \frac{f'(x)}{f(x)}\,dx$ with $f(x) = x^2 + 1$ (see Section 18.3). So

$$\ln|y| = \tfrac{1}{2}\ln(x^2 + 1) + k.$$

(a) Substituting $x = 1$, $y = 2$ in this equation gives $\ln 2 = \tfrac{1}{2}\ln 2 + k$, so the required solution has $k = \tfrac{1}{2}\ln 2$, and is

$$\ln|y| = \tfrac{1}{2}\ln(x^2 + 1) + \tfrac{1}{2}\ln 2.$$

This equation can be written without logarithms, as $|y| = \sqrt{2(x^2 + 1)}$.

In this form the equation represents not one, but two solution curves, with equations $y = \pm\sqrt{2(x^2 + 1)}$. Since the square root on the right is positive, the curve which passes through $(1,2)$ has the equation with the positive sign,

$$y = \sqrt{2(x^2 + 1)}.$$

(b) You already have one form of the general solution in the equation found above, but you should try to rearrange it in a simpler form. For the particular solution through $(1,2)$, the constant k came out as a logarithm. Similarly, in the general solution it helps to write the arbitrary constant k as $\ln A$, where A is a positive number. Then

$$\ln|y| = \tfrac{1}{2}\ln(x^2 + 1) + \ln A,$$

which can be written without logarithms as

$$|y| = A\sqrt{x^2 + 1}.$$

Now $|y|$ is positive, and so is $A\sqrt{x^2 + 1}$. But y might be negative, so

$$y = \pm A\sqrt{x^2 + 1}.$$

Finally, instead of writing the constant as $\pm A$, where A is positive, it is easier to write it as c, where c can be positive or negative. The solution can then be expressed as

$$y = c\sqrt{x^2 + 1}.$$

There are a number of points to notice about the solution to the last example:

- Although when integrating you need to put the modulus sign in $\ln|y|$, it is not needed in $\ln(x^2 + 1)$, because $x^2 + 1$ is always positive.
- When integration introduces logarithms into the equation for the general solution, it is often worth adding the arbitrary constant in the form $+\ln A$, rather than $+k$.
- What about the value $c = 0$? You can't include $A = 0$ in the solution because $\ln 0$ has no meaning. But obviously $y = 0$ (the x-axis) is a solution of the original differential equation, since $\dfrac{dy}{dx} = 0$ at every point. This solution in fact got lost at the very first step, dividing by y to separate the variables; you can't do this if $y = 0$. But now that this special case has been checked, you can say that the general solution of the differential equation is $y = c\sqrt{x^2 + 1}$, where the constant c can be any number, positive, negative or zero.

Check this solution for yourself by finding $\dfrac{dy}{dx}$ and showing that it does satisfy the differential equation for any value of c.

Example 19.5.3
For a certain period of about 12 years, the rate of growth of a country's gross national product (GNP) is predicted to vary between $+5\%$ and -1%. This variation is modelled by the formula $\left(2 + 3\cos\tfrac{1}{2}t\right)\%$, where t is the time in years. Find a formula for the GNP during the 12-year period.

Denote the GNP after t years by P. The rate of growth $\dfrac{dP}{dt}$ is given as a percentage of its current value, so

$$\frac{dP}{dt} = \frac{2 + 3\cos\tfrac{1}{2}t}{100}P.$$

The variables can be separated by dividing by P:

$$\frac{1}{P}\frac{dP}{dt} = \frac{2+3\cos\frac{1}{2}t}{100}.$$

Integrating,

$$\ln P = \frac{2t+6\sin\frac{1}{2}t}{100} + k.$$

(Note that the GNP is always positive, so by definition $P>0$.)

If P has the value P_0 when $t=0$, then

$$\ln P_0 = 0 + k,$$

and the equation can be written

$$\ln P = \frac{2t+6\sin\frac{1}{2}t}{100} + \ln P_0, \quad \text{or} \quad P = P_0 e^{\frac{1}{100}\left(2t+6\sin\frac{1}{2}t\right)}.$$

Exercise 19C

1 Find the general solution of each of the following differential equations.

(a) $\dfrac{dy}{dx} = \dfrac{x^2}{y^2}$ (b) $\dfrac{dy}{dx} = \dfrac{x}{y}$ (c) $\dfrac{dy}{dx} = xy$ (d) $\dfrac{dy}{dx} = \dfrac{1}{xy}$

2 Find the equation of the curve which satisfies the differential equation $\dfrac{dy}{dx} = \dfrac{y}{x(x+1)}$ and passes through the point $(1,2)$.

3 Find the general solution of the differential equation $\dfrac{dy}{dx} = -\dfrac{x}{y}$. Describe the solution curves, and find the equation of the curve which passes through $(-4,3)$.

4 Solve the differential equation $\dfrac{dy}{dx} = \dfrac{x+1}{2-y}$, and describe the solution curves.

5 Find the equations of the curves which satisfy the given differential equations and pass through the given points.

(a) $\dfrac{dy}{dx} = \dfrac{3y}{2x}$ $(2,4)$ (b) $\dfrac{dy}{dx} = -\dfrac{3y}{2x}$ $(2,4)$

(c) $\dfrac{dy}{dx} = \dfrac{\sin x}{\cos y}$ $\left(\frac{1}{3}\pi,0\right)$ (d) $\dfrac{dy}{dx} = \dfrac{\tan x}{\tan y}$ $\left(\frac{1}{3}\pi,0\right)$

6 Solve the equation $v\dfrac{dv}{dx} = -\omega^2 x$, where ω is a constant. Find the particular solution for which $v=0$ when $x=a$.

7 Find the general solution of the equations

(a) $\dfrac{dy}{dx} = \dfrac{2x(y^2+1)}{y(x^2+1)}$,   (b) $\dfrac{dy}{dx} = \tan x \cot y$.

8 Find the equations of the curves which satisfy the following differential equations and pass through the given points.

(a) $\dfrac{dy}{dx} = \dfrac{y(y-1)}{x}$ $(1,2)$ (b) $\dfrac{dy}{dx} = \cot x \cot y$ $\left(\tfrac{1}{6}\pi, 0\right)$

(c) $\dfrac{dy}{dx} = \dfrac{1+y^2}{y(1-x^2)}$ $\left(\tfrac{3}{2}, 2\right)$ (d) $\dfrac{dy}{dx} = y \tan x$ $(0,2)$

9 Find the general solution of the differential equations

(a) $4 + x\dfrac{dy}{dx} = y^2$,   (b) $e^y \dfrac{dy}{dx} - 1 = \ln x$,   (c) $y \cos x \dfrac{dy}{dx} = 2 - y\dfrac{dy}{dx}$.

10 The gradient at each point of a curve is n times the gradient of the line joining the origin to that point. Find the general equation of the curve.

11 The size of an insect population n, which fluctuates during the year, is modelled by the equation $\dfrac{dn}{dt} = 0.01\,n\,(0.05 - \cos 0.02t)$, where t is the number of days from the start of observations. The initial number of insects is 5000.

   (a) Solve the differential equation to find n in terms of t.

   (b) Show that the model predicts that the number of insects will fall to a minimum after about 76 days, and find this minimum value.

12 The velocity v m s^{-1} of a spacecraft moving vertically x metres above the centre of the earth can be modelled by the equation $v\dfrac{dv}{dx} = -\dfrac{10R^2}{x^2}$, where R metres is the radius of the earth. The inital velocity at blast-off, when $x = R$, is V m s^{-1}.

Find an expression for v^2 in terms of V, x and R, and show that, according to this model, if the spacecraft is to be able to escape from the earth, then $V^2 \geqslant 20R$.

<hr>

Miscellaneous exercise 19

1 Find the solution of the differential equation $x\dfrac{dy}{dx} = 2x^2 + 7x + 3$ for which $y = 10$ when $x = 1$.   (OCR)

2 In a chemical reaction, the amount z grams of a substance after t hours is modelled by the differential equation $\dfrac{dz}{dt} = 0.005(20 - z)^2$. Initially $z = 0$. Find an expression for t in terms of z, and show that $t = 15$ when $z = 12$.   (OCR)

3 The gradient of a curve is given by $\dfrac{dy}{dx} = 3x^2 - 8x + 5$. The curve passes through the point $(0,3)$. Find the equation of the curve. Find the coordinates of the two stationary points. State, with a reason, the nature of each stationary point. (MEI)

4 The area of a circle of radius r metres is A m^2.

(a) Find $\dfrac{dA}{dr}$ and write down an expression, in terms of r, for $\dfrac{dr}{dA}$.

(b) The area increases with time t seconds in such a way that $\dfrac{dA}{dt} = \dfrac{2}{(t+1)^3}$. Find an expression, in terms of r and t, for $\dfrac{dr}{dt}$.

(c) Solve the differential equation $\dfrac{dA}{dt} = \dfrac{2}{(t+1)^3}$ to obtain A in terms of t, given that $A = 0$ when $t = 0$.

(d) Show that, when $t = 1$, $\dfrac{dr}{dt} = 0.081$ correct to 2 significant figures. (OCR)

5 The rate of destruction of a drug by the kidneys is proportional to the amount of drug present in the body. The constant of proportionality is denoted by k. At time t the quantity of drug in the body is x. Write down a differential equation relating x and t, and show that the general solution is $x = Ae^{-kt}$, where A is an arbitrary constant.

Before $t = 0$ there is no drug in the body, but at $t = 0$ a quantity Q of the drug is administered. When $t = 1$ the amount of drug in the body is $Q\alpha$, where α is a constant such that $0 < \alpha < 1$. Show that $x = Q\alpha^t$.

When $t = 1$ and again when $t = 2$ another dose Q is administered. Show that the amount of drug in the body immediately after $t = 2$ is $Q(1 + \alpha + \alpha^2)$.

If the drug is administered at regular intervals for an indefinite period, and if the greatest amount of the drug that the body can tolerate is T, show that Q should not exceed $T(1 - \alpha)$. (OCR, adapted)

6 (a) The number of people, x, in a queue at a travel centre t minutes after it opens is modelled by the differential equation $\dfrac{dx}{dt} = 1.4t - 4$ for values of t up to 10. Interpret the term '-4' on the right side of the equation. Solve the differential equation, given that $x = 8$ when $t = 0$.

(b) An alternative model gives the differential equation $\dfrac{dx}{dt} = 1.4t - 0.5x$ for the same values of t. Verify that $x = 13.6e^{-0.5t} + 2.8t - 5.6$ satisfies this differential equation. Verify also that when $t = 0$ this function takes the value 8. (OCR)

7 (a) Two quantities x and y are related to each other by the differential equation $y\dfrac{dy}{dx} = -16x$. Solve this equation to get an implicit equation of the solution curve for which $y = 0$ when $x = 0.1$.

(b) Sketch your solution curve from part (a), showing the values of x and y at which the curve cuts the coordinate axes. (OCR)

8 At time $t = 0$ there are 8000 fish in a lake. At time t days the birth-rate of fish is equal to one-fiftieth of the number N of fish present. Fish are taken from the lake at the rate of 100 per day. Modelling N as a continuous variable, show that $50 \dfrac{\mathrm{d}N}{\mathrm{d}t} = N - 5000$.

Solve the differential equation to find N in terms of t. Find the time taken for the population of fish in the lake to increase to $11\,000$.

When the population of fish has reached $11\,000$, it is decided to increase the number of fish taken from the lake from 100 per day to F per day. Write down, in terms of F, the new differential equation satisfied by N. Show that if $F > 220$, then $\dfrac{\mathrm{d}N}{\mathrm{d}t} < 0$ when $N = 11\,000$. For this range of values of F, give a reason why the population of fish in the lake continues to decrease.   (OCR)

9 A metal rod is 60 cm long and is heated at one end. The temperature at a point on the rod at distance x cm from the heated end is denoted by $T\ ^{\circ}\mathrm{C}$. At a point halfway along the rod, $T = 290$ and $\dfrac{\mathrm{d}T}{\mathrm{d}x} = -6$.

(a) In a simple model for the temperature of the rod, it is assumed that $\dfrac{\mathrm{d}T}{\mathrm{d}x}$ has the same value at all points on the rod. For this model, express T in terms of x and hence determine the temperature difference between the ends of the rod.

(b) In a more refined model, the rate of change of T with respect to x is taken to be proportional to x. Set up a differential equation for T, involving a constant of proportionality k. Solve the differential equation and hence show that, in this refined model, the temperature along the rod is predicted to vary from $380\ ^{\circ}\mathrm{C}$ to $20\ ^{\circ}\mathrm{C}$.   (OCR)

10 A battery is being charged. The charging rate is modelled by $\dfrac{\mathrm{d}q}{\mathrm{d}t} = k(Q - q)$, where q is the charge in the battery (measured in ampere-hours) at time t (measured in hours), Q is the maximum charge the battery can store and k is a constant of proportionality. The model is valid for $q \geqslant 0.4Q$.

(a) It is given that $q = \lambda Q$ when $t = 0$, where λ is a constant such that $0.4 \leqslant \lambda < 1$. Solve the differential equation to find q in terms of t. Sketch the graph of the solution.

(b) It is noticed that the charging rate halves every 40 minutes. Show that $k = \frac{3}{2}\ln 2$.

(c) Charging is always stopped when $q = 0.95Q$. If T is the time until charging is stopped, show that $T = \dfrac{2\ln(20(1 - \lambda))}{3\ln 2}$ for $0.4 \leqslant \lambda \leqslant 0.95$.   (MEI)

11 Find the general solution of the differential equation $\dfrac{\mathrm{d}y}{\mathrm{d}x} = \dfrac{x(y^2 + 1)}{(x - 1)y}$, expressing y in terms of x.   (OCR, adapted)

12 Solve the differential equation $\dfrac{\mathrm{d}y}{\mathrm{d}x} = xy\,\mathrm{e}^{2x}$, given that $y = 1$ when $x = 0$.

13 The rate at which the water level in a cylindrical barrel goes down is modelled by the equation $\dfrac{dh}{dt} = -\sqrt{h}$, where h is the height in metres of the level above the tap and t is the time in minutes. When $t = 0$, $h = 1$. Show by integration that $h = \left(1 - \tfrac{1}{2}t\right)^2$. How long does it take for the water flow to stop?

An alternative model would be to use a sine function, such as $h = 1 - \sin kt$. Find the value of k which gives the same time before the water flow stops as the previous model. Show that this model satisfies the differential equation $\dfrac{dh}{dt} = -k\sqrt{2h - h^2}$. (OCR, adapted)

14 A tropical island is being set up as a nature reserve. Initially there are 100 nesting pairs of fancy terns on the island. In the first year this increases by 8. In one theory being tested, the number N of nesting pairs after t years is assumed to satisfy the differential equation $\dfrac{dN}{dt} = \dfrac{1}{5000} N(500 - N)$.

(a) Show that, according to this model, the rate of increase of N is 8 per year when $N = 100$. Find the rate of increase when $N = 300$ and when $N = 450$. Describe what happens as N approaches 500, and interpret your answer.

(b) Use your answers to part (a) to sketch the solution curve of the differential equation for which $N = 100$ when $t = 0$.

(c) Obtain the general solution of the differential equation, and the solution for which $N = 100$ when $t = 0$. Use your answer to predict after how many years the number of pairs of nesting fancy terns on the island will first exceed 300. (OCR)

15 A biologist is researching the population of a species. She tries a number of different models for the rate of growth of the population and solves them to compare with observed data. Her first model is $\dfrac{dp}{dt} = kp\left(1 - \dfrac{p}{m}\right)$ where p is the population at time t years, k is a constant and m is the maximum population sustainable by the environment. Find the general solution of the differential equation.

Her observations suggest that $k = 0.2$ and $m = 100\,000$. If the initial population is $30\,000$, estimate the population after 5 years to 2 significant figures.

She decides that the model needs to be refined. She proposes a model $\dfrac{dp}{dt} = kp\left(1 - \left(\dfrac{p}{m}\right)^{\alpha}\right)$ and investigates suitable values of α. Her observations lead her to the conclusion that the maximum growth rate occurs when the population is 70% of its maximum. Show that $(\alpha + 1)0.7^{\alpha} = 1$, and that an approximate solution of this equation is $\alpha \approx 5$. Express the time that it will take the population to reach $54\,000$ according to this model as a definite integral, and use the trapezium rule to find this time approximately. (MEI, adapted)

16 Obtain the general solution of the differential equation $y\dfrac{dy}{dx}\tan 2x = 1 - y^2$.

(OCR, adapted)

17 Find the general solution of the differential equation $\dfrac{dy}{dx} = \dfrac{y^2}{x^2 - x - 2}$ in the region $x > 2$.

Find also the particular solution which satisfies $y = 1$ when $x = 5$. (OCR)

18 Find the solution of the differential equation $\dfrac{dy}{dx} = \dfrac{\sin^2 x}{y^2}$ which also satisfies $y = 1$ when $x = 0$. (OCR)

19 Solve the differential equation $\dfrac{dy}{dx} = \dfrac{x}{y} e^{x+y}$, in the form $f(y) = g(x)$, given that $y = 0$ when $x = 0$.

20 To control the pests inside a large greenhouse, 600 ladybirds were introduced. After t days there are P ladybirds in the greenhouse. In a simple model, P is assumed to be a continuous variable satisfying the differential equation $\dfrac{dP}{dt} = kP$, where k is a constant.

Solve the differential equation, with initial condition $P = 600$ when $t = 0$, to express P in terms of k and t.

Observations of the number of ladybirds (estimated to the nearest hundred) were made as follows:

t	0	150	250
P	600	1200	3100

Show that $P = 1200$ when $t = 150$ implies that $k \approx 0.00462$. Show that this is not consistent with the observed value when $t = 250$.

In a refined model, allowing for seasonal variations, it is assumed that P satisfies the differential equation $\dfrac{dP}{dt} = P(0.005 - 0.008 \cos 0.02t)$ with initial condition $P = 600$ when $t = 0$. Solve this differential equation to express P in terms of t, and comment on how well this fits with the data given above.

Show that, according to the refined model, the number of ladybirds will decrease initially, and find the smallest number of ladybirds in the greenhouse. (MEI)

21 The organiser of a sale, which lasted for 3 hours and raised a total of £1000, attempted to create a model to represent the relationship between s and t, where £s is the amount which had been raised at time t hours after the start of the sale. In the model s and t were taken to be continuous variables. The organiser assumed that the rate of raising money varied directly as the time remaining and inversely as the amount already raised. Show that, for this model, $\dfrac{ds}{dt} = k\dfrac{3-t}{s}$, where k is a constant. Solve the differential equation, and show that the solution can be written in the form $\dfrac{s^2}{1000^2} + \dfrac{(3-t)^2}{3^2} = 1$. Hence

(a) find the amount raised during the first hour of the sale,

(b) find the rate of raising money one hour after the start of the sale. (OCR)

22 A biologist studying fluctuations in the size of a particular population decides to investigate
    a model for which $\dfrac{dP}{dt} = kP \cos kt$, where P is the size of the population at time t days and
    k is a positive constant.

    (a) Given that $P = P_0$ when $t = 0$, express P in terms of k, t and P_0.

    (b) Find the ratio of the maximum size of the population to the minimum size.   (OCR)

23 For $x > 0$ and $0 < y < \frac{1}{2}\pi$, the variables y and x are connected by the differential
    equation $\dfrac{dy}{dx} = \dfrac{\ln x}{\cot y}$, and $y = \frac{1}{6}\pi$ when $x = e$.

    Find the value of y when $x = 1$, giving your answer to 3 significant figures. Use the
    differential equation to show that this value of y is a stationary value, and determine its
    nature.   (MEI)

Revision exercise 3

1 Express $\dfrac{1}{r(r+2)}$ in partial fractions. (MEI, adapted)

2 Find the factors of x^3+8, and hence split $\dfrac{6(x-2)}{x^3+8}$ into partial fractions.

Show that $\displaystyle\int_0^1 \dfrac{6(x-2)}{x^3+8}\,dx = a\ln b$, where a is a negative integer and b is a positive integer.

Find the values of a and b.

How do you interpret the fact that the value of the integral is negative?

3 Find

(a) $\displaystyle\int_{\frac{1}{4}\pi}^{\frac{1}{3}\pi} \cot^4 x\,dx$, (b) $\displaystyle\int_0^{\frac{1}{2}\pi} \dfrac{\cos x}{\sqrt{1+\sin x}}\,dx$, (c) $\displaystyle\int_0^{\frac{1}{4}\pi} x\tan^2 x\,dx$, (d) $\displaystyle\int_0^{\frac{1}{2}\pi} e^{3x}\cos 4x\,dx$.

4 Find the complex numbers which satisfy the following equations.

(a) $(1+i)z = 1+3i$

(b) $z^2+4z+13=0$

(c) $(1-i)z^2-4z+(1+3i)=0$

(d) $\begin{cases} (1-i)z+(1+i)w = 2, \\ (1+3i)z-(4+i)w = 3i \end{cases}$

5 If $z = \cos\theta + i\sin\theta$, where $-\pi < \theta \leqslant \pi$, find the modulus and argument of
(a) z^2 and (b) $1+z^2$, distinguishing the cases

(i) $\theta = 0$, (ii) $\theta = \frac{1}{2}\pi$, (iii) $\theta = \pi$, (iv) $\theta = -\frac{1}{2}\pi$,

(v) $0 < \theta < \frac{1}{2}\pi$, (vi) $\frac{1}{2}\pi < \theta < \pi$, (vii) $-\frac{1}{2}\pi < \theta < 0$, (viii) $-\pi < \theta < -\frac{1}{2}\pi$.

6 (a) If $z = x+yi$, sketch in an Argand diagram the curves given by $\operatorname{Re} z^2 = a$ and $\operatorname{Im} z^2 = b$, where a and b are positive constants. Show that, where the two curves intersect, their tangents are in perpendicular directions.

(b) Repeat part (a) for the curves given by $\operatorname{Re}\dfrac{1}{z} = a$ and $\operatorname{Im}\dfrac{1}{z} = b$.

7 (a) Write x^3-1 and x^3+1 as products of real factors.

(b) Write z^3-1 and z^3+1 as products of complex factors.

(c) Solve the equations $x^6-1=0$, $x^6+1=0$ and $x^{12}-1=0$ in real numbers, and illustrate your answers with graphs of $y=x^6-1$, $y=x^6+1$ and $y=x^{12}-1$.

(d) Solve the equations $z^6-1=0$, $z^6+1=0$ and $z^{12}-1=0$ in complex numbers, and illustrate your answers using Argand diagrams.

8 Write down the first five terms in the expansion of $(1+2x)^{\frac{1}{2}}$. Show that, when differentiated, the result is zero plus the first four terms in the expansion of $(1+2x)^{-\frac{1}{2}}$.

Investigate similarly the effect of differentiation on the terms of $(1+3x)^{\frac{1}{3}}$ and $(1+x)^{-3}$.

9 Determine whether or not the point $(1,2,-1)$ lies on the line passing through $(3,1,2)$ and $(5,0,5)$.

10 A, B and C are points in an Argand diagram representing the complex numbers $-1+0\,\mathrm{i}$, $1+0\,\mathrm{i}$ and $0+\mathrm{i}$ respectively, and P is the point representing the complex number z (with $\mathrm{Im}(z)>0$). The displacements $\overrightarrow{AP}$ and $\overrightarrow{BP}$ make angles α and β with the x-axis, and the angle $APB=\frac{1}{4}\pi$.

(a) Show that $\arg(z-1)-\arg(z+1)=\frac{1}{4}\pi$.

(b) Show that $\dfrac{z+1}{z^*+1}=\cos 2\alpha+\mathrm{i}\sin 2\alpha$ and write down a similar expression for $\dfrac{z-1}{z^*-1}$.

(c) Show that $\dfrac{(z-1)(z^*+1)}{(z^*-1)(z+1)}=\mathrm{i}$ and deduce that $zz^*+\mathrm{i}(z-z^*)=1$.

(d) Show that the equation in part (c) can be written as $(z-\mathrm{i})(z^*+\mathrm{i})=2$ and deduce that $|z-\mathrm{i}|=\sqrt{2}$.

(e) State in words what geometrical property is established by combining the results of parts (a) to (d).

11 Find the following integrals

(a) $\displaystyle\int \sin\left(2x+\tfrac{1}{6}\pi\right)\mathrm{d}x$ (b) $\displaystyle\int \sin^2 3x\,\mathrm{d}x$ (c) $\displaystyle\int \sin^2 2x\cos 2x\,\mathrm{d}x$

12 Find a vector equation of the line which passes through $(1,4,2)$ and $(-2,3,3)$, and find the coordinates of its point of intersection with the line with vector equation

$$\mathbf{r}=\begin{pmatrix}1\\0\\2\end{pmatrix}+t\begin{pmatrix}3\\-1\\-1\end{pmatrix}.$$

13 Expand $\left(1-x+x^2\right)^{\frac{1}{2}}$ as a series in ascending powers of x up to and including the term in x^3.

14 By identifying the series $1-\dfrac{1}{4}+\dfrac{1\times 3}{4\times 8}-\dfrac{1\times 3\times 5}{4\times 8\times 12}+\ldots$ as a binomial series of the form $(1+x)^n$ and finding the values of x and n, find the sum to infinity of the series

$$1-\dfrac{1}{4}+\dfrac{1\times 3}{4\times 8}-\dfrac{1\times 3\times 5}{4\times 8\times 12}+\ldots.$$

15 Find the vector equation of the straight line parallel to $\mathbf{r}=\begin{pmatrix}-2\\1\\3\end{pmatrix}+s\begin{pmatrix}1\\-1\\1\end{pmatrix}$ through the point $(2,-1,4)$.

16 A straight line has vector equation $\mathbf{r}=\begin{pmatrix}1\\2\end{pmatrix}+t\begin{pmatrix}3\\4\end{pmatrix}$. Find its cartesian equation.

17 $1+ax+bx^2$ are the first three terms of a binomial expansion for $(1+cx)^n$. Write two equations involving n and c, and hence express n and c in terms of a and b.

Prove that the next term of the expansion is $\dfrac{b}{3a}\left(4b-a^2\right)x^3$.

18 By squaring both sides of the expansion $(1+x)^{-1} = 1 - x + x^2 - x^3 + \dots$, obtain the expansion of $(1+x)^{-2}$. Then use $(1+x)^{-2}(1+x)^{-1} \equiv (1+x)^{-3}$ to obtain the expansion of $(1+x)^{-3}$.

19 (a) Find the area of the region enclosed by the curve with equation $y = \tan x$, the x-axis and the lines $x = 0$ and $x = \frac{1}{3}\pi$.

    (b) Find the volume generated when this area is rotated about the x-axis.

20 Express $\dfrac{1}{(1+x)(3-x)}$ as the sum of partial fractions. Hence express $\dfrac{1}{(1+x)^2(3-x)^2}$ as the sum of partial fractions.

    A region is bounded by parts of the x- and y-axes, the curve $y = \dfrac{1}{(1+x)(3-x)}$ and the line $x = 2$. Find the area of the region, and the volume of the solid of revolution formed by rotating it about the x-axis.   (OCR)

21 (a) Use the substitution $y = \frac{1}{2}\pi - x$ to show that $\displaystyle\int_0^{\frac{1}{2}\pi} \sin^2 x \, dx = \int_0^{\frac{1}{2}\pi} \cos^2 y \, dy$.

    (b) Show that $\displaystyle\int_0^{\frac{1}{2}\pi} \cos^2 y \, dy = \int_0^{\frac{1}{2}\pi} \cos^2 x \, dx$ and $\displaystyle\int_0^{\frac{1}{2}\pi} \sin^2 x \, dx = \int_0^{\frac{1}{2}\pi} \cos^2 x \, dx$.

    (c) Find $\displaystyle\int_0^{\frac{1}{2}\pi} \left(\sin^2 x + \cos^2 x\right) dx$, and hence show that $\displaystyle\int_0^{\frac{1}{2}\pi} \sin^2 x \, dx = \frac{1}{4}\pi$.

22 An anthropologist is modelling the population of the island of A. In the model, the population at the start of the year t is P. The birth rate is 10 births per 1000 population per year. The death rate is m deaths per 1000 population per year.

    (a) Show that $\dfrac{dP}{dt} = \dfrac{(10-m)P}{1000}$.

    (b) At the start of year 0 the population was 108 000. Find an expression for P in terms of t.

    (c) State one assumption about the population of A that is required for this model to be valid.

    (d) If the population is to double in 100 years, find the value of m.

    (e) Explain why the population cannot double in less than 69 years.   (OCR)

23 A model for the way in which a population of animals in a closed environment varies with time is given, for $P > \frac{1}{3}$, by $\dfrac{dP}{dt} = \frac{1}{2}\left(3P^2 - P\right)\sin t$, where P is the size of the population in thousands at time t. Given that $P = \frac{1}{2}$ when $t = 0$, show that $\ln\dfrac{3P-1}{P} = \frac{1}{2}(1 - \cos t)$.

    Rearrange this equation to show that $P = \dfrac{1}{3 - e^{\frac{1}{2}(1-\cos t)}}$.

    Calculate the smallest positive value of t for which $P = 1$, and find the two values between which the number of animals in the population oscillates.   (MEI, adapted)

24 Two small insects A and B are crawling on the walls of a room, with A starting from the ceiling. The floor is horizontal and forms the xy-plane, and the z-axis is vertically upwards. Relative to the origin O, the position vectors of the insects at time t seconds $(0 \leqslant t \leqslant 10)$ are $\overrightarrow{OA} = \mathbf{i} + 3\mathbf{j} + \left(4 - \frac{1}{10}t\right)\mathbf{k}$, $\overrightarrow{OB} = \left(\frac{1}{5}t + 1\right)\mathbf{i} - 3\mathbf{j} + 2\mathbf{k}$, where the unit of distance is the metre.

(a) Write down the height of the room.

(b) Show that the insects move in such a way that angle $BOA = 90°$.

(c) For each insect, write down a vector to represent its displacement between $t = 0$ and $t = 10$, and show that these displacements are perpendicular to each other.

(d) Write down expressions for the vector $\overrightarrow{AB}$ and for $\left|\overrightarrow{AB}\right|$, and hence find the minimum distance between the insects, correct to 3 significant figures.   (OCR)

25 (a) Differentiate $x\sqrt{2-x}$ with respect to x.

(b) Find $\displaystyle\int x\sqrt{2-x}\,dx$

    (i) by using the substitution $2 - x = u$,   (ii) by integration by parts.

26 The lines l_1 and l_2 intersect at the point C with position vector $\mathbf{i} + 5\mathbf{j} + 11\mathbf{k}$. The equations of l_1 and l_2 are $\mathbf{r} = \mathbf{i} + 5\mathbf{j} + 11\mathbf{k} + \lambda(3\mathbf{i} + 2\mathbf{j} - 2\mathbf{k})$ and $\mathbf{r} = \mathbf{i} + 5\mathbf{j} + 11\mathbf{k} + \mu(8\mathbf{i} + 11\mathbf{j} + 6\mathbf{k})$, where λ and μ are real parameters. Find, in the form $ax + by + cz = d$, an equation of the plane Π which contains l_1 and l_2.

The point A has position vector $4\mathbf{i} - \mathbf{j} + 5\mathbf{k}$ and the line through A perpendicular to Π meets Π at B. Find

(a) the length of AB,

(b) the perpendicular distance of B from l_1, giving your answer correct to 3 significant figures.

27 With respect to an origin O, the point A has position vector $30\mathbf{i} - 3\mathbf{j} - 5\mathbf{k}$. The line l passes through O and is parallel to the vector $4\mathbf{i} - 5\mathbf{j} - 3\mathbf{k}$. The point B on l is such that AB is perpendicular to l. In either order,

(a) find the length of AB,   (b) find the position vector of B.

The plane Π passes through A and is parallel to both l and the vector $-2\mathbf{i} + 2\mathbf{j} + \mathbf{k}$. The point Q on AB is such that $AQ = \frac{1}{4}QB$. Find, correct to 2 decimal places, the perpendicular distance from Q to Π.   (OCR)

28 The plane π has equation $\mathbf{r} \cdot (2\mathbf{i} - 3\mathbf{j} + 6\mathbf{k}) = 0$, and P and Q are the points with position vectors $7\mathbf{i} + 6\mathbf{j} + 5\mathbf{k}$ and $\mathbf{i} + 3\mathbf{j} - \mathbf{k}$ respectively. Find the position vector of the point in which the line passing through P and Q meets the plane π.

Find, in the form $ax + by + cz = d$, the equation of the plane which contains the line PQ and which is perpendicular to π.   (OCR)

29 The region bounded by the curve with equation $y = \sqrt{\dfrac{10x}{(x+4)(x^2+4)}}$, the x-axis and the lines with equations $x = 0$ and $x = 2$ is rotated through 2π radians about the x-axis. Calculate the volume of the solid of revolution formed.

Practice examination 1 for P3

Time 1 hour 45 minutes

Answer all the questions.
The use of an electronic calculator is expected, where appropriate.

1   By writing $\cot x$ in the form $\dfrac{\cos x}{\sin x}$, show that the result of differentiating $\cot x$ with respect to x is $-\csc^2 x$. [4]

2   The variables x and y are related by the equation $2^y = 3^{1-x}$.

Show that the graph of y against x is a straight line, and state the values of the gradient and the intercept on the y-axis. [4]

3   The cubic polynomial $f(x) = x^3 + 3x^2 + ax + b$, where a and b are constants, has a factor $x+1$. The remainder when $f(x)$ is divided by $x+2$ is the same as the remainder when $f(x)$ is divided by $x-2$. Find this remainder. [6]

4   The population of a community with finite resources is modelled by the differential equation

$$\frac{dn}{dt} = 0.01n\,e^{-0.01t},$$

where n is the population at time t. At time $t = 0$ the population is 5000.

(i)   Solve the differential equation, expressing $\ln n$ in terms of t. [5]

(ii)   What happens to the population as t becomes large? [2]

5   Two planes have equations $x + 2y - z = 3$ and $2x - z = 0$. Find

(i)   the acute angle between the planes, [3]

(ii)   the coordinates of two points on the line of intersection, l, of the planes, [2]

(iii)   the equation of l, giving your answer in the form $\mathbf{r} = \mathbf{a} + t\mathbf{b}$. [2]

6   The angle $\theta°$ satisfies the equation $\tan 2\theta° = \sin\theta°$.

(i)   Show that either $\sin\theta° = 0$ or $2\cos\theta° = 2\cos^2\theta° - 1$. [5]

(ii)   Hence find the smallest positive value of θ. [3]

7   (i)   Express $\dfrac{2x+1}{(x-2)(x^2+1)}$ in partial fractions. [4]

(ii)   Hence show that

$$\int_0^1 \frac{2x+1}{(x-2)(x^2+1)}\,dx = -\frac{3}{2}\ln 2.$$

[4]

8 (a) Find $\displaystyle\int x \ln x \, dx$. [3]

(b) Show that $\dfrac{1}{\sin x \cos x} \equiv \dfrac{\sec^2 x}{\tan x}$, and hence evaluate

$$\int_{\frac{1}{6}\pi}^{\frac{1}{3}\pi} \frac{1}{\sin x \cos x} \, dx.$$ [5]

9 The parametric equations of a curve are

$$x = t^3 - e^{-t}, \quad y = t^2 - e^{-2t}.$$

(i) Find the equation of the tangent to the curve at the point where $t = 0$. [5]

The curve cuts the y-axis at the point A.

(ii) Show that the value of t at A lies between 0 and 1. [2]

(iii) Use the iteration $t_{n+1} = \sqrt[3]{e^{-t_n}}$ to find this value of t correct to 2 significant figures, and hence determine the y-coordinate of A, correct to 1 significant figure. [4]

10 One root of the cubic equation $z^3 + az + 10 = 0$ is $1 + 2i$.

(i) Find the value of the real constant a. [4]

(ii) Show all three roots of the equation on an Argand diagram. [4]

(iii) Show that all three roots satisfy the equation $|6z - 1| = 13$, and show the locus represented by this equation on your diagram. [4]

Practice examination 2 for P3

Time 1 hour 45 minutes

Answer all the questions.
The use of an electronic calculator is expected, where appropriate.

1 Solve the inequality $2 \times 3^{-x} < 5 \times 10^{-2}$. [4]

2 (i) Use the trapezium rule, with three intervals each of width $\frac{1}{18}\pi$, to estimate the value of

$$\int_0^{\frac{1}{6}\pi} \sec x \, dx.$$ [3]

(ii) State with a reason whether the trapezium rule gives an overestimate or an underestimate of the true value of the integral in this case. [2]

3 (i) Show that the iteration given by

$$x_{n+1} = \frac{2 + k x_n}{x_n^2 + k}$$

corresponds to the equation $x^3 = 2$, whatever the value of the constant k. [2]

(ii) Taking $k = 3$, use the iteration, with $x_1 = 1$, to find the value of $\sqrt[3]{2}$ correct to 2 decimal places. [3]

4 The equation of a curve is $\sin y = x \cos 2x$. Find $\dfrac{dy}{dx}$ in terms of x and y, and hence find the gradient of the curve at the point $\left(\frac{1}{4}\pi, 0\right)$. [5]

5 (a) Find the quotient and remainder when $4x^2$ is divided by $2x + 1$. [4]

(b) Use the binomial series to show that, when x is small,

$$\frac{1}{\sqrt{1 + 2x}} \approx 1 - x + k x^2,$$

where the value of the constant k is to be stated. [3]

6 (i) Show that the equation $2 \sec \theta^\circ - \tan \theta^\circ = 3$ can be expressed in the form $R \cos(\theta - \alpha)^\circ = 2$, where the values of R and α (with $0 < \alpha < 90$) are to be stated. [4]

(ii) Hence solve the equation $2 \sec \theta^\circ - \tan \theta^\circ = 3$, giving all values of θ such that $0 < \theta < 360$. [4]

7 (i) Show that the substitution $y = e^{-x}$ transforms the integral

$$\int_0^{\ln 2} \frac{1}{1+e^{-x}}\, dx \quad \text{to} \quad \int_{\frac{1}{2}}^1 \frac{1}{y(1+y)}\, dy.$$ [4]

(ii) Hence, or otherwise, evaluate

$$\int_0^{\ln 2} \frac{1}{1+e^{-x}}\, dx.$$ [4]

8 (i) Find the modulus and argument of the complex number $2 + 2\sqrt{3}\,i$. [2]

(ii) Hence, or otherwise, find the two square roots of $2 + 2\sqrt{3}\,i$, giving your answers in the form $a + ib$. [3]

(iii) Find the exact solutions of the equation

$$i z^2 - 2\sqrt{2}\,z - 2\sqrt{3} = 0,$$

giving your answers in the form $a + ib$. [4]

9

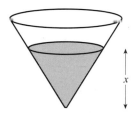

Water is flowing out of a small hole at the bottom of a conical container whose axis is vertical. At time t, the depth of the water in the container is x and the volume of the water in the container is V (see diagram). You are given that V is proportional to x^3, and that the rate at which V decreases is proportional to $\sqrt{x}$.

(i) Express $\dfrac{dV}{dt}$ in terms of x, $\dfrac{dx}{dt}$ and a constant. [2]

(ii) Show that x satisfies a differential equation of the form

$$\frac{dx}{dt} = -\frac{A}{x^{\frac{3}{2}}},$$

where A is a positive constant. [2]

(iii) Find the general solution of the differential equation in part (ii). [3]

(iv) Given that $x = 4$ when $t = 0$ and that $x = 1$ when $t = 1$, find the value of t when $x = 0$. [4]

10 The lines l_1 and l_2 have equations

$$\mathbf{r} = \begin{pmatrix} 3 \\ 5 \\ 1 \end{pmatrix} + s \begin{pmatrix} 1 \\ 2 \\ -4 \end{pmatrix} \quad \text{and} \quad \mathbf{r} = \begin{pmatrix} 0 \\ 2 \\ 4 \end{pmatrix} + t \begin{pmatrix} 1 \\ -1 \\ 5 \end{pmatrix}$$

respectively.

(i)   Show that l_1 and l_2 intersect, and find the position vector of the point of intersection. [4]

The plane p passes through the point with position vector $\begin{pmatrix} 3 \\ 5 \\ 1 \end{pmatrix}$ and is perpendicular to l_1.

(ii)   Find the equation of p, giving your answer in the form $ax + by + cz = d$. [3]

(iii)   Find the position vector of the point of intersection of l_2 and p. [3]

(iv)   Find the acute angle between l_2 and p. [3]

Answers to P2&3

1 Polynomials

Exercise 1A (page 5)

1. (a) 3   (b) 1   (c) 4
   (d) 0   (e) 1   (f) 0

2. (a) $4x^2 + 7x + 6$
   (b) $5x^3 + 3x^2 - 6x - 3$
   (c) $8x^4 - 3x^3 + 7x^2 - 3x + 1$
   (d) $2x^5 + 2x^4 - 5x^2 + 3$
   (e) $-x^3 - 2x^2 - 5x + 4$

3. (a) $2x^2 + x - 8$
   (b) $3x^3 + 7x^2 - 8x + 9$
   (c) $-2x^4 - x^3 + 7x^2 + 3x - 3$
   (d) $2x^5 - 2x^4 - 6x^3 + 5x^2 + 1$
   (e) $-x^3 - 6x^2 + 9x + 2$

4. (a) $2x^3 - 3x^2 + 9x - 2$
   (b) $3x^3 - 7x^2 + 16x - 13$
   (c) $x^3 - 4x^2 + 7x - 11$
   (d) $3x^3 - 8x^2 + 17x - 17$

5. (a) $6x^2 - 7x - 3$
   (b) $x^3 + x^2 - 7x + 2$
   (c) $2x^3 + 5x^2 - 3x - 9$
   (d) $12x^3 - 13x^2 + 9x - 2$
   (e) $x^4 + 2x^3 - 2x^2 + 2x - 3$
   (f) $8x^4 - 6x^3 - 15x^2 + 18x - 5$
   (g) $x^4 + 5x^3 + 5x^2 + 3x + 18$
   (h) $x^5 + 5x^4 + 3x^3 + 10x^2 - 8x + 5$
   (i) $3 + 8x - 4x^2 + 13x^3 - 4x^4 + 4x^5$
   (j) $8 - 22x + 19x^2 - 3x^3 - 3x^4 + x^5$
   (k) $6x^3 + 29x^2 - 7x - 10$
   (l) $2x^5 - 7x^4 + 6x^3 - 10x^2 + 4x - 3$

6. (a) $0, -1$ (b) $-11, -1$
   (c) $-3, -9$ (d) $25, -8$
   (e) $20, -21$ (f) $-16, 13$
   (g) $-17, -1$ (h) $-5, 11$
   (i) $5, -6$ (j) $-11, 8$

7. (a) $4, 1$ (b) $2, -3$
   (c) $2, 1$ (d) $3, -2$
   (e) $1, 2$ (f) $2, -3$
   (g) $2, 3$ (h) $2, -1$

Exercise 1B (page 8)

1. (a) $x - 3$ (b) $x + 17$
   (c) $3x + 11$ (d) $7x - 3$
   (e) $x - 3$ (f) $7x - 2$

2. (a) $1, -5, 22$ (b) $1, 8, -11$
   (c) $3, -4, 0$ (d) $3, -1, -4$
   (e) $4, 1, 4$ (f) $7, 1, 8$

3. (a) $1, -3, 5, 2$ (b) $1, -2, 4, 22$
   (c) $1, 1, -1, 3$ (d) $4, 1, -3, 11$
   (e) $2, 7, -1, 0$ (f) $3, 0, -5, 10$

4. (a) $2, -3, 4, -1, -2$ (b) $4, 1, 0, -2, 3$
   (c) $3, 1, -1, 2, 0$ (d) $1, -2, 5, -3, 2$

Exercise 1C (page 12)

1. (a) $x - 2, -4$ (b) $x + 1, -7$
   (c) $2x + 7, 13$ (d) $x + 2, 3$
   (e) $2x - 1, -1$ (f) $x, 0$

2. (a) $x^2 - 3, 7$
   (b) $x^2 + 2x + 15, 71$
   (c) $2x^2 - 6x + 22, -71$
   (d) $5x^2 + 20x + 77, 315$
   (e) $x^2 - x - 1, -6$
   (f) $2x^2 + 7x - 1, 3$

3. (a) $x^2 - 4x + 2, -x + 7$
   (b) $x^2 - 2x + 3, 2x + 5$
   (c) $2x - 6, 9x^2 + 4x + 11$
   (d) $3x^2 + 2x - 5, 30$

4. (a) -5 (b) 13 (c) 50 (d) -355
   (e) $\frac{7}{8}$ (f) $7\frac{13}{27}$ (g) 0 (h) 279

5. -1

6. -2

7. -5

8. 3

9. $5, -3$

10. $4, -3$

11. $2, 1$

12. $5, 3$

Exercise 1D (page 15)

1. (a) $(x+1)(x-2)(x+3)$ $-3, -1, 2$
   (b) $(x-1)(x-3)(x+1)$ $-1, 1, 3$
   (c) $(x-1)(x-5)(x+3)$ $-3, 1, 5$
   (d) $(x+1)^2(x-5)$ $-1, 5$
   (e) $(x-2)(x+2)(x+3)$ $-3, -2, 2$
   (f) $(2x+1)(x-1)(x+4)$ $-4, -\frac{1}{2}, 1$
   (g) $(3x-1)(x-2)(x+2)$ $-2, \frac{1}{3}, 2$
   (h) $(x+1)(2x-1)(3x+2)$ $-1, -\frac{2}{3}, \frac{1}{2}$
   (i) $(x-1)(x^2 + 3x - 1)$ $1, \frac{1}{2}(\pm\sqrt{13} - 3)$

2 (a) $(x-3)(x-1)(x+1)(x+2)$ $-2,-1,1,3$
    (b) $(x-2)(x+1)(x+2)(x+3)$ $-3,-2,-1,2$
    (c) $(x-3)(x-1)(x+2)(2x+1)$ $-2,-\frac{1}{2},1,3$
    (d) $(x+1)(x-2)(2x+1)(3x+2)$ $-1,-\frac{2}{3},-\frac{1}{2},2$
    (e) $(x-1)^3(x+1)$ $-1,1$
    (f) $(x-2)^2(2x+1)^2$ $-\frac{1}{2},2$

3 (a) $(x-2)\left(x^2+2x+4\right)$
    (b) $(x+2)\left(x^2-2x+4\right)$
    (c) $(x-a)\left(x^2+ax+a^2\right)$
    (d) $(x+a)\left(x^2-ax+a^2\right)$
    (e) $(x-a)(x+a)\left(x^2+a^2\right)$
    (f) $(x+a)\left(x^4-ax^3+a^2x^2-a^3x+a^4\right)$

4 (b) n must be odd;
    $x^{n-1}-ax^{n-2}+a^2x^{n-3}-\ldots+a^{n-1}$

Miscellaneous exercise 1 (page 16)

1 $a=-3, b=1$

2 1

3 $3x+4$, $2x+3$

4 $a=-1, b=-10$

5 -6

6 $-1,\frac{1}{2},-\frac{3}{2}$

7 (a) 2
    (b) $2,\sqrt{2},-\sqrt{2}$

8 $(x-3)\left(x^2+x+2\right)$; one root only as the
discriminant of the quadratic is negative; one
point only, as the equation for the intersections is
the given cubic.

9 $x^2+4x+6=(x+2)^2+2$

10 $x^2-3; x^2+x+2$

11 (b) $p=-\sqrt{5}, q=\sqrt{5}$

12 $4x+7$

13 $x^2+2x+2, 0$

14 (a) $84, 0$; $x-2$ is not a factor of p(x), but
$x+2$ is.
    (b) $-2,-1\frac{1}{2},\frac{1}{2}$

15 (a) -15
    (b) $(x+2)(x-5)\left(x+\sqrt{5}\right)\left(x-\sqrt{5}\right)$
    (c) $x<-\sqrt{5},-2<x<\sqrt{5},x>5$

16 $A\left(-\sqrt{3},0\right), B\left(\sqrt{3},0\right)$; $(x-2)(x+1)^2$
    (a) 2
    (b) They touch at $(-1,-2)$.

17 (b) $6x-4$
    (d) $1,1-\sqrt{2},1+\sqrt{2}$

2 The modulus function

Exercise 2A (page 23)

3 (a) $2<x<4$ (b) $-2.1\leqslant x\leqslant-1.9$
    (c) $1.4995\leqslant x\leqslant1.5005$
    (d) $-1.25\leqslant x\leqslant2.75$

4 (a) $|x-1.5|\leqslant0.5$ (b) $|x-1|<2$
    (c) $|x+3.65|\leqslant0.15$ (d) $|x-2.85|<0.55$

5 $|a+b|\leqslant|a|+|b|$; $|a+b|\geqslant||a|-|b||$

Exercise 2B (page 28)

1 (a) $3,-7$ (b) $8,-6$ (c) $0,3$
    (d) $3,-3\frac{2}{3}$ (e) $4,\frac{2}{3}$ (f) $-2,\frac{1}{2}$
    (g) $-2,-8$ (h) $-4,1\frac{3}{7}$

2 (a) $-3<x<-1$ (b) $x<-2$ or $x>8$
    (c) $-5\leqslant x\leqslant-2$ (d) $x\leqslant-3\frac{1}{3}$ or $x\geqslant2$
    (e) $x<-\frac{3}{4}$ or $x>\frac{1}{2}$ (f) $x<-3$ or $x>2\frac{1}{3}$
    (g) $1<x<3$ (h) $x\leqslant0$

3 (a) $-1\leqslant x\leqslant1$ (b) $x\geqslant1$ (c) $x\leqslant-1$

4 (a) $x\geqslant1$ (b) $x\leqslant0$ (c) $0\leqslant x\leqslant1$

5 (a) True   (b) False

Miscellaneous exercise 2 (page 28)

1 $x<\frac{1}{2}$

2 3 and -2 respectively

3 $-10<x<\frac{2}{3}$

4 $-\frac{1}{3},-1$

5 $x<1$

6 $-\frac{1}{3},\frac{1}{5}$; $x<-\frac{1}{3}$ or $x>\frac{1}{5}$

7 Any value of x such that $-3\leqslant x\leqslant3$

8 (a) (ii) $0\leqslant$ f$(x)\leqslant2$, $-1\leqslant$ g$(x)\leqslant1$
    (b) f(x) is periodic, with period 180; g(x) is
not periodic.

9 $x<2.4$ or $x>4$

10 (a) $2x-1$ (b) 7 (c) $1-2x$

11 (a) -2, $\frac{4}{3}$ (b) No solution

12 (a) 8 (b) -3

13 (a) $y=3x-1$ (b) $y=1-x$

15 (a) $x<1$ or $x>1\frac{2}{3}$
    (b) $x<-\frac{5}{3}$ or $-1<x<1$ or $x>\frac{5}{3}$
    (c) $x<\frac{3}{5}$ or $x>\frac{5}{3}$

3 Exponential and logarithmic functions

Exercise 3A (page 34)

1 (a) 800 (b) 141 (c) 336

2 (a) 1.059 (b) 262 (c) between D and D#

3 (a) 45.5 °C (b) 13.6 minutes

4 (a) 5.42 (b) 5.42

6 (a) $8 = 2^3$ (b) $81 = 3^4$
(c) $0.04 = 5^{-2}$ (d) $x = 7^4$
(e) $5 = x^t$ (f) $q = p^r$

7 (a) $3 = \log_2 8$ (b) $6 = \log_3 729$
(c) $-3 = \log_4 \frac{1}{64}$ (d) $8 = \log_a 20$
(e) $9 = \log_h g$ (f) $n = \log_m p$

8 (a) 4 (b) 2 (c) -2
(d) 0 (e) 1 (f) $-\frac{1}{3}$
(g) $\frac{3}{4}$ (h) $\frac{3}{2}$ (i) 7

9 (a) 7 (b) $\frac{1}{64}$ (c) 4
(d) $\frac{1}{10}$ (e) $4\sqrt{2}$ (f) 6
(g) $\frac{1}{256}$ (h) -10 (i) $\frac{1}{3}\sqrt{3}$

Exercise 3B (page 36)

1 (a) $\log p + \log q + \log r$
(b) $\log p + 2\log q + 3\log r$
(c) $2 + \log p + 5\log r$
(d) $\frac{1}{2}(\log p - 2\log q - \log r)$
(e) $\log p + \log q - 2\log r$
(f) $-(\log p + \log q + \log r)$
(g) $\log p - \frac{1}{2}\log r$
(h) $\log p + \log q + 7\log r - 1$
(i) $\frac{1}{2}(1 + 10\log p - \log q + \log r)$

2 (a) 2 (b) -1 (c) $\log 30\,575$ (d) 0
(e) 3 (f) -3 (g) $\log 8$ (h) 0

3 (a) $r - q$ (b) $2p + q$ (c) $p + \frac{1}{2}r$
(d) $-q$ (e) $p + 2q + r$ (f) $p - q + 2r$
(g) $q - p - r$ (h) $4p + q - 2r$ (i) $p + q - 2r$

Exercise 3C (page 40)

1 (a) 1.46 (b) 1.56 (c) 1.14
(d) 1.22 (e) 3.58 (f) 1.71
(g) -2.21 (h) 3 (i) -0.202

2 (a) $x > 1.89$ (b) $x < 1.43$ (c) $x \leqslant -1.68$
(d) $x > 9.97$ (e) $x > 8.54$ (f) $x < -2$
(g) $x \geqslant -2$ (h) $x \leqslant -5.61$ (i) $x \geqslant 3.77$

3 37

4 14

5 28

6 7

7 9.56

8 71

9 (a) 0.891 (b) 12 days (c) 19.9 days

10 9.49 a.m. Tuesday

11 389 years

12 (a) 1.79 (b) 2.37 (c) 0.486
(d) -7.97 (e) 1.04 (f) 2.32

Exercise 3D (page 46)

1 (a) $y = 2.51 \times 3.98^x$ (b) $y = 10^{12} \times 0.001^x$
(c) $y = 5.01 \times 50.1^x$ (d) $y = 5.01x^2$
(e) $y = \dfrac{0.316}{x^5}$

2 (a) $y = 1.49 \times 1.82^x$
(b) $y = 1.63 \times 10^5 \times 20.1^{-x}$
(c) $y = 2.01 \times 5.47^x$
(d) $y = 2.01x^2$ (e) $y = \dfrac{0.607}{x^5}$

3 $p = 39.7 \times 1.022^x$ gives $p = 39.7, 49.4, 61.3,$ 76.3, 94.8. The exponential model does not fit so well in this period.

4 61, 30, 19, 13

5 $a \approx 1, n \approx 1.5$

6 Investment = £850, interest at 7.5%

7 (a) 3, 2 (b) 3.15×10^6 (c) 3, ln 2

8 (a) Between 1050 and 1170
(b) 3 (c) 843

Miscellaneous exercise 3 (page 48)

1 (a) $\dfrac{1}{\log a - 2}$ (b) $\sqrt{\dfrac{5a}{2}}$

2 0.774

3 $f^{-1}: x \mapsto e^x - 1, \quad x \in \mathbb{R}$

4 $2 - \log 2$

6 $\log 3 - \frac{1}{6}$

11 $\log_2 \dfrac{x+2}{x}, \frac{2}{7}$

12 $10\,000, 451$

13 (a) $0.202, 0.4$ (b) 1.84 tonnes

14 (a) 15.3, 2.7; 39.6 mm (or 39.3 mm if you use the exact values $a = \frac{46}{3}$, $b = \frac{8}{3}$)
(b) 12.5, 1.46
(c) Both models give a reasonable fit, but (b) is slightly better.

4 Differentiating exponentials and logarithms

Exercise 4A (page 54)

1 (a) $3e^{3x}$ (b) $-e^{-x}$ (c) $6e^{2x}$
(d) $16e^{-4x}$ (e) $3e^{3x+4}$ (f) $-2e^{3-2x}$
(g) $-e^{1-x}$ (h) $12e^{3+4x}$ (i) $3x^2 e^{x^3}$
(j) $-xe^{-\frac{1}{2}x^2}$ (k) $-\dfrac{1}{x^2}e^{\frac{1}{x}}$ (l) $\dfrac{1}{2\sqrt{x}}e^{\sqrt{x}}$

2 (a) $3e^2$ 　　(b) $-2e$ 　　(c) -1
　　(d) -2

3 (a) $ey = x + 2$
　　(b) $y = 3x - 1$
　　(c) $y = (4 + 4e^4)x - (4 + 6e^4)$
　　(d) $4y = -2x + 1 + 2\ln 2$

4 (a) $2e^{x^2+x+1}(2x+1)$ 　　(b) $-15e^{-x}(e^{-x}+1)^4$

　　(c) $-x(1-x^2)^{-\frac{1}{2}}e^{\sqrt{1-x^2}}$

5 $-3\frac{3}{4}$

6 (a) No stationary points
　　(b) $(\ln 2, 2\ln 2 - 2)$, maximum
　　(c) $(0,1)$, minimum; $(\pm 1, 2 + e^{-1})$, maximum

7 (a) $\frac{1}{3}e^{3x} + k$ 　　(b) $-e^{-x} + k$

　　(c) $\frac{3}{2}e^{2x} + k$ 　　(d) $e^{-4x} + k$

　　(e) $\frac{1}{3}e^{3x+4} + k$ 　　(f) $-\frac{1}{2}e^{3-2x} + k$

　　(g) $-e^{1-x} + k$ 　　(h) $\frac{3}{4}e^{3+4x} + k$

8 (a) $\frac{1}{2}(e^4 - e^2)$ (b) $e - \frac{1}{e}$ (c) $e^5 - e$

　　(d) $e^{10} - e^8$ (e) 6 (f) $\frac{3}{4}e$

　　(g) $\dfrac{1}{\ln 2}$ 　　(h) $\dfrac{1}{\ln 3}(3^9 - \frac{1}{27})$

9 $\frac{1}{2}(e^4 - 1)$

10 $1 - e^{-N}$; 1

11 $\frac{1}{2}$

Exercise 4B (page 57)

1 (a) $\dfrac{1}{x}$ (b) $\dfrac{2}{2x-1}$ (c) $\dfrac{-2}{1-2x}$

　　(d) $\dfrac{2}{x}$ (e) $\dfrac{b}{a+bx}$ (f) $\dfrac{-1}{x}$

　　(g) $\dfrac{-3}{3x+1}$ (h) $\dfrac{2}{2x+1} - \dfrac{3}{3x-1}$

　　(i) $-\dfrac{6}{x}$ (j) $\dfrac{1}{x} + \dfrac{1}{x+1}$ (k) $\dfrac{2}{x} + \dfrac{1}{x-1}$

　　(l) $\dfrac{1}{x-1} + \dfrac{1}{x+2}$

2 (a) $y = 2x - \ln 2 - 1$
　　(b) $y = 2x - 1$
　　(c) $y = -3x - \ln 3 - 1$
　　(d) $ey = x + e\ln 3$

3 (a) 1, minimum
　　(b) $\frac{1}{2} - \ln 2$, minimum
　　(c) 1, minimum
　　(d) 1, minimum

4 (a) $\dfrac{3x^2}{1+x^3}$ (b) $\dfrac{2x^3}{2+x^4}$ (c) $\dfrac{3x^2+4}{x^3+4x}$

6 $2y = 2 - x$

7 $x > 6$; $\dfrac{1}{x-2} + \dfrac{1}{x-6}$ 　(a) $x > 6$ 　(b) None

8 (i) $2 < x < 6$; $\dfrac{1}{x-2} - \dfrac{1}{6-x}$
　　　(a) $2 < x < 4$ 　(b) $4 < x < 6$
　　(ii) There are no points in the domain.

Exercise 4C (page 58)

1 (a) $y = \frac{1}{2}\ln x + k$, 　$x > 0$
　　(b) $y = \ln(x-1) + k$, 　$x > 1$
　　(c) $y = -\ln(1-x) + k$, 　$x < 1$
　　(d) $y = \frac{1}{4}\ln(4x+3) + k$, 　$x > -\frac{3}{4}$
　　(e) $y = -2\ln(1-2x) + k$, 　$x < \frac{1}{2}$
　　(f) $y = 2\ln(1+2x) + k$, 　$x > -\frac{1}{2}$
　　(g) $y = -2\ln(-1-2x) + k$, 　$x < -\frac{1}{2}$
　　(h) $y = 2\ln(2x-1) + k$, 　$x > \frac{1}{2}$

2 (a) $\ln 2$ (b) $\ln 2$ (c) $\ln 2$ (d) $\ln 2$

3 (a) $\ln 2$ 　　(b) $\frac{1}{2}\ln 3$ (c) $\frac{2}{3}\ln\frac{13}{7}$

　　(d) $\ln\dfrac{5c-7}{4e-7}$ (e) $\ln 2$ (f) $8 + \ln 5$

4 $2\ln\frac{3}{2}$

5 $\pi\ln\frac{5}{2}$

6 $\frac{1}{2}\pi\ln 3$

7 $y = \frac{3}{2}\ln\left(\frac{1}{3}(2x+1)\right)$

8 $y = 2\ln(4x-3) + 2$

9 $2\pi\ln 2$

Exercise 4D (page 62)

1 (a) $-\ln 4$ 　　(b) $-\frac{1}{2}\ln 3$ 　　(c) $\frac{2}{3}\ln 5 - \ln 4$

　　(d) $\ln\dfrac{7-2e}{7-e}$ (e) $-\ln\frac{5}{3}$ (f) $2 - \ln 2$

2 $\ln 7$, $-\frac{2}{7}$

3 $2^x \ln 2, 3^x \ln 3, 10^x \ln 10, -\left(\frac{1}{2}\right)^x \ln 2$

Miscellaneous exercise 4 (page 62)

1 (a) $\dfrac{3}{3x-4}$ (b) $\dfrac{-3}{4-3x}$ (c) $3e^{3x}$

　　(d) $-e^{-x}$ (e) $\dfrac{1}{3-x} - \dfrac{1}{2-x}$ (f) $\dfrac{-6}{3-2x}$

2 E.g., 13

3 $\left(\frac{1}{2}, 8\right), (4,1)$; $15\frac{3}{4} - 12\ln 2$

4 (a) $\frac{2}{3}e - \frac{1}{6}e^{-2} - \frac{1}{2}$ 　　(b) Minimum

5 $2x - e^{-x} + k$

6 $x = \frac{1}{2}$, minimum

8 $y = x + 4$

9 $y = -\frac{1}{2}x + \frac{1}{4}$

10 (a) $\dfrac{2}{x-1}$ (b) $3\ln 2$

11 (3, 0)

13 (a) -1 (b) 1

14 (a) $2 - e^{-a} = a$ (b) $\displaystyle\int_0^a \left(2 - e^{-x} - x\right)dx$

15 $\left(-3, e^{-81}\right), (0, 1), \left(3, e^{-81}\right)$

16 (a) $e^{-2a} - 3a = 0$

(d) $\frac{1}{2} - \frac{3}{2}a^2 - \frac{1}{2}e^{-2a}$

17 $\frac{1}{2}e$

18 (a) 3.00 (b) 1.22×10^6

5 Trigonometry

Exercise 5A (page 68)

1 (a) -0.675 (b) 1.494 (c) 1.133

2 (a) $\operatorname{cosec} x$ (b) $\cot x$ (c) $\sec x$
(d) $\sec^2 x$ (e) $\cot x$ (f) $-\operatorname{cosec} x$

3 (a) $\sqrt{2}$ (b) 1 (c) $-\sqrt{3}$
(d) $-\sqrt{2}$ (e) $-\frac{1}{3}\sqrt{3}$ (f) $\frac{2}{3}\sqrt{3}$
(g) 0 (h) $-\frac{2}{3}\sqrt{3}$

4 (a) 0.951 (b) 1.05 (c) 3.73
(d) 2 (e) -0.924 (f) 3.73
(g) -1.04 (h) $-\sqrt{3}$

5 (a) $\frac{5}{4}$ (b) $\frac{4}{3}$ (c) $-\frac{1}{3}\sqrt{3}$
(d) $\frac{2}{3}\sqrt{3}$

6 $\pm 4\sqrt{3}$, $\pm\sqrt{2}$, $\pm\frac{1}{2}\sqrt{5}$

7 (a) $|\tan\phi|$ (b) $\sin\phi\cos\phi$ (c) $\cot\phi$
(d) $|\sin\phi|$ (e) $|\tan\phi|$ (f) $\cot^2\phi$

8 (a) $3\sec^2\theta - \sec\theta - 3$ (b) 0.72, π, 5.56

9 1.11, 2.82, 4.25, 5.96

10 $\frac{1}{4}\pi, \frac{1}{2}\pi, \frac{3}{4}\pi, \frac{5}{4}\pi, \frac{3}{2}\pi, \frac{7}{4}\pi$

Exercise 5B (page 73)

1 $\frac{1}{4}\left(\sqrt{6} + \sqrt{2}\right)$, $2 + \sqrt{3}$

2 (a) $\frac{1}{4}\left(\sqrt{2} - \sqrt{6}\right)$ (b) $\frac{1}{4}\left(\sqrt{6} + \sqrt{2}\right)$
(c) $-2 - \sqrt{3}$

3 $\frac{1}{2}\cos x - \frac{1}{2}\sqrt{3}\sin x$

4 $-\cos\phi$, $-\sin\phi$

5 $\dfrac{\sqrt{3} + \tan x}{1 - \sqrt{3}\tan x}$, $\dfrac{1 + \sqrt{3}\tan x}{\tan x - \sqrt{3}}$

7 (a) $\frac{4}{3}$ (b) $\frac{7}{25}$ (c) $\frac{4}{5}$ (d) $\frac{117}{44}$

8 $-\frac{63}{65}$, $-\frac{33}{56}$

Exercise 5C (page 75)

1 $-\frac{1}{3}\sqrt{5}$, $-\frac{4}{9}\sqrt{5}$, $-4\sqrt{5}$

2 $\frac{1}{8}$, $\pm\frac{1}{4}\sqrt{14}$

3 $\sin 3A \equiv 3\sin A - 4\sin^3 A$

4 $\cos 3A \equiv 4\cos^3 A - 3\cos A$

5 $\tan^2 \frac{1}{2}x$

7 $\pm\frac{5}{6}$, $\pm\frac{1}{6}\sqrt{11}$

8 $\frac{2}{3}$, $-\frac{3}{2}$

9 $\pm\sqrt{2} - 1$, $\sqrt{2} - 1$

10 (a) π (b) $\frac{1}{2}\pi$, 3.99, 5.44
(c) 0, 0.87, 2.27, π, 4.01, 5.41, 2π

Exercise 5D (page 78)

1 0.588

2 53.1

3 $\sqrt{34}$

4 $\sqrt{37}$, 0.165

5 (a) $\sqrt{5}$, 1.11 (b) $\sqrt{5}$, 0.464
(c) $\sqrt{5}$, 1.11 (d) $\sqrt{5}$, 0.464

6 $\sqrt{61}\cos(\theta - 0.876)$
(a) $\sqrt{61}$ when $\theta = 0.876$
(b) $-\sqrt{61}$ when $\theta = 4.018$

7 $10\sin(x + 36.9)°$ (a) 1 (b) 0

8 0.87 or 3.45, correct to 2 decimal places

Miscellaneous exercise 5 (page 78)

1 (b) $-\frac{25}{24}$

2 $60, 120, 240, 300$

3 (a) 1 (b) $\frac{4}{5}$

4 (b) $60, 300$

5 $\frac{1}{4}\left(\sqrt{6} - \sqrt{2}\right)$

7 (a) $-\frac{4}{5}$ (b) $-\frac{24}{25}$, $\frac{7}{25}$

8 73.9

9 $26.6, 90, 206.6, 270$

12 $2\sin(\theta + 60)°$; $90, 330$

13 (a) $2\cos(x + 60)°$ (b) $0, 60, 180, 240, 360$

14 (a) $15\cos(x - 0.644)$ (b) 0.276

15 $\sqrt{5}\cos(x - 26.6)°$
(a) $90, 323.1$ (b) $-\sqrt{5} \leqslant k \leqslant \sqrt{5}$

16 -0.5

17 (a) $13\sin(x + 67.4)°$
(b) 7.5 when $x = 202.6$, 1 when $x = 22.6$

18 $5\cos(x+53.1)°$ (a) $13.3, 240.4$
  (b) $\frac{1}{3}, \frac{1}{13}$
19 (a) $\frac{1}{6}\pi$ (b) $\dfrac{x+y}{1-xy}$
20 (b) $\dfrac{1}{x}$, $1-2x^2$

21 (a) $x+2y=3$
  (b) $\sqrt{5}\sin(\theta+63.4)°$
  (c) $(2\sin\theta°, 2\cos\theta°), 74.4$
23 (a) $(56.8,0)$, $(123.2,0)$
  (b) 53.1, 120; $(53.1, 0.6)$, $(120, \pm 0.5)$

6 Differentiating trigonometric functions

Exercise 6A (page 88)

2 (a) $-\cos x$ (b) $\sin x$
  (c) $4\cos 4x$ (d) $-6\sin 3x$
  (e) $\frac{1}{2}\pi\cos\frac{1}{2}\pi x$ (f) $-3\pi\sin 3\pi x$
  (g) $-2\sin(2x-1)$ (h) $15\cos\left(3x+\frac{1}{4}\pi\right)$
  (i) $5\cos 5x$ (j) $2\cos\left(\frac{1}{4}\pi-2x\right)$
  (k) $2\cos 2x$ (l) $-\pi\sin\pi x$

3 (a) $2\sin x\cos x$ (b) $-2\cos x\sin x$
  (c) $-3\cos^2 x\sin x$ (d) $5\sin\frac{1}{2}x\cos\frac{1}{2}x$
  (e) $-8\cos^3 2x\sin 2x$ (f) $2x\cos x^2$
  (g) $-42x^2\sin 2x^3$
  (h) $\sin\left(\frac{1}{2}x-\frac{1}{3}\pi\right)\cos\left(\frac{1}{2}x-\frac{1}{3}\pi\right)$
  (i) $-6\pi\cos^2 2\pi x\sin 2\pi x$
  (j) $6x\sin^2 x^2\cos x^2$
  (k) 0 (l) $-\cos\frac{1}{2}x\sin\frac{1}{2}x$

4 (a) $2\sec 2x\tan 2x$ (b) $-3\csc 3x\cot 3x$
  (c) $-3\csc\left(3x+\frac{1}{5}\pi\right)\cot\left(3x+\frac{1}{5}\pi\right)$
  (d) $\sec\left(x-\frac{1}{3}\pi\right)\tan\left(x-\frac{1}{3}\pi\right)$
  (e) $8\sec^2 x\tan x$ (f) $-3\csc^3 x\cot x$
  (g) $-12\csc^4 3x\cot 3x$
  (h) $10\sec^2\left(5x-\frac{1}{4}\pi\right)\tan\left(5x-\frac{1}{4}\pi\right)$

5 (a) $2\cot 2x$ (b) $-3\tan 3x$
  (c) $-\cot x$ (d) $4\tan 4x$
  (e) $2\cot x$ (f) $-6\tan 2x$

6 (a) $\cos x\,e^{\sin x}$ (b) $-3\sin 3x\,e^{\cos 3x}$
  (c) $10\sin x\cos x\,e^{\sin^2 x}$

8 (a) $2y-x=\sqrt{3}-\frac{1}{3}\pi$
  (b) $3y=\sqrt{2}\left(x-\frac{3}{2}-\frac{1}{4}\pi\right)$
  (c) $\sqrt{2}y+x=2+\frac{1}{4}\pi$

(d) $y=x+\frac{1}{2}\ln 2-\frac{1}{4}\pi$
(e) $y=3$

9 (a) $\left(\frac{1}{4}\pi, \sqrt{2}\right)$, maximum; $\left(\frac{5}{4}\pi, -\sqrt{2}\right)$, minimum
  (b) (π, π), neither
  (c) $(0,2)$, maximum; $(\pi, -2)$, minimum
  (d) $\left(\frac{1}{12}\pi, \frac{1}{2}\sqrt{3}+\frac{1}{12}\pi\right)$, maximum;
    $\left(\frac{5}{12}\pi, -\frac{1}{2}\sqrt{3}+\frac{5}{12}\pi\right)$, minimum;
    $\left(\frac{13}{12}\pi, \frac{1}{2}\sqrt{3}+\frac{13}{12}\pi\right)$, maximum;
    $\left(\frac{17}{12}\pi, -\frac{1}{2}\sqrt{3}+\frac{17}{12}\pi\right)$, minimum
  (e) $\left(\frac{1}{4}\pi, 2\sqrt{2}\right)$, minimum;
    $\left(\frac{5}{4}\pi, -2\sqrt{2}\right)$, maximum
  (f) $\left(\frac{1}{2}\pi, -3\right)$, minimum; $\left(\frac{7}{6}\pi, \frac{3}{2}\right)$, maximum;
    $\left(\frac{3}{2}\pi, 1\right)$, minimum; $\left(\frac{13}{6}\pi, \frac{3}{2}\right)$, maximum

10 $\cos(a+x)$, $-\sin a\sin x+\cos a\cos x$

11 $\sin\left(\frac{3}{2}\pi-x\right)$; $-\sin x$, $-\cos x$

12 As $2\cos^2 x-1=1-2\sin^2 x=\cos 2x$, they all differ by only a constant, and therefore have the same derivative.

13 $\cos 2x$

14 Above, at $y=\cos\frac{5}{6}\pi+\frac{5}{12}\pi\approx 0.443$

15 (a) $\dfrac{\cos\sqrt{x}}{2\sqrt{x}}$ (b) $\dfrac{-\sin x}{2\sqrt{\cos x}}$ (c) $-\dfrac{1}{x^2}\cos\dfrac{1}{x}$

16 The curve bends downwards when $y>0$, and upwards when $y<0$; $y=\sin(nx+\alpha)$.

17 $\dfrac{1+\sin^2 x}{\cos^3 x}=\sec^3 x+\tan^2 x\sec x$

19 (a) Growing at 50 million dollars per year
  (b) Falling at 9.7 million dollars per year

20 (a) 55.3 mm s^{-1} (b) 153 m s^{-2}

Exercise 6B (page 92)

1 (a) $\frac{1}{2}\sin 2x+k$ (b) $-\frac{1}{3}\cos 3x+k$
  (c) $\frac{1}{2}\sin(2x+1)+k$ (d) $-\frac{1}{3}\cos(3x-1)+k$
  (e) $\cos(1-x)+k$ (f) $-2\sin\left(4-\frac{1}{2}x\right)+k$
  (g) $-2\cos\left(\frac{1}{2}x+\frac{1}{3}\pi\right)+k$
  (h) $\frac{1}{3}\sin\left(3x-\frac{1}{4}\pi\right)+k$ (i) $2\cos\frac{1}{2}x+k$

2 (a) 1   (b) $\frac{1}{2}\sqrt{2}$ (c) $\frac{1}{2}$

    (d) $-\frac{1}{6}\sqrt{2}$ (e) $\frac{1}{6}\left(\sqrt{3}-1\right)$ (f) 0

    (g) $\sin 1$ (h) $2\cos 1 - 2\cos\frac{5}{4}$ (i) 4

3 (a) $\frac{1}{2}\ln\sec 2x + k$ (b) $\frac{1}{5}\ln\sin 5x + k$

    (c) $\frac{1}{3}\sec 3x + k$ (d) $-\frac{1}{4}\operatorname{cosec} 4x + k$

    (e) $\ln\cos\left(\frac{1}{4}\pi - x\right) + k$

    (f) $-\frac{1}{2}\ln\sin\left(\frac{1}{3}\pi - 2x\right) + k$

    (g) $2\sec\left(\frac{1}{2}x + 1\right) + k$ (h) $\frac{1}{2}\operatorname{cosec}(1 - 2x) + k$

    (i) $\frac{1}{2}\sec 2x + k$

4 (a) $\frac{1}{2}\ln 2$ (b) $\frac{1}{6}\ln 2$

    (c) $\frac{1}{3}\left(1 - \sqrt{2}\right)$ (d) $\ln 2$

    (e) $\frac{1}{2}\left(\operatorname{cosec}\frac{1}{2} - \operatorname{cosec} 1\right)$

    (f) $4(\sec 0.075 - \sec 0.025)$

5 (a) $\frac{1}{2}\left(x + \frac{1}{2}\sin 2x\right) + k$ (b) $\frac{1}{2}(x + \sin x) + k$

    (c) $\frac{1}{2}\left(x - \frac{1}{4}\sin 4x\right) + k$ (d) $\frac{1}{4}\sin^4 x + k$

    (e) $\frac{1}{3}\sec^3 x + k$ (f) $-\frac{1}{10}\operatorname{cosec}^5 2x + k$

    (g) $\frac{1}{2}\cos x - \frac{1}{14}\cos 7x + k$

    (h) $\frac{1}{3}\cos^3 x - \cos x + k$

    (i) $-\cos x + \frac{2}{3}\cos^3 x - \frac{1}{5}\cos^5 x + k$

    (j) $\frac{1}{8}\sin 4x - \frac{1}{16}\sin 8x + k$

6 (a) $2\sec^2 x\tan x$ (b) $\sec^2 x$

    (c) $\tan x + k$, $\tan x - x + k$

    (d) $-\cot x + k$, $-x - \cot x + k$

7 1; $\frac{1}{4}\pi^2$

8 $\pi + 2$, $\frac{1}{2}\pi(8 + 3\pi)$

9 (a) $2 - \sqrt{2}$ (b) $\frac{1}{4}\pi(\pi - 2)$

10 $(0,1), \left(\frac{3}{4}\pi, 0\right)$; $1 + \sqrt{2}$; $\frac{1}{4}\pi(3\pi + 2)$

Miscellaneous exercise 6 (page 94)

1 (a) $2\cot 2x$

    (b) $-\frac{1}{4}(\cos 2x + 2\cos x) + k$

    (c) $-2\cos x\sin x$

    (d) $3t^2\cos(t^3 + 4)$

    (e) $\frac{1}{2}x + \frac{1}{12}\sin 6x + k$

    (f) $-\dfrac{1}{2\sqrt{x}}\sin\sqrt{x}$

    (g) $\frac{1}{2}x - \frac{3}{4}\sin\frac{2}{3}x + k$

2 (a) $\frac{1}{2}(1 - \cos 2x)$

3 (a) $\frac{1}{4}\sqrt{2}\left(\sqrt{3} - 1\right)$, $\dfrac{\sqrt{3} - 1}{\sqrt{3} + 1}$

    (c) 3.106, 3.215

4 $\dfrac{\sec^2 x\tan x}{\sqrt{\sec^2 x - 1}} = \sec^2 x$; $\sec^2 x$

5 $(0, -0.404)$; $1.404, 1.360, 1.622$

6 $\frac{1}{6}$; better, values are 0.5236 and 0.4997 approximating to 0.5.

7 (a) $v = 11\left(1 - \cos\left(\dfrac{\pi}{45}t\right)\right)$, 90 seconds

    (b) 990 metres, $11\ \mathrm{m\ s^{-1}}$

    (c) $0.665\ \mathrm{m\ s^{-2}}$

8 (b) $\frac{1}{2}\pi, \frac{5}{6}\pi$

    (c) The model suggests that the motion will continue indefinitely, but in practice it will gradually die out because of friction. In the new model the motion will die out.

10 (a) $\frac{1}{2}\cos\frac{1}{2}x - \frac{1}{3}\sin\frac{1}{3}x$ (b) $1, \frac{1}{2}$

    (c) $4\pi, 6\pi$

    (d) Any integer multiple of 12π

11 (a) $\frac{2}{3}\pi, \pi, \frac{4}{3}\pi, 2\pi$

Revision exercise 1
(page 97)

1 (a) -2 (b) 1   (c) 3   (d) ± 3

3 (a) $3e^{3x-1}$ (b) $\dfrac{2x}{x^2 - 1}$

4 (a) $-7, 25$ (b) $-5, 5$

    (c) $-7 \leqslant x \leqslant 25$ (d) $-5 \leqslant x \leqslant 5$

5 (a) $\frac{1}{2}e^6 - \frac{1}{2}$ (b) $\frac{1}{4}\pi e^{12} - \frac{1}{4}\pi$

6 $\sqrt{37}\cos(x + 0.165\ldots)$; 0.441, -0.771

7 $\dfrac{300}{\pi}$, $240\cos 600t$ (a) 0   (b) ± 0.4

8 (a) $-\frac{1}{2}\cos\left(2x + \frac{1}{6}\pi\right) + k$

    (b) $\frac{1}{2}x - \frac{1}{12}\sin 6x + k$

    (c) $\frac{1}{6}\sin^3 2x + k$

9 (a) $2, 2$

    (b) $(x + 2)(x + 1)(x - 1)(x - 2)$; $-2, -1, 1, 2$

10 $(x + 4)(x - 2)(x - 3)$

11 (a) At 3.49 hours   (b) After 15 hours

    (c) 2.51×10^5

12 (a) $(6x-6)e^{3x^2-6x}$ (b) $(1,e^{-3})$, minimum
   (c) $x+6y=8$

13 (a) $e-1$ (b) $\frac{1}{2}(e-(e-1)\ln(e-1))$

14 (b) $\sin\alpha$ (c) $\left(\alpha,\frac{1}{2}\sin\alpha\right)$

15   1.04, 3.03, 4.18, 6.17

7 Differentiating products

Exercise 7A (page 101)

1 (a) $2x$ (b) $3x^2+4x$
   (c) $5x^4+9x^2+8x$ (d) $63x^2+100x+39$
   (e) $3x^2$ (f) $(m+n)x^{m+n-1}$

2 (a) $(x+1)e^x$ (b) $x(2\ln x+1)$
   (c) $3x^2(\sin x+1)+x^3\cos x$
   (d) $\cos^2 x-\sin^2 x$ (e) $\cos x-x\sin x$
   (f) $e^{-x}(\cos x-\sin x)$

3 (a) $(x^2+2x+3)e^x$
   (b) $2x(\sin x+\cos x)+x^2(\cos x-\sin x)$
   (c) $\sin^2 x+2x\sin x\cos x$

4 (a) $4x+(2x+x^2)e^x$ (b) $x^2(3+2x)e^{2x}$
   (c) $6x\ln x+\dfrac{4}{x}+3x$

5 (a) $-3e^{-4}$ (b) $e^2(\sin 2+\cos 2)$
   (c) $1+\ln 6$

6 (a) $y=-\pi x+\pi^2$ (b) $y=x-1$
   (c) $8y=47x-75$ (d) $y=0$

7 $(2,4e^{-2})$, $(0,0)$

8 (a) $2x\sin^3 2x+6x^2\sin^2 2x\cos 2x$
   (b) $\dfrac{(5x^2+5x+2)e^x}{\sqrt{5x^2+2}}$
   (c) $8\sin^3 2x\cos 2x\cos^3 5x$
           $-15\sin^4 2x\cos^2 5x\sin 5x$
   (d) $12(4x+1)^2\ln 3x+\dfrac{(4x+1)^3}{x}$
   (e) $\dfrac{\ln 2x+2}{2\sqrt{x}}$ (f) $-e^{ax}(a\sin bx+b\cos bx)$

9 $\sin 2+2\cos 2$

10 $x+2y=1$

11 $(1,-3\sqrt{3})$

12 $V\approx 51.8$, $x=6.4$

13   When n is even, there is a maximum at $x=n$,
   and a minimum at $x=0$. When n is odd, there is
   a maximum at $x=n$; if $n>1$, there is also a
   point of inflexion at $x=0$.

14 (a) $(\sin x+x\sin x+x\cos x)e^x$
   (b) $(2x\cos 4x-3x^2\cos 4x-4x^2\sin 4x)e^{-3x}$

Exercise 7B (page 105)

1 (a) $\dfrac{1}{(1+5x)^2}$ (b) $\dfrac{3x^2-4x}{(3x-2)^2}$
   (c) $\dfrac{2x}{(1+2x^2)^2}$ (d) $\dfrac{(12x-13)e^{3x}}{(4x-3)^2}$
   (e) $\dfrac{1-2x^3}{(1+x^3)^2}$ (f) $\dfrac{(x-1)^2 e^x}{(x^2+1)^2}$

2 $-\operatorname{cosec}^2 x$

3 (a) $\dfrac{x\cos x-\sin x}{x^2}$ (b) $\dfrac{\sin x-x\cos x}{\sin^2 x}$
   (c) $\dfrac{2x(\sin x-x\cos x)}{\sin^3 x}$

4 (a) $\dfrac{x+2}{2(x+1)^{\frac{3}{2}}}$ (b) $\dfrac{10-x}{2x^2\sqrt{x-5}}$
   (c) $-\dfrac{3x+4}{4x^2\sqrt{3x+2}}$

5 (a) $-\dfrac{2x\sin x+\cos x}{2x\sqrt{x}}$ (b) $\dfrac{3e^x-10-5xe^x}{(e^x-2)^2}$
   (c) $\dfrac{-1}{(1+x)^{\frac{3}{2}}(1-x)^{\frac{1}{2}}}$

6 (a) $\dfrac{1-\ln x}{x^2}$ (b) $\dfrac{2}{x^2+4}-\dfrac{\ln(x^2+4)}{x^2}$
   (c) $\dfrac{3}{(3x+2)(2x-1)}-\dfrac{2\ln(3x+2)}{(2x-1)^2}$

7 $4y=x+3$

8 (a) $\dfrac{(2x-1)e^x}{(2x+1)^2}$ (b) $\left(\frac{1}{2},\frac{1}{2}e^{\frac{1}{2}}\right)$

9 $14y=8x-37$

10 $\left(2(\sqrt{2}-1),2(1+\sqrt{2})\right)$, $\left(-2(1+\sqrt{2}),-2(\sqrt{2}-1)\right)$

11 (a) $\dfrac{x^2+2x-3}{(x+1)^2}$
   (b) $-3\leqslant x<-1$, $-1<x\leqslant 1$

12 (a) 2 (b) $\frac{2}{3}$

Miscellaneous exercise 7 (page 106)

1 (a) $3x^2\sin x+x^3\cos x$ (b) $\dfrac{x+6}{2(x+3)^{\frac{3}{2}}}$

2 $(1-3x)e^{-3x}$; $\left(\frac{1}{3},\frac{1}{3}e^{-1}\right)$

3 (a) $e^x(\cos x - \sin x)$ (b) $\frac{1}{4}\pi < x < \frac{5}{4}\pi$

(c) The gradient is negative for these values of x.

4 $-\dfrac{1}{\pi^2}$

5 (a) $2\cos 2x \cos 4x - 4\sin 2x \sin 4x$

(b) $\dfrac{3x(2\ln x - 1)}{(\ln x)^2}$ (c) $-2\left(1 - \dfrac{x}{5}\right)^9$

6 $\left(\frac{1}{2}, 4\sqrt{2}\right)$

7 $\dfrac{x\cos x - \sin x}{x^2}$

11 (a) $\dfrac{2}{a}$

8 Solving equations numerically

Exercise 8A (page 112)

2 2

3 $1.91, -29.6$; as the graph is continuous there is at least one root in the interval $3 < x < 4$.

4 (a) (i) -2 (ii) -1.71
(b) (i) 2 (ii) 2.63
(c) (i) 1 (ii) 1.33
(d) (i) 6 (ii) 6.30
(e) (i) -8 (ii) -7.86
(f) (i) 4 (ii) 4.53

5 The graph of the function f may have a break in it. $f(x) = \dfrac{1}{x}$ has $f(-1)f(1) < 0$, but $f(x) = 0$ has no root between -1 and 1.

Exercise 8B (page 114)

1 Here are three possible examples for each part.

(a) $x = \sqrt[5]{5x - 6}; x = \frac{1}{5}(x^5 + 6); x = \sqrt{\dfrac{5x - 6}{x^3}}$

(b) $x = 5e^{-x}; x = \ln 5 - \ln x; x = \sqrt{5xe^{-x}}$

(c) $x = \sqrt[5]{1999 - x^3}; x = \sqrt[3]{1999 - x^5}; x = \sqrt[3]{\dfrac{1999}{x^2 + 1}}$

2 (a) (i) $f(x) = x^{11} - x^7 + 6 = 0$ (ii) -1.1769, -1.2227, -1.2338, -1.2367, -1.2375
(iii) Converging to a root
(iv) x_5 is an approximate root of $f(x) = 0$

(b) (i) $f(x) = x^4 - x^3 - 34x^2 + 289 = 0$
(ii) 7.1111, 22.283, 463.11, $214\,440$, 4.598×10^{10} (iii) Diverging

(c) (i) $f(x) = x^4 - 500x - 10 = 0$ (ii) 7.9446, 7.9437, 7.9437, 7.9437, 7.9437
(iii) Converging to a root
(iv) x_5 is an approximate root of $f(x) = 0$

3 $1.6198, 1.8781, 1.6932, 1.8208, 1.7304, 1.7933$; converging

4 (b) 0.7895; it is converging to another root.

5 $x_{n+1} = \sqrt[3]{e^{x_n} + 2}$ with $x_0 = 2$ converges to 2.27 in 11 steps;

$x_{n+1} = \dfrac{e^{x_n} + 2}{x_n^2}$ with $x_0 = 2$ converges to 2.27 in 3 steps

6 (a) 9
(b) 5 steps; x_4 and x_5 are the same to 4 significant figures; 9.725

7 One root; $x_{r+1} = \cos x_r$ with $x_0 = \frac{1}{4}\pi$ converges to 0.739

Exercise 8C (page 119)

1 (a) No convergence; $F^{-1}(x) = \dfrac{3}{x+1}, 1.303$

(b) No convergence; $F^{-1}(x) = \frac{1}{3}\ln(5 - x), 0.501$

(c) No convergence; $F^{-1}(x) = \tan^{-1}(2x)$, 1.1656

(d) No convergence; $F^{-1}(x) = -\sqrt[6]{300 - 10x}$, -2.6237

(e) Converges to 1.8955

2 (a) $5, 1.179$ (b) $-13, 6.730$
(c) $10, -1.896$ (d) $-3.485, 12.87$

Miscellaneous exercise 8 (page 119)

1 $5, 13$ (a) There exists a root between $x = 1$ and $x = 2$. (b) $1\frac{5}{8}$

2 2

3 2.15

4 $0.381 < x < 0.382$

5 5.0

6 (b) 0.77

7 (a) Two roots

8 233.1

9 (a) $x + 15y + 30 = 0$
(b) One graph is increasing from $-\infty$ to ∞, and the other is decreasing in $-90 < x < 90$. Hence there is a point of intersection, and only one.
(c) -35

10 2.13

11 1.210

12 6.72

13 (a) $1, 1.389$. There could be an even number of roots in the interval.
  (b) 0.310 or 0.756 (c) 0.62 or 1.51

14 (b) 3.11111, $3.103\,32$, $3.103\,84$, $3.103\,80$; 3.104

15 (b)   The second, as its derivative is numerically less than 1 between $x = 2$ and $x = 3$; 2.93

16 (a) 3 (b) $F(t) = 1 + \dfrac{\ln(9t)}{\ln 4}$, $F^{-1}(t) = \frac{1}{9} 4^{t-1}$, or vice versa; $t = F(t)$; 3.486

17 (a) -5 (b) $k = -63$, $\alpha = -4.4188$

18 (a) $\left(\pm\sqrt{5}, \pm\sqrt{5}\right)$
  (c)   The function g is self-inverse.
  (d) $2.236\,068$

19 (b) $30, 40$ (c) $30, 35$ (d) $32.5, 35$
  (e) $a_r = 30$, 30, 32.5, 33.75, 34.375, 34.375, $34.531\,25$, $34.609\,375$, $34.648\,437\,5$, $34.667\,968\,75$
     $b_r = 40$, 35, 35, 35, 35, 34.6875, 34.6875, 34.6875, 34.6875, 34.6875
     $\alpha = 34.7$ correct to 1 d.p.

20 (a) $1.324\,72$ (b) $0.111\,83$, $3.577\,15$

9 The trapezium rule

Exercise 9 (page 125)

1 (a) 2.12 (b) 0.75

2 5.73

3 3.09

4 (a) 1.26 (b) 0.94 (c) 0.8

5 8.04

6 1.034

8 2.86

9 3.14

10 (a) 19.40 (b)   Underestimates   (c) 6.25π
  (d) 3.10

Miscellaneous exercise 9 (page 126)

1 10.6

2 3.28; overestimate

3 1.701

4 0.52

5 0.70

6 (a) 1.41

7 51

8 1140 m

9 8.15 km

10 (a) $0, 1.708, 2.309, 2.598, 2.582, 2.141, 0$
  (b) 34.0 m^2 (c) 3400 m^3

11 0.55

12 (a) 0.622 m^2 (b) 12.4 m^3 (c) 3.9 m^3
  (d)   (b) overestimate, (c) underestimate

13 (a) $\frac{592}{3}$ (b) $98 + 70\sqrt{2}$

14 $a = 10$, $(x+1)(x-2)(2x-5)$; 16; 18

15 0.535, 0.5; $T = \frac{3}{8}h - \frac{1}{4} + \frac{1}{2}h^{-2}$; $h = 2 \times 3^{-\frac{1}{3}}$, or 1.39

16 (a) $\frac{1}{3}$ (b) $\frac{1}{2}, \frac{3}{8}, \frac{11}{32}, \frac{43}{128}$
  (c) $-\frac{1}{6}, -\frac{1}{24}, -\frac{1}{96}, -\frac{1}{384}$
  (d) $E_n = -\dfrac{1}{6n^2}$ (e) 409, or more

10 Parametric equations

Exercise 10A (page 133)

1 (a) $(180, 60)$ (b) $(5, -10)$

2 (a) $\left(\frac{2}{3}, \frac{4}{3}\right)$ (b) $(2, 0)$

3 $\frac{1}{2}\pi$

4 $\frac{2}{3}\pi$

9 (a) $y^2 = \dfrac{1}{x}$ (b) $y^2 = 12x$
  (c) $x^2 + y^2 = 4$

10 (a) $x + y = 1$, for $0 \leqslant x \leqslant 1$
  (b) $x^{\frac{2}{3}} + y^{\frac{2}{3}} = 1$
  (c) $x + y = 2$, excluding $(1, 1)$
  (d) $4x^3 = 27y^2$

Exercise 10B (page 136)

1 (a) $\dfrac{2}{3t^2}$ (b) $-\tan t$
  (c) $-\frac{3}{2}\cot t$ (d) $\dfrac{2t - 1}{3t^2 + 1}$

2 (a) 2 (b) $\frac{1}{3}$ (c) -1 (d) $-\frac{1}{54}$

3 (a) -3 (b) 1 (c) $\sqrt{3}$ (d) -8

5 (a) $\frac{1}{3}$ (b) $3y = x - 1$

6 $x + y = 1 + \pi$

7 (a) $3y = x + 9$ (b) $5y = 3\sqrt{3}x - 30$

8 (a) $3x + y = 165$ (b) $y = -\sqrt{3}x$

9 (a) $y = 4x - 30$ (b) $\left(-\frac{1}{2}, -32\right)$

10 (a) $y = 2x - 36$ (b) $(27, 18)$

Exercise 10C (page 138)

7 (b)   The point N always lies on the circle with centre at the origin and radius $\sqrt{2}$.

Miscellaneous exercise 10 (page 139)

1 (a) $\frac{1}{4}\pi$ (b) $2x+y=2\sqrt{2}$

2   $6\cos t$

3 (a) $\dfrac{t^2-1}{t^2+1}$ (c) $y^2-x^2=4$

4 (a) $\dfrac{2t}{3t^2+1}$ (b) $2x+y=6$

5 (a) $-\sqrt{3}\tan t$ (b) $x+y=2$

6   $\dfrac{1}{e^t-1}$, $\ln 2$

7 (b) $y=2x-2$ (c) $(0,-2)$

8 (a) $2x+y=9$ (b) $y=4x-x^2$

10 (a) The half-line of gradient 1 through $(0,0)$ for which $x\geqslant 0$

     (b) $y=x$; the straight line of gradient 1 through $(0,0)$

     (c) Each point of the curve given by the parametric equations lies on the curve given by the cartesian equation, but the reverse is not necessarily true, as this example shows.

11 Curves defined implicitly

Exercise 11A (page 147)

1 (a) $\frac{3}{4}$ (b) 6   (c) $\frac{3}{4}$ (d) $-\frac{1}{2}$

2   $-\frac{1}{3}$

3   $3x-2y=8$

4   $x-2y=7$, $-\frac{86}{13}$

5 (a) $(\pm 1,0),\left(0,\pm\frac{1}{2}\right)$

     (b) $-1\leqslant x\leqslant 1$, $-\frac{1}{2}\leqslant y\leqslant\frac{1}{2}$

     (d) $2x+8y\dfrac{dy}{dx}=0$; the gradient is zero where the graph cuts the y-axis.

     (e) $2x\dfrac{dx}{dy}+8y=0$; the tangent is vertical where the graph cuts the x-axis.

6 (a) $(\pm 1,0)$

     (b) $x\geqslant 1$ or $x\leqslant -1$; y can take any value.

     (d) $2x-2y\dfrac{dy}{dx}=0$; x is never zero, so the gradient is never zero.

     (e) $2x\dfrac{dx}{dy}-2y=0$; the tangent is vertical where the graph cuts the x-axis.

7 (a) $(1,0)$, $(0,1)$ (b) $y<0$

     (c) $\dfrac{2(x-1)}{3y^2}$, $-\frac{2}{3}$

     (d) The modulus of the gradient becomes very large.

9 (b) $\dfrac{dy}{dx}=-e^{x-y}$

     (c) Both x and y are less than $\ln 2$.

10 (a) The curve is symmetrical about the y-axis.

     (b) $\dfrac{dy}{dx}=-\dfrac{2x}{3y^2}$; when x is positive $\dfrac{dy}{dx}$ is negative, and vice versa.

     (c) Maximum

11   $(0,0)$, $(-1,0)$; 0, 1

12   1, -2 and 1 at $(0,0)$, $(0,1)$ and $(0,2)$ respectively

Exercise 11B (page 150)

1 (a) $y+x\dfrac{dy}{dx}$ (b) $y^2+2xy\dfrac{dy}{dx}$

     (c) $2xy^2+2x^2y\dfrac{dy}{dx}$ (d) $\dfrac{2xy-x^2\dfrac{dy}{dx}}{y^2}$

2 (a) $\dfrac{y+x\dfrac{dy}{dx}}{2\sqrt{xy}}$ (b) $\left(2xy+x^2\dfrac{dy}{dx}\right)\cos(x^2y)$

     (c) $\dfrac{1}{x}+\dfrac{1}{y}\dfrac{dy}{dx}$ (d) $\left(y+x\dfrac{dy}{dx}+\dfrac{dy}{dx}\right)e^{xy+y}$

3 (a) $-\frac{4}{3}$ (b) $\frac{20}{11}$

4 (a) $-\frac{1}{3}\sqrt{3}$ (b) 0   (c) -2 (d) $-\frac{1}{3}\pi$

5   $3y=x+5$

6   $x=1$

7   $(3,1)$, $(-3,-1)$

8 (a) -1 (b) $\frac{1}{3}$

     (c) $(0,0)$, where there are two branches, one parallel to each axis; $\left(2^{\frac{1}{3}},2^{\frac{2}{3}}\right)$, $\left(2^{\frac{2}{3}},2^{\frac{1}{3}}\right)$

9 (a) $\left(x^2+y^2\right)^2\geqslant 0$ so $x^2-y^2\geqslant 0$

     (b) $\left(\pm\frac{1}{4}\sqrt{6},\pm\frac{1}{4}\sqrt{2}\right)$, $(\pm 1,0)$

Miscellaneous exercise 11 (page 151)

1   $5x-13y+3=0$

2 (a) $\dfrac{x-y}{x-4y}$, $(2,2)$, $(-2,-2)$

     (b) $2x-y=3\sqrt{3}$

3 $3x - 7y = 13$

4 (a) $\dfrac{2 + x - 2xy}{2 + x^2}$ (b) $(2,1), \left(-1, \tfrac{1}{2}\right)$

5 (a) $\dfrac{a^2}{x^2} + \dfrac{b^2}{y^2} = 4$

6 3, maximum; −3, minimum

7 4

8 (a) $\dfrac{1}{1 - 8y^3}$

(c) 1; since $\dfrac{d^2 y}{dx^2}$ is zero at the origin, but positive close to the origin, the curve has a point of inflexion there of the same orientation as $y = x^3$.

Revision exercise 2
(page 152)

1 $-1.17, 0.69, 2.48$

2 (a) $3\tfrac{1}{2}$ (b) 3.21

3 (a) $2x \ln x + x$ (b) $\dfrac{1 - 2\ln x}{x^3}$

(c) $\dfrac{e^x \left(1 - 2xe^x\right)}{\left(x^2 e^x + 1\right)^2}$ (d) $\left(1 + 2x^2\right) e^{x^2}$

4 25.5

5 2.74

6 4.4934

7 (a) 1.9113 (b) $\tfrac{1}{3}\pi + \tfrac{1}{2}\sqrt{3}$ (c) 0.1%

8 (33,27)

9 5

10 $t^2 y + x = 2ct$

11 (a) $-\sin\theta\cos^2\theta$ (b) $3x + 8y = 15$

12 $(4,-2), (-4,2); \left(2\sqrt{5}, -\tfrac{4}{5}\sqrt{5}\right), \left(-2\sqrt{5}, \tfrac{4}{5}\sqrt{5}\right)$

13 (a) $e^{2x}(7 + 6x)$ (b) $-\dfrac{x\sin x + 2\cos x}{x^3}$

(c) $\dfrac{2(\cos 2x - \sin 2x)}{e^{2x}}$ (d) $\dfrac{2x\sec^2 2x - \tan 2x}{x^2}$

(e) $e^{-x}(\sin x - x\sin x + x\cos x)$

(f) $\dfrac{e^{2x}(x\sin 2x + x\cos 2x - x\sin x)}{2x^3}$

14 (a) $-\dfrac{e^{-x}(x+1)^2}{\left(1 + x^2\right)^2}$

15 (a) $3x + 2y = 5$ (b) $\left(8\tfrac{1}{3}, -10\right)$

16 (a) $\tfrac{1}{2}\pi < t \leqslant \pi$

(b) $x + 4y\cos T = \cos T + 4\cos T\cos 2T$

(c) $1 + 3\cos T = 4\cos T\cos 2T$

(d) $8X^3 - 7X - 1 = 0; X = 1, T = 0$

(e) 1.72, 2.59

Practice examinations for P2

Practice examination 1 (page 155)

1 $1, \tfrac{5}{3}$

2 (ii) 2.61

4 (a) $\ln 3$ (b) 2.17

6 (i) 9.51 units (ii) 69.3
(iii) 0.0951 units per second

7 (i)

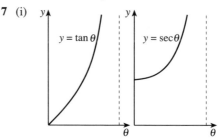

(iv) 0.644

Practice examination 2 (page 157)

1 $2x - 1, 3$

2 (i) $1 < x < 5$ (ii) $0 < y < \dfrac{\ln 5}{\ln 2}$

3 7.5, 0.5

4 (a) $\tfrac{1}{12}\pi - \tfrac{1}{16}\sqrt{3}$ (b) $\tfrac{1}{2}\pi + 1$

5 (i) $1 - e^{-a}$ (iii) 2.3

6 (i) $2x(x+1)e^{2x}$ (ii) $\dfrac{1 - \ln 2x}{x^2}$

(iii) $\dfrac{1}{x(x+1)}$

7 (i) $\dfrac{1 + \cos\theta}{2 - \sin\theta}$ (ii) $\tan^{-1}\tfrac{4}{3} \approx 0.927$

(iii) 1.80, 5.77

Answers to P3

12 Vectors: lines in two and three dimensions

Since vector equations are not unique, other correct answers are sometimes possible.

Exercise 12A (page 164)

1. (a) $\mathbf{r} = \begin{pmatrix} 2 \\ -3 \end{pmatrix} + t\begin{pmatrix} 1 \\ 2 \end{pmatrix}, y = 2x - 7$

   (b) $\mathbf{r} = \begin{pmatrix} 4 \\ 1 \end{pmatrix} + t\begin{pmatrix} -3 \\ 2 \end{pmatrix}, 2x + 3y = 11$

   (c) $\mathbf{r} = \begin{pmatrix} 5 \\ 7 \end{pmatrix} + t\begin{pmatrix} 1 \\ 0 \end{pmatrix}, y = 7$

   (d) $\mathbf{r} = t\begin{pmatrix} 2 \\ -1 \end{pmatrix}, x + 2y = 0$

   (e) $\mathbf{r} = \begin{pmatrix} a \\ b \end{pmatrix} + t\begin{pmatrix} 0 \\ 1 \end{pmatrix}, x = a$

   (f) $\mathbf{r} = \begin{pmatrix} \cos\alpha \\ \sin\alpha \end{pmatrix} + t\begin{pmatrix} -\sin\alpha \\ \cos\alpha \end{pmatrix}, x\cos\alpha + y\sin\alpha = 1$

2. (a) $\mathbf{r} = \begin{pmatrix} 2 \\ 0 \end{pmatrix} + t\begin{pmatrix} 0 \\ 1 \end{pmatrix}$ (b) $\mathbf{r} = \begin{pmatrix} 1 \\ 2 \end{pmatrix} + t\begin{pmatrix} 3 \\ -1 \end{pmatrix}$

   (c) $\mathbf{r} = \begin{pmatrix} -1 \\ -1 \end{pmatrix} + t\begin{pmatrix} 5 \\ 2 \end{pmatrix}$

3. (a) $(7,3)$ (b) $(8,-5)$ (c) No common points
   (d) $(4.76, 3.68)$
   (e) The lines coincide, $2x + 3y = 17$
   (f) $(-1,1)$

4. $x = 2 + t, y = -1 + 3t;\ (4,5)$

5. $(3,1)$

6. (a), (d), (e)

7. (a) $\mathbf{r} = \begin{pmatrix} 3 \\ 7 \end{pmatrix} + t\begin{pmatrix} 2 \\ -3 \end{pmatrix}$ (b) $\mathbf{r} = \begin{pmatrix} 2 \\ 3 \end{pmatrix} + t\begin{pmatrix} 0 \\ 1 \end{pmatrix}$

   (c) $\mathbf{r} = \begin{pmatrix} -1 \\ 2 \end{pmatrix} + t\begin{pmatrix} 2 \\ -1 \end{pmatrix}$ (d) $\mathbf{r} = \begin{pmatrix} -3 \\ -4 \end{pmatrix} + t\begin{pmatrix} 2 \\ 3 \end{pmatrix}$

   (e) $\mathbf{r} = \begin{pmatrix} -2 \\ 7 \end{pmatrix} + t\begin{pmatrix} 1 \\ 0 \end{pmatrix}$ (f) $\mathbf{r} = \begin{pmatrix} 1 \\ 3 \end{pmatrix} + t\begin{pmatrix} 1 \\ 1 \end{pmatrix}$

8. (a) $\mathbf{r} = \begin{pmatrix} 4 \\ -1 \end{pmatrix} + s\begin{pmatrix} 3 \\ 1 \end{pmatrix}, \mathbf{r} = \begin{pmatrix} -3 \\ 2 \end{pmatrix} + t\begin{pmatrix} 1 \\ -1 \end{pmatrix}; (1,-2)$

   (b) $\left(13\tfrac{1}{3}, -5\right), \left(4, 11\tfrac{4}{5}\right)$

9. (a) Yes   (b) Yes
   (c) Meaningless, since $\mathbf{0}$ has no direction

   $\mathbf{r} = \begin{pmatrix} 1 \\ 2 \end{pmatrix} + t\begin{pmatrix} -4 \\ 3 \end{pmatrix}$

10. $\begin{pmatrix} 1 \\ 3 \end{pmatrix}; \mathbf{r} = \begin{pmatrix} 1 \\ 5 \end{pmatrix} + t\begin{pmatrix} -3 \\ 1 \end{pmatrix}; (4,4)$

11. $(2,0)$

12. $\mathbf{r} = \begin{pmatrix} -1 \\ 1 \end{pmatrix} + t\begin{pmatrix} 1 \\ 2 \end{pmatrix}; x = -1 + t, y = 1 + 2t;$
    $(-1,1), (3,9)$

13. $(1,8), (-7,-4)$

Exercise 12B (page 168)

1. (a) $\mathbf{r} = \begin{pmatrix} 1 \\ 2 \\ 3 \end{pmatrix} + t\begin{pmatrix} 0 \\ 1 \\ 2 \end{pmatrix}$ (b) $\mathbf{r} = t\begin{pmatrix} 0 \\ 0 \\ 1 \end{pmatrix}$

   (c) $\mathbf{r} = \begin{pmatrix} 2 \\ -1 \\ 1 \end{pmatrix} + t\begin{pmatrix} 3 \\ -1 \\ 1 \end{pmatrix}$ (d) $\mathbf{r} = \begin{pmatrix} 3 \\ 0 \\ 2 \end{pmatrix} + t\begin{pmatrix} 4 \\ -2 \\ 3 \end{pmatrix}$

2. (a) $\mathbf{r} = \begin{pmatrix} 2 \\ -1 \\ 2 \end{pmatrix} + t\begin{pmatrix} 1 \\ 0 \\ 2 \end{pmatrix}$ (b) $\mathbf{r} = \begin{pmatrix} 1 \\ 2 \\ 2 \end{pmatrix} + t\begin{pmatrix} 1 \\ -4 \\ 0 \end{pmatrix}$

   (c) $\mathbf{r} = \begin{pmatrix} 3 \\ 1 \\ 4 \end{pmatrix} + t\begin{pmatrix} -4 \\ 1 \\ -1 \end{pmatrix}$

3. They all represent the same straight line.
4. The point lies on line (a) only.
5. Set (a) lies on a straight line, but set (b) does not.
6. (a) $2, -3$ (b) $3,1$ (c) No solution
7. (a) $(-3,1,5)$ (b) $(3,-5,4)$
8. Any multiple of $1, 2, -3$; the translations are all parallel to the same plane.
9. $(4,-3,0)$
10. All $\tfrac{1}{4}(\mathbf{a}+\mathbf{b}+\mathbf{c}+\mathbf{d})$; the lines joining the mid-points of opposite edges of a tetrahedron meet and bisect one another.
11. $\tfrac{1}{4}\mathbf{e} + \tfrac{3}{4}\mathbf{f}; \tfrac{1}{3}(\mathbf{a}+\mathbf{b}+\mathbf{c}), \tfrac{1}{4}(\mathbf{a}+\mathbf{b}+\mathbf{c}+\mathbf{d})$
12. (a) Intersect at $(1,-1,0)$
    (b) Parallel   (c) Skew
13. 0.4 m

Exercise 12C (page 172)

1. 5
2. $\tfrac{1}{13}\sqrt{69}$
3. 3
4. 5
5. $\mathbf{r} = \mathbf{i} + 3\mathbf{j} + \mathbf{k} + t(-6\mathbf{j} - 2\mathbf{k}), \tfrac{21}{5}\sqrt{15}$

Miscellaneous exercise 12 (page 172)

1. $\begin{pmatrix} 9 \\ -1 \\ 4 \end{pmatrix}$

2. (a) $\mathbf{r} = \begin{pmatrix} 2 \\ 1 \end{pmatrix} + t\begin{pmatrix} 3 \\ -2 \end{pmatrix}$ (b) $(2,1)$

3 (a) $\mathbf{r} = \begin{pmatrix} 2 \\ 3 \\ 5 \end{pmatrix} + \lambda \begin{pmatrix} 1 \\ 1 \\ -0.5 \end{pmatrix}$ (b) 25 m

4 $(\cos 2\alpha, \sin 2\alpha)$; for all α, the intersection lies on the circle with $(-1,0)$ and $(1,0)$ at ends of a diameter.

5 1.68 km, 20 seconds

6 (a) (i) $\begin{pmatrix} -2 \\ 6 \\ 2 \end{pmatrix}$ (iii) $1, 2\sqrt{11}$ (b) $47.9°$ (c) 21

7 (a) Above $(61,77)$ on the ground   (b) 4200 m
  (c) 384 km h^{-1}, $38.7°$
  (d) 386 km h^{-1}, $5.35°$

8 (a) (i) $3\mathbf{i} + 3\mathbf{j} + 4\mathbf{k}$ (ii) $25°$
  (b) $5\mathbf{i} + 9\mathbf{j} + 12\mathbf{k}$, $-\frac{1}{13}(15\mathbf{i} + 123\mathbf{j} + 164\mathbf{k})$
  (c) $\frac{5}{2}\mathbf{i} + \frac{3}{2}\mathbf{j} + 2\mathbf{k}$

9 $(85,-10)$; 2200 m, 2 minutes

10 $u < -1$ or $u > 0.5$; $0.5 < u < 0.753$

13 Vectors: planes in three dimensions

Since vector equations are not unique, other correct answers are sometimes possible.

Exercise 13A (page 179)

1 $5x - 8y + 4z = 1$

2 $2x - y - z = 4$

3 There are many possibilities. Two are $(2,0,0)$ and $(0,0,1)$.

5 $5x + y + 7z = 0$

6 $\mathbf{r} = \begin{pmatrix} 4 \\ 2 \\ -1 \end{pmatrix} + t \begin{pmatrix} 3 \\ 4 \\ -1 \end{pmatrix}$

7 $2, \frac{4}{3}; \frac{2}{3}$

Exercise 13B (page 185)

1 (a) $\begin{pmatrix} 2 \\ 7 \\ 5 \end{pmatrix}$ (b) $\begin{pmatrix} 14 \\ -13 \\ 23 \end{pmatrix}$ (c) $\begin{pmatrix} 0 \\ -3 \\ 0 \end{pmatrix}$ or $\begin{pmatrix} 0 \\ 1 \\ 0 \end{pmatrix}$

2 (a) $-6\mathbf{j}$ or $\mathbf{j}$ (b) $-10\mathbf{i} + 5\mathbf{j}$ or $-2\mathbf{i} + \mathbf{j}$
  (c) $-2\mathbf{k}$ or $\mathbf{k}$

3 $\mathbf{i} - 4\mathbf{j} - 7\mathbf{k}$, $x - 4y - 7z = 23$

4 $\begin{pmatrix} 1 \\ 1 \\ 1 \end{pmatrix}$, $x + y + z = -3$

5 (a) $z = 0$ (b) $x + y + z = 0$
  (c) $11x + 2y + 5z = 30$ (d) $5x + 13y + 7z = 21$

6 $(0,0,0), 7$

7 1

8 $5x + y + 7z = 0$

9 $\mathbf{r} = 4\mathbf{i} + 2\mathbf{j} - \mathbf{k} + t(3\mathbf{i} + 4\mathbf{j} - \mathbf{k})$

10 $53°$

11 They lie in the plane $5x + y = 10$.

12 $\mathbf{r} = -7\mathbf{i} + 3\mathbf{j} + t(33\mathbf{i} - 9\mathbf{j} + \mathbf{k})$

13 $\mathbf{r} = 4\mathbf{i} + 2\mathbf{j} - 3\mathbf{k} + t(2\mathbf{i} + 3\mathbf{j} - 4\mathbf{k})$

14 $2x + y - z = 0$

15 $5x - y + 7z = 20$

16 $5x - y + 3z = 12$

17 (a) $\mathbf{r} = \begin{pmatrix} 2 \\ -1 \\ 1 \end{pmatrix} + s \begin{pmatrix} -1 \\ 3 \\ -4 \end{pmatrix}$, $\mathbf{r} = \begin{pmatrix} 1 \\ -1 \\ 1 \end{pmatrix} + t \begin{pmatrix} 0 \\ -3 \\ 4 \end{pmatrix}$

  (b) $(1,2,-3)$ (c) $\cos^{-1}\dfrac{5}{\sqrt{26}}$ (d) $\begin{pmatrix} 0 \\ 4 \\ 3 \end{pmatrix}$

Miscellaneous exercise 13 (page 186)

1 $8x + 3y + 4z = -11$

2 (a) $63.1°$ (b) $\mathbf{r} = \begin{pmatrix} 0 \\ -2 \\ 8 \end{pmatrix} + t \begin{pmatrix} 3 \\ 2 \\ -7 \end{pmatrix}$

3 (a) $\begin{pmatrix} -6 \\ 5 \\ 13 \end{pmatrix}$ (b) $\mathbf{r} = \begin{pmatrix} 1 \\ 2 \\ -3 \end{pmatrix} + t \begin{pmatrix} -6 \\ 5 \\ 13 \end{pmatrix}$

4 $\left| \dfrac{ap + bq + cr - d}{\sqrt{a^2 + b^2 + c^2}} \right|$

5 $2x - 3y - 8z = 28$

7 (a) $(5,1,-1)$ is a point on the line, so the position vector of $(5,1,-1)$ is suitable for **a**. The vector $2\mathbf{i} - 3\mathbf{j} + \mathbf{k}$ is normal to the plane, and therefore lies along the line, so it is a suitable choice for **b**.
  (b) (i) $69°$ or $111°$

8 (a) $\mathbf{r} = 2\mathbf{i} + \mathbf{j} - \mathbf{k} + s(\mathbf{i} - \mathbf{j})$,
     $\mathbf{r} = 5\mathbf{i} - 2\mathbf{j} - \mathbf{k} + t(\mathbf{j} + 2\mathbf{k})$
  (c) $71.6°$ or $108.4°$
  (e) $2x + 2y - z = 7$

9 $3x - 2y - z = 16$

10 (a) 3 (b) $\begin{pmatrix} -17 \\ -10 \\ 14 \end{pmatrix}$ (c) $\dfrac{21}{\sqrt{65}}$

14 The binomial expansion

Exercise 14 (page 193)

1 (a) $1 - 3x + 6x^2$ (b) $1 - 5x + 15x^2$
  (c) $1 + 4x + 10x^2$ (d) $1 + 6x + 21x^2$

2 (a) $1-4x+16x^2$ (b) $1+6x+24x^2$
    (c) $1+12x+90x^2$ (d) $1-x+\frac{3}{4}x^2$

3 (a) 84   (b) -8 (c) -270 (d) 256
    (e) $\frac{56}{27}$ (f) $-20a^3$ (g) $20b^3$
    (h) $\frac{1}{6}n(n+1)(n+2)c^3$

4 (a) $1+\frac{1}{3}x-\frac{1}{9}x^2$ (b) $1+\frac{3}{4}x-\frac{3}{32}x^2$
    (c) $1-\frac{3}{2}x+\frac{3}{8}x^2$ (d) $1+\frac{1}{2}x+\frac{3}{8}x^2$

5 (a) $1+2x-2x^2$ (b) $1-x+2x^2$
    (c) $1-8x+8x^2$ (d) $1+\frac{1}{8}x+\frac{5}{128}x^2$

6 (a) $-\frac{1}{2}$ (b) $\frac{625}{16}$ (c) $\frac{5}{24}$ (d) $-\frac{5}{2}$
    (e) 20   (f) $\frac{1}{8}\sqrt{2}$ (g) $-\frac{1}{16}a^3$
    (h) $\frac{1}{48}n(n+2)(n+4)b^3$

7 (a) $1-\frac{1}{8}x-\frac{1}{128}x^2$ (b) $1+\frac{1}{8}x^2-\frac{1}{128}x^4$
    (c) $2+\frac{1}{4}x-\frac{1}{64}x^2$ (d) $6+\frac{3}{4}x-\frac{3}{64}x^2$

8 $|x|<\frac{2}{3}$
    (a) $4+12x+27x^2+54x^3$
    (b) $\frac{1}{4}+\frac{3}{4}x+\frac{27}{16}x^2+\frac{27}{8}x^3$

9 (a) $1-3x-\frac{9}{2}x^2-\frac{27}{2}x^3, |x|<\frac{1}{6}$
    (b) $1-5x+25x^2-125x^3, |x|<\frac{1}{5}$
    (c) $1-3x+18x^2-126x^3, |x|<\frac{1}{9}$
    (d) $1+8x+40x^2+160x^3, |x|<\frac{1}{2}$
    (e) $1+x^2-\frac{1}{2}x^4+\frac{1}{2}x^4, |x|<\frac{1}{2}\sqrt{2}$
    (f) $2-\frac{4}{3}x-\frac{8}{9}x^2-\frac{80}{81}x^3, |x|<\frac{1}{2}$
    (g) $10-4x+\frac{6}{5}x^2-\frac{8}{25}x^3, |x|<5$
    (h) $1+\frac{1}{2}x+\frac{1}{4}x^2+\frac{1}{8}x^3, |x|<2$
    (i) $\frac{1}{8}-\frac{3}{16}x+\frac{3}{16}x^2-\frac{5}{32}x^3, |x|<2$
    (j) $2x-\frac{1}{4}x^4+\frac{3}{64}x^7-\frac{5}{512}x^{10}, |x|<\sqrt[3]{4}$
    (k) $1+2x-6x^2+28x^3, |x|<\frac{1}{8}$
    (l) $\frac{4}{3}-\frac{16}{9}\sqrt{3}x+\frac{40}{9}x^2+\frac{80}{27}\sqrt{3}x^3, |x|<\sqrt{3}$

10 $1+4x-8x^2+32x^3, 1.039\,232$
    (a) $10.392\,32$ (b) $1.732\,05$

11 $1+\frac{4}{3}x-\frac{16}{9}x^2$ (a) $5.065\,78$ (b) $9.996\,67$

12 6

13 15

14 $1+3x+\frac{15}{2}x^2, |x|<\frac{1}{4}, 4.123$

15 $4, 6, -100$

16 $1+x+2x^2+3x^3+5x^4, 1.001\,002\,003\,005$

17 (a) $2-2x+\frac{7}{2}x^2$ (b) $1+5x+6x^2$

18 $-\frac{56}{27}$

Miscellaneous exercise 14 (page 195)

1 $1+5x+\frac{15}{2}x^2+\frac{5}{2}x^3$

2 $1-2x-2x^2-4x^3$

3 $1-6x+24x^2-80x^3$

4 $1-4x^2+12x^4-32x^6$

5 $2+\frac{1}{4}x-\frac{1}{64}x^2, |x|<4$

6 $1-\frac{3x^2}{2a^2}+\frac{15x^4}{8a^4}$

7 $1+\frac{1}{2}x-\frac{1}{8}x^2, a=2, b=-\frac{3}{4}$

8 $1+2x+3x^2+4x^3, a=5, b=7$

9 $4+x-\frac{1}{16}x^2, |x|<\frac{8}{3}$

10 $1-\frac{1}{3}x-\frac{1}{9}x^2-\frac{5}{81}x^3$

11 $1+\frac{1}{4}x-\frac{3}{32}x^2$

12 $\frac{3}{2}-\frac{3}{4}x+3x^2$

13 $n=15, 1-\frac{1}{2}x-\frac{1}{8}x^2, \frac{1351}{780}$

15 $|x|<\frac{1}{4}$

16 (a) $1-2x+3x^2-4x^3$
    (b) $1+2x^2+3x^4+4x^6$
    (c) $1-4x^2+12x^4-32x^6$

17 $A=1, B=-1, C=0, D=1, E=-1$
    (a) $0.999\,700\,000\,026\,991\,9$

18 $1+2x+3x^2+4x^3, 1.000\,200\,030\,004$

19 $1-\frac{1}{4}x+\frac{5}{32}x^2, 1.495\,35$

20 $1-\frac{1}{2}x+\frac{3}{8}x^2+\frac{5}{16}x^3$

23 $1+\frac{3}{2}x+\frac{3}{8}x^2, 3.605\,525$

24 $1-2x+4x^2-8x^3+16x^4, 0.346\,056, 0.69$

25 $3+\frac{7}{2}x+\frac{1}{2}x^2, 3\frac{1}{24}$

26 $1-2x^2+15x^4, 0.531$

15 Rational functions

Exercise 15A (page 201)

1 (a) $2x-4$ (b) $3x+2$
    (c) x^2-3x+6 (d) $\dfrac{1}{3x+2}$
    (e) $(x+3)(x-2)$ (f) $\dfrac{1}{x^2+x+1}$

2 (a) 5   (b) $\frac{1}{4}$ (c) 1
    (d) -1 (e) 1   (f) $\frac{2}{3}$

3 (a) $x+4$ (b) $\dfrac{1}{x+7}$ (c) $\dfrac{3x+2}{2x+1}$
    (d) $\dfrac{x+6}{x-3}$ (e) $\dfrac{x+4}{x-4}$ (f) $\dfrac{-2(4x+5)}{3x+1}$

4 (a) $\dfrac{5x}{12}$ (b) $\dfrac{25x}{12}$ (c) $\dfrac{7x+11}{12}$
    (d) $\dfrac{x-7}{15}$ (e) $\dfrac{x^2+4x+2}{4}$ (f) $\dfrac{13x+14}{5}$

5 (a) $\dfrac{8+3x}{4x}$ (b) $\dfrac{5}{2x}$
    (c) $\dfrac{7}{12x}$ (d) $\dfrac{3x-5}{2x}$
    (e) $\dfrac{5x-2-x^2}{2x}$ (f) $\dfrac{x^2+2x+1}{x^2}$

6 (a) $\dfrac{6x+10}{(x+1)(x+3)}$ (b) $\dfrac{13x-1}{(x-2)(2x+1)}$

(c) $\dfrac{2x+10}{(x+3)(x+4)}$ (d) $\dfrac{5x+13}{(x-3)(x+1)}$

(e) $\dfrac{22x+19}{(2x+3)(3x+1)}$ (f) $\dfrac{26x-20}{(2x+1)(5x-3)}$

7 (a) $\dfrac{4x+7}{(3x-1)(2x+1)}$ (b) $\dfrac{-3}{2x(4x+1)}$

(c) $\dfrac{8x^2+13x}{(x+2)(x+1)}$ (d) $\dfrac{6x^2+17x}{(2x-1)(x+2)}$

(e) $\dfrac{2x^2+6x+5}{(x+1)(x+2)}$ (f) $\dfrac{x^2-2x+18}{(x+4)(x-2)}$

8 (a) $\dfrac{4x+5}{(x+1)(x+3)}$ (b) $\dfrac{6x-1}{(x+2)(x-1)}$

(c) $\dfrac{6x+2}{x(x-3)}$ (d) $\dfrac{-4x}{(x+2)(x-2)}$

(e) $\dfrac{1}{x-3}$ (f) $\dfrac{5}{2x-1}$

9 (a) 6 (b) $3x$ (c) $\dfrac{3x+15}{x+3}$ (d) $\dfrac{x+3}{x+1}$

(e) $\dfrac{2x+2}{x+3}$ (f) $\dfrac{(2x+3)(3x-2)}{(2x-3)(3x+2)}$

10 (a) 2 (b) $-\tfrac{1}{3}$ (c) $\dfrac{x+4}{x+2}$

(d) $\dfrac{(5x-1)(x+2)}{x-1}$ (e) 1 (f) -1

11 $a=5, b=10, c=4$

12 x^2-9

13 (a) $\dfrac{16x-6}{(x+4)(x-3)}$ (b) $\dfrac{7x-77}{(x+4)(x-3)}$

14 (a) $\dfrac{3}{x(x-1)(x-3)}$ (b) $\dfrac{1}{(2x+1)(3x-1)}$

Exercise 15B (page 206)

1 (a) $\dfrac{1}{x+5}+\dfrac{1}{x+3}$ (b) $\dfrac{3}{x-1}+\dfrac{7}{x+5}$

(c) $\dfrac{-4}{x-4}+\dfrac{5}{x-5}$ (d) $\dfrac{8}{2x-1}-\dfrac{4}{x+3}$

2 (a) $\dfrac{5}{x+2}+\dfrac{3}{x-1}$ (b) $\dfrac{5}{x-4}-\dfrac{5}{x+1}$

(c) $\dfrac{4}{x-3}+\dfrac{6}{x+3}$ (d) $\dfrac{3}{x}-\dfrac{6}{2x+1}$

3 (a) $\dfrac{3}{x+2}-\dfrac{5}{x-1}+\dfrac{2}{x-3}$

(b) $\dfrac{9}{x+3}-\dfrac{2}{x+1}+\dfrac{1}{x-1}$ (c) $\dfrac{3}{x}+\dfrac{5}{x-6}+\dfrac{7}{x+4}$

4 (a) $10\ln|x-3|-3\ln|x-1|+k$

(b) $\ln|x-2|-\ln|x+2|+k$

(c) $7\ln|x|+\tfrac{1}{2}\ln|2x+5|+k$

(d) $\tfrac{5}{3}\ln|3x+1|-\tfrac{3}{2}\ln|2x-1|+k$

5 (a) $\ln 18$ (b) $\ln 40$ (c) $3\ln\tfrac{16}{7}$ (d) $\ln\tfrac{54}{125}$

6 $\dfrac{1}{1+x}+\dfrac{1}{1-2x}$, $2+x+5x^2+7x^3$

7 $\dfrac{6}{1+4x}-\dfrac{3}{1+2x}$, $3-18x+84x^2-360x^3$; $|x|<\tfrac{1}{4}$

8 $\dfrac{3a}{x+2a}+\dfrac{a}{x-a}$

9 $2\ln\tfrac{3}{2}$

Exercise 15C (page 211)

1 (a) $\dfrac{1}{x-1}-\dfrac{1}{x-3}+\dfrac{2}{(x-3)^2}$

(b) $\dfrac{5}{x+2}+\dfrac{2}{(x+2)^2}+\dfrac{1}{x-1}$

(c) $\dfrac{3}{2x}-\dfrac{3}{2(x-2)}+\dfrac{3}{(x-2)^2}$

(d) $\dfrac{2}{2x-1}-\dfrac{1}{x+1}-\dfrac{5}{(x+1)^2}$

2 (a) $4\ln|x+1|+2\ln|x+2|-\dfrac{5}{x+2}+k$

(b) $\ln|2x-3|-\ln|5x+2|-\dfrac{1}{5(5x+2)}+k$

4 $\tfrac{5}{2}+2\ln 2-\tfrac{3}{2}\ln\tfrac{7}{5}$

5 $1-3x+9x^2$

Exercise 15D (page 215)

1 (a) $\dfrac{1}{x-1}+\dfrac{1}{x^2+1}$ (b) $\dfrac{1}{x+1}-\dfrac{x}{x^2+4}$

(c) $\dfrac{1}{x+3}+\dfrac{x-2}{x^2+4}$ (d) $\dfrac{4}{2x-3}-\dfrac{x-4}{x^2+1}$

(e) $\dfrac{1}{3x+2}-\dfrac{1}{x^2+16}$ (f) $\dfrac{1}{1+4x}-\dfrac{1-4x}{4+x^2}$

(g) $\dfrac{2}{1+2x}-\dfrac{2+x}{4+x^2}$ (h) $\dfrac{2}{x}-\dfrac{x}{x^2+9}$

(i) $\dfrac{3}{x+4}-\dfrac{6x+1}{2x^2+7}$

2 (a) $1-3x+x^2+x^3$ (b) $-2x+\tfrac{17}{4}x^2-8x^3$

(c) $3-9\tfrac{2}{25}x+27x^2-80\tfrac{623}{625}x^3$

3 (a) $-\dfrac{3}{(x+2)^2}+\dfrac{4x}{\left(x^2+1\right)^2}$ (b) $-\dfrac{1}{x^2}+\dfrac{2\left(4-x^2\right)}{\left(x^2+4\right)^2}$

(c) $-\dfrac{1}{\left(x+4\right)^2}+\dfrac{8x}{\left(x^2+16\right)^2}$

Exercise 15E (page 219)

1 (a) $1+\dfrac{1}{x}$ (b) $1-\dfrac{1}{x+1}$ (c) $1+\dfrac{2}{x-1}$

(d) $1+\dfrac{2}{x^2-1}$, $1+\dfrac{1}{x-1}-\dfrac{1}{x+1}$

(e) $3-\dfrac{x}{(2x-3)(x-2)}$, $3+\dfrac{3}{2x-3}-\dfrac{2}{x-2}$

(f) $4-\dfrac{x+9}{(2x+5)(3x+1)}$, $4+\dfrac{1}{2x+5}-\dfrac{2}{3x+1}$

2 (a) $1-\dfrac{1}{x^2}+\dfrac{1}{x}-\dfrac{2}{x+1}$ (b) $1+\dfrac{1}{x}-\dfrac{x-1}{x^2+1}$

(c) $1-\dfrac{2}{x+4}+\dfrac{x-4}{x^2+1}$ (d) $2+\dfrac{2}{x-2}-\dfrac{3}{(x+2)^2}$

(e) $3+\dfrac{5}{x-2}-\dfrac{2}{x+2}-\dfrac{3}{2x-1}$

(f) $-1+\dfrac{2}{x}+\dfrac{1}{2x-5}+\dfrac{3}{2x+5}$

(g) $1+\dfrac{2}{x-1}-\dfrac{1}{x}-\dfrac{1}{x+1}$

(h) $1+\dfrac{3}{x^2}-\dfrac{4}{x+2}$ (i) $3-\dfrac{2}{x-1}+\dfrac{4}{4x^2+9}$

Miscellaneous exercise 15 (page 220)

1 $\dfrac{1}{x-3}-\dfrac{1}{x+1}$

2 $-\dfrac{2}{x}+\dfrac{1}{x-1}+\dfrac{1}{x+1}$

3 $\dfrac{1}{x}+\dfrac{1}{x-1}+\dfrac{3}{(x-1)^2}$

4 $\dfrac{1}{x+2}+\dfrac{1}{(x+2)^2}-\dfrac{2}{3x-1}$

5 $\ln|x|-\ln|x+1|+k$

6 $\dfrac{1}{x^2}+\dfrac{1}{x}+\dfrac{1}{3-x}$

7 $-\ln|x+1|+2\ln|x+2|+k$

8 $3x-\dfrac{3}{x}+\ln\left|x^2(x-1)\right|+k$

9 $\dfrac{7}{4}+\dfrac{7}{4}x-\dfrac{63}{16}x^2$

10 $-\dfrac{1}{x}-\dfrac{1}{x^2}+\dfrac{1}{x-1}$, $-\ln|x|+\dfrac{1}{x}+\ln|x-1|+k$

11 $\left|\dfrac{3x-1}{4x-3}\right|$

12 $\dfrac{2x}{(x-3)(x+3)}$

13 $1-\dfrac{2}{(2x-3)^2}+\dfrac{\frac{1}{3}}{2x-3}-\dfrac{\frac{1}{3}}{2x+3}$

14 (a) $\dfrac{1}{x}-\dfrac{x+1}{2x^2+3}$ (b) $\dfrac{2}{x-1}+\dfrac{x-1}{3x^2+2}$

15 $-\dfrac{2}{2x+1}+\dfrac{1}{x-1}$, $\ln\frac{10}{7}$

16 $\dfrac{1}{7(x+3)}+\dfrac{1}{7(4-x)}$, $\frac{1}{7}\ln\frac{10}{3}$

17 $\dfrac{2}{1-x}-\dfrac{2}{2-x}$, $1+\frac{3}{2}x+\frac{7}{4}x^2+\frac{15}{8}x^3$; $|x|<1$

18 $\dfrac{1}{2+x}+\dfrac{1}{1-2x}$, $\frac{3}{2}+\frac{7}{4}x+\frac{33}{8}x^2$; $|x|<\frac{1}{2}$

19 $B=1, C=3$

20 $A=2, B=2, C=-1$; $2+\frac{1}{2}x-\frac{3}{4}x^2+\frac{17}{8}x^3$; $\frac{1}{2}$

21 $\dfrac{1}{1-x}+\dfrac{2}{(1-x)^2}+\dfrac{3}{4-x}$, $2\ln 2+1$

22 $-\dfrac{1}{30(x+3)}+\dfrac{1}{30(x-3)}-\dfrac{1}{20(x-2)}+\dfrac{1}{20(x+2)}$

23 $-\dfrac{1}{1+x}-\dfrac{2}{1-x}+\dfrac{3}{(1-x)^2}$, $c_0=0, c_1=5, c_2=6$;

$3r+1-(-1)^r$

24 (a) $2x^3-3x^2-11x+6$ (b) -14

(c) $\dfrac{2}{5(2x-1)}-\dfrac{1}{5(x+2)}$

25 $\pi(2-\ln 2)$

26 (a) $A=1, B=4, C=2$

27 (a) $\dfrac{1}{x-3}-\dfrac{1}{x-1}$,

$\dfrac{1}{(x-3)^2}-\dfrac{1}{x-3}+\dfrac{1}{(x-1)^2}+\dfrac{1}{x-1}$

(c) $\left(\frac{25}{48}-\ln\frac{3}{2}\right)\pi$

16 Complex numbers

Exercise 16A (page 226)

1 (a) 4 (b) 6i (c) 13 (d) 24i
(e) 24i (f) −10 (g) 16 (h) −36

2 (a) $4-i$ (b) $2+3i$ (c) $7+0i$
(d) $5+2i$ (e) $5-5i$ (f) $8+6i$
(g) $\frac{1}{5}(1+7i)$ (h) $\frac{1}{10}(1-7i)$ (i) $1-3i$
(j) $2+4i$ (k) $\frac{1}{2}(-1-3i)$ (l) $\frac{1}{5}(3+i)$

3 $2x-y=1$, $x+2y=3$, $x=1$, $y=1$; $1+i$

4 (a) 3 (b) −3 (c) 3 (d) −6
(e) $\frac{1}{2}$ (f) −1

5 (a), (b), (d)

Exercise 16B (page 230)

1 (a) $-2-5i$ (b) $\frac{1}{3}(7-2i)$

  (c) $\frac{1}{2}(-5+i)$ (d) $\frac{1}{25}(7+24i)$

2 (a) $z=2, w=i$ (b) $z=1+i, w=2i$

3 (a) $\pm 3i$ (b) $-2\pm i$
  (c) $3\pm 4i$ (d) $\frac{1}{2}(-1\pm 5i)$

4 (a) $1-7i$ (i) 2 (ii) $14i$ (iii) 50
    (iv) $\frac{1}{25}(-24+7i)$

  (b) $-2-i$ (i) -4 (ii) $2i$ (iii) 5
    (iv) $\frac{1}{5}(3-4i)$

  (c) 5 (i) 10 (ii) 0 (iii) 25
    (iv) 1

  (d) $-3i$ (i) 0 (ii) $6i$ (iii) 9
    (iv) -1

5 (a) $(z-5i)(z+5i)$
  (b) $(3z-1-2i)(3z-1+2i)$
  (c) $(2z+3-2i)(2z+3+2i)$
  (d) $(z-2)(z+2)(z-2i)(z+2i)$
  (e) $(z-3)(z+3)(z-i)(z+i)$
  (f) $(z-2)(z+1-2i)(z+1+2i)$
  (g) $(z+1)(z-2-i)(z-2+i)$
  (h) $(z-1)^2(z+1-i)(z+1+i)$

6 $1-i, -1+2i, -1-2i$

7 $-2-i, 2+\sqrt{7}i, 2-\sqrt{7}i$

10 $a^5-10a^3b^2+5ab^4, 5a^4b-10a^2b^3+b^5;$
   $a^5-10a^3b^2+5ab^4, -5a^4b+10a^2b^3-b^5$

Exercise 16C (page 235)

3 (a) $\pm 1, \pm i$ (b) $-1, \frac{1}{2}(1\pm\sqrt{3}i)$

  (c) $-2, 1\pm 3i$ (d) $-3, 1, -1\pm\sqrt{2}i$

  (e) $\pm 2i, \frac{1}{2}(-1\pm\sqrt{3}i)$

4 $-1, 1, \frac{1}{2}(1\pm\sqrt{3}i)$

5 (a) Circle centre O radius 5, $x^2+y^2=25$
  (b) Line $x=3$ (c) Line $x=3$
  (d) Line $y=1$
  (e) Circle centre $2+0i$ radius 2,
    $x^2+y^2-4x=0$
  (f) Line $x=2$ (g) Line $y=2x+3$
  (h) Circle centre $\frac{1}{2}+0i$ radius $1\frac{1}{2}$,
    $x^2+y^2-x=2$
  (i) Parabola $y^2=4x$

6 (a) Exterior of circle centre O radius 2,
    $x^2+y^2>4$
  (b) Interior and boundary of circle centre $0+3i$
    radius 1, $x^2+y^2-6y+8\leq 0$
  (c) Half-plane including boundary, $x+y\leq 0$
  (d) Interior of circle centre $-1+0i$ radius 2,
    $x^2+y^2+2x-3<0$

11 If in a triangle ABC, O is the mid-point of BC,
then $AB^2+AC^2=2OA^2+2OC^2$ (this is
Apollonius' theorem).

Exercise 16D (page 239)

1 (a) $1-i, -1+i$ (b) $1+2i, -1-2i$
  (c) $3+2i, -3-2i$ (d) $3-i, -3+i$

2 (a) $i, -1-i$ (b) $2, -3+i$
  (c) $-1-i, -3+i$ (d) $-2i, -1+i$
  (e) $1, -2-i$

3 (a) $2+2i, -2-2i, 2-2i, -2+2i$
  (b) $\frac{1}{2}\sqrt{2}(3+i), \frac{1}{2}\sqrt{2}(-3-i), \frac{1}{2}\sqrt{2}(1-3i),$
    $\frac{1}{2}\sqrt{2}(-1+3i)$

4 $-2i, \sqrt{3}+i, -\sqrt{3}+i$

5 $-1-i; z^2-(1+i)z+2i=0;$
  $\frac{1}{2}(1+\sqrt{3})+\frac{1}{2}(1-\sqrt{3})i, \frac{1}{2}(1-\sqrt{3})+\frac{1}{2}(1+\sqrt{3})i$

Miscellaneous exercise 16 (page 239)

1 $3-4i$

2 $-3i, \frac{5}{3}$

3 $1-3i; z^3-4z^2+14z-20=0$

4 (a) $3,3$ (b) $4i$

5 $5-2i, -5+2i$

6 $3-i, -3+3i$

7 $-2i, -2-2i, -4$
  (a) $p=-4, q=2$ (b) $1-i, 1+i, -1, -4$

8 2

9 $z_3=3z_1, z_4=4z_1; z_n=nz_1$, correct for $n=5$.

10 (a) $kz^2+2iz-k=0; a=1, b=-1, c=0,$
    $d=1$
  (b) α is real and negative.
  (c) $\beta-\frac{1}{2}ki=\dfrac{k}{2(k^2+1)}\left(2k+(1-k^2)i\right)$

17 Complex numbers in polar form

Exercise 17A (page 242)

1 (a) $1+\sqrt{3}i$ (b) $-5\sqrt{2}+5\sqrt{2}i$
  (c) $0-5i$ (d) $-3+0i$
  (e) $-4.16+9.09i$ (f) $-0.99-0.14i$

2 $r(\cos\theta+i\sin\theta)$ where
  (a) $r=2.24, \theta=1.11$ (b) $r=5, \theta=-0.93$
  (c) $r=7.81, \theta=2.27$
  (d) $r=10.63, \theta=-2.29$
  (e) $r=1, \theta=0$ (f) $r=2, \theta=\frac{1}{2}\pi$

(g) $r=3, \theta=\pi$ (h) $r=4, \theta=-\frac{1}{2}\pi$

(i) $r=2, \theta=-\frac{1}{4}\pi$ (j) $r=2, \theta=\frac{2}{3}\pi$

5 (a) $-2+2\sqrt{3}\,i$ (b) $1+2i$

   (c) $-3-6i$ (d) $\sqrt{3}-i$

   (e) $2-2\sqrt{3}+i$ (f) $\left(1-\sqrt{3}\right)+\left(1+\sqrt{3}\right)i$

6 (a) $\sqrt{3}+i$ (b) $3+4i$

   (c) $-3+4i$, $-9+4i$

   (d) $\mp\frac{1}{2}\sqrt{2}+\left(2\pm\frac{1}{2}\sqrt{2}\right)i$

7   $\frac{1}{5}\pi$

8   $-\frac{5}{6}\pi$

9   $\frac{1}{4}\pi$

10 $\frac{1}{3}\pi$

11 $\sqrt{5}$

12 $\frac{1}{3}\pi$

Exercise 17B (page 247)

1 $r(\cos\theta+i\sin\theta)$, where

   (a) $r=2, \theta=\frac{7}{12}\pi$ (b) $r=2, \theta=\frac{1}{12}\pi$

   (c) $r=\frac{1}{2}, \theta=-\frac{1}{12}\pi$ (d) $r=8, \theta=-\frac{1}{2}\pi$

   (e) $r=2, \theta=\frac{5}{6}\pi$ (f) $r=\frac{1}{4}, \theta=-\frac{11}{12}\pi$

   (g) $r=4, \theta=\frac{2}{3}\pi$ (h) $r=16, \theta=\frac{1}{3}\pi$

   (i) $r=8, \theta=0$ (j) $r=1, \theta=\frac{11}{12}\pi$

   (k) $r=2, \theta=-\frac{1}{3}\pi$ (l) $r=2, \theta=\frac{1}{12}\pi$

   (m) $r=2, \theta=-\frac{1}{12}\pi$ (n) $r=4, \theta=\frac{1}{3}\pi$

   (o) $r=1, \theta=-\frac{1}{4}\pi$ (p) $r=1, \theta=-\frac{1}{3}\pi$

   (q) $r=1, \theta=-\frac{1}{3}\pi$ (r) $r=2, \theta=-\frac{1}{6}\pi$

2 $2\left(\cos\frac{4}{5}\pi+i\sin\frac{4}{5}\pi\right)$

5 (a) $\cos\theta-i\sin\theta$ (b) $\cos\theta-i\sin\theta$

   (c) $r(\cos\theta-i\sin\theta)$ (d) $\frac{1}{r}(\cos\theta-i\sin\theta)$

6 $2\left(\cos\frac{1}{3}\pi+i\sin\frac{1}{3}\pi\right)$,

   $\sqrt{2}\left(\cos\left(-\frac{1}{4}\pi\right)+i\sin\left(-\frac{1}{4}\pi\right)\right)$; $-\sqrt{3}+i$

7 (a) $29+278\,i$ (b) $-122-597\,i$

   (c) $(-8.432+5.376\,i)\times10^{-5}$

8 (a) $-\frac{1}{6}\pi$ (b) $\frac{1}{6}\pi$ (c) $\frac{1}{3}\pi$ (d) $\frac{1}{3}\pi$

9 (a) $\frac{1}{2}\pi$ (b) $-\frac{1}{2}\pi$ (c) $-\frac{1}{2}\pi$

10 The semicircle in the first quadrant of the circle with 3 and 4i at ends of a diameter

11 The line segment AB

12 The major arc of the circle with centre -1 passing through i and $-i$

13 (a) $\sec\theta, \theta$ (b) $-\sec\theta, \theta-\pi$

    (c) $-\sec\theta, \theta-\pi$ (d) $\sec\theta, \theta-2\pi$

14 (a) $2\sin\theta, \theta-\frac{1}{2}\pi$ (b) $-2\sin\theta, \theta+\frac{1}{2}\pi$

Exercise 17C (page 250)

1 $-2+4i$, $4+2i$

2 $\left(3+2\sqrt{3}\right)+\left(3\sqrt{3}-2\right)i$, $-\left(2\sqrt{3}-3\right)-\left(3\sqrt{3}+2\right)i$

3 $\pm2\sqrt{3}+\left(3\pm\sqrt{3}\right)i$

4 $\frac{8}{13}(1-5i)$

7 (a) $6+3i$ (b) $\frac{17}{16}(6+3i)$ (c) $\frac{16}{15}(6+3i)$

8 (a) $10(1+i)$ (b) $\frac{15}{2}(1+i)$ (c) $8(1+i)$

11 $(1-i)a+ic$, $(1+i)b-ic$, $\frac{1}{2}(a+b)+\frac{1}{2}(b-a)i$; M is the third vertex of an isosceles right-angled triangle having AB as hypotenuse.

12 $3-3i$, $5+3i$, $4i$, -2

Exercise 17D (page 254)

1 (a) $\pm2\left(\cos\frac{1}{5}\pi+i\sin\frac{1}{5}\pi\right)$

   (b) $\pm3\left(\cos\frac{2}{7}\pi-i\sin\frac{2}{7}\pi\right)$ (c) $\pm(1-i)$

   (d) $\pm(2+5i)$ (e) $+(1.098\ldots+0.455\ldots i)$

   (f) $\pm(3-2i)$

2 (a) $\pm\left(2+2\sqrt{3}\right)i$ (b)$\pm\left(\sqrt{3}+i\right)$, $\pm\left(1-\sqrt{3}\,i\right)$

3 e^z-e^{z*} is imaginary, $e^z\div e^{z*}$ has modulus 1.

5 (a) $e^{\frac{1}{3}\pi i}$, $e^{\frac{4}{3}\pi i}$ (b) $\pm\sqrt{e}(\cos 1+i\sin 1)$

6 $i\tan y$

10 $e^{\cos\theta}$, $\sin\theta$

Miscellaneous exercise 17 (page 255)

1 (a) $\sec\alpha$ (b) $4\sec\alpha$ (c) $\frac{1}{2}\pi-\alpha$

   (d) $\frac{2}{5}\pi-\alpha$

2 $5, -0.927$ (a) $\frac{5}{3}$ (b) 0.120

3 6

4 Circle with centre $3+4i$ and radius 2

   (a) 7 (b) $2\sin^{-1}0.4\approx0.823$

6 Interior and boundary of the circle with centre $-2+2\sqrt{3}\,i$ and radius 2   (a) 2 (b) $\frac{5}{6}\pi$

7 (a) $-3\pm5i$ (b) $\sqrt{34}$, ±2.11 (c) 10

8 (a) $\frac{1}{4}\pi$, $\frac{5}{6}\pi$ (b) 8, $-\frac{11}{12}\pi$

   (c) Perpendicular bisector of line segment joining points representing α and β

   (d) $\frac{13}{24}\pi$

9 (a) $-3-4i$, $11-2i$; $-1-2i$, -5

   (b) $\sqrt{5}$, ±2.03; 5, π

10 (a) 2, $\frac{1}{8}\pi$; $4\sqrt{2}$, $\frac{5}{8}\pi$; $8\sqrt{2}$, $\frac{3}{4}\pi$;

    $\frac{1}{4}\sqrt{2}$, $-\frac{1}{2}\pi$

    (b) $-8+8i$

11 (a) $1-\sqrt{3}\,i,\ -\frac{2}{3}$

   (b) $2,\ \frac{1}{3}\pi;\ 2,\ -\frac{1}{3}\pi;\ 4, 0;\ 1,\ \frac{2}{3}\pi$

   (c) Circle with centre $1+\sqrt{3}\,i$ and radius $\sqrt{3}$

   (d) $1-\sqrt{3}\,i$

12 (a) (i) Line through A parallel to OB

       (ii) Line AB

       (iii) Circle having AB as a diameter

13 $|f(1)-f(0)|=1$

If $g(z)=z$ and $g(w)=w*$,
then $|z-w*|=|z-w|$, $(z-z*)(w-w*)=0$,
$\operatorname{Im}z\times\operatorname{Im}w=0$, so $\operatorname{Im}z=0$ or $\operatorname{Im}w=0$;
$g(z)=z$ or $g(z)=z*$;
$f(z)=\alpha+\beta z$ with $|\beta|=1$ (rotation then
translation) or $f(z)=\alpha+\beta z*$ with $|\beta|=1$
(reflection in real axis then rotation then
translation)

18 Integration

Exercise 18A (page 261)

1 (a) $2\ln\left|\sqrt{x}-2\right|+k$ (b) $-\dfrac{1}{3(3x+4)}+k$

   (c) $2\cos\left(\frac{1}{3}\pi-\frac{1}{2}x\right)+k$

   (d) $\frac{1}{6}(x-1)^6+\frac{1}{7}(x-1)^7+k$

      $\equiv\frac{1}{42}(6x+1)(x-1)^6+k$

   (e) $\ln(1+e^x)+k$ (f) $\frac{1}{2}\ln(3+4\sqrt{x})+k$

   (g) $\frac{6}{5}(x+2)^{\frac{5}{2}}-4(x+2)^{\frac{3}{2}}+k$

      $\equiv\frac{2}{5}(3x+4)(x+2)^{\frac{3}{2}}+k$

   (h) $6(x-3)^{\frac{1}{2}}+\frac{2}{3}(x-3)^{\frac{3}{2}}+k$

      $\equiv\frac{2}{3}(x+6)\sqrt{x-3}+k$

   (i) $\ln(\ln x)+k$ (j) $\sin^{-1}\left(\frac{1}{2}x\right)+k$

2 (a) $\frac{1}{20}(2x+1)^5-\frac{1}{16}(2x+1)^4+k$

      $\equiv\frac{1}{80}(8x-1)(2x+1)^4+k$

   (b) $\frac{1}{28}(2x-3)^7+\frac{7}{24}(2x-3)^6+k$

      $\equiv\frac{1}{168}(12x+31)(2x-3)^6+k$

   (c) $\frac{1}{6}(2x-1)^{\frac{3}{2}}+\frac{1}{10}(2x-1)^{\frac{5}{2}}+k$

      $\equiv\frac{1}{15}(3x+1)(2x-1)^{\frac{3}{2}}+k$

   (d) $4(x-4)^{\frac{1}{2}}+\frac{2}{3}(x-4)^{\frac{3}{2}}+k$

      $\equiv\frac{2}{3}(x+2)\sqrt{x-4}+k$

   (e) $\ln|x+1|+\dfrac{1}{x+1}+k$

   (f) $\frac{1}{2}x-\frac{3}{4}\ln|2x+3|+k$

3 (a) $\frac{1}{3}\sin^{-1}3x+k$

   (b) $\frac{8}{3}\sin^{-1}\frac{3}{4}x+\frac{1}{2}x\sqrt{16-9x^2}+k$

   (c) $\frac{1}{2}\ln(2e^x+1)+k$

   (d) $\frac{3}{5}(x+1)^{\frac{5}{3}}-\frac{3}{2}(x+1)^{\frac{2}{3}}+k$

      $\equiv\frac{3}{10}(2x-3)(x+1)^{\frac{2}{3}}+k$

   (e) $\dfrac{x}{\sqrt{1-x^2}}+k$

   (f) $-4\ln\left|2-\sqrt{x}\right|-2\sqrt{x}+k$

4 (b) $\tan^{-1}e^x+k$

5 (a) $\frac{1}{2}\ln\left|\dfrac{e^x-1}{e^x+1}\right|+k$ (b) $2\ln\dfrac{\sqrt{x}}{\sqrt{x}+1}+k$

Exercise 18B (page 263)

1 (a) $\ln\left(\frac{1}{2}(1+e)\right)$ (b) $2\ln 2$ (c) $\frac{13}{42}$

   (d) $1\frac{1}{15}$ (e) $\frac{1}{6}\pi$ (f) $109\frac{1}{15}$

   (g) 8π (h) $\frac{1}{2}\left(\tan^{-1}3-\tan^{-1}\frac{1}{2}\right)=\frac{1}{8}\pi$

   (i) $\frac{1}{2}$ (j) $\frac{1}{3}\sqrt{3}$ (k) $3\ln\frac{4}{3}$

2 $\frac{1}{4}\pi-\frac{1}{2}$

3 (a) $\frac{1}{4}\pi$ (b) $\frac{1}{2}\pi$ (c) $\frac{1}{6}\pi$

   (d) π (e) $1-\frac{1}{2}\sqrt{2}$ (f) $\frac{1}{2}\pi$

4 (a) $\dfrac{1}{2(\ln 2)^2}$ (b) $\ln 2$

   (c) $2(1-\ln 2)$ (d) 1

7 π

Exercise 18C (page 266)

1 (a) $\frac{1}{4}(x^2+1)^4+k$ (b) $\frac{1}{3}(4+x^2)^{\frac{3}{2}}+k$

   (c) $\frac{1}{6}\sin^6 x+k$ (d) $\frac{1}{4}\tan^4 x+k$

   (e) $-\sqrt{1-x^4}+k$ (f) $-\frac{1}{8}\cos^4 2x+k$

2 (a) $-\frac{1}{12}(1-x^2)^6+k$ (b) $-\frac{1}{6}(3-2x^2)^{\frac{3}{2}}+k$

   (c) $-\frac{1}{63}(5-3x^3)^7+k$ (d) $\frac{2}{3}\sqrt{1+x^3}+k$

   (e) $\frac{1}{4}\sec^4 x+k$ (f) $\frac{1}{16}\sin^4 4x+k$

3 (a) $\ln(1+\sin x)+k$ (b) $\frac{1}{3}\ln\left|1+x^3\right|+k$

   (c) $\ln|\sin x|+k$ (d) $\ln(4+e^x)+k$

   (e) $-\frac{2}{3}\ln\left|5-e^{3x}\right|+k$ (f) $\frac{1}{3}\ln|\sec 3x|+k$

4 (a) $\ln(e+1)$ (b) $\frac{1}{2}\ln 2$ (c) $\frac{1}{2}\ln\frac{4}{3}$

5 (a) $\frac{2}{3}$ (b) 78   (c) $\frac{1}{12}(4-\sqrt{2})$

(d) $\frac{860}{3}\sqrt{2}$ (f) $\frac{1}{6}$ (f) $\frac{2}{3}(10\sqrt{10}-1)$

(g) $\frac{1}{3}$ (h) $\dfrac{1}{n+1}$ (i) $\frac{7}{3}$

6 $\dfrac{1}{2(n-1)}\left(1-\dfrac{1}{\left(1+a^2\right)^{n-1}}\right)$ if $n\neq 1$,

$\frac{1}{2}\ln\left(1+a^2\right)$ if $n=1$; $n>1$; $\dfrac{1}{2(n-1)}$

Exercise 18D (page 270)

1 (a) $\sin x - x\cos x + k$ (b) $3(x-1)e^x + k$

(c) $(x+3)e^x + k$

2 (a) $\frac{1}{4}(2x-1)e^{2x} + k$

(b) $\frac{1}{4}x\sin 4x + \frac{1}{16}\cos 4x + k$

(c) $\frac{1}{4}x^2(2\ln 2x - 1) + k$

3 (a) $\frac{1}{36}x^6(6\ln 3x - 1) + k$

(b) $\frac{1}{4}(2x-1)e^{2x+1} + k$

(c) $x(\ln 2x - 1) + k$

4 (a) $\frac{1}{4}(e^2+1)$ (b) $\frac{1}{2}\sqrt{2}(4-\pi)$

(c) $\dfrac{ne^{n+1}+1}{(n+1)^2}$

5 $(2-x^2)\cos x + 2x\sin x + k$

(a) $\frac{1}{4}(2x^2 - 2x + 1)e^{2x} + k$

(b) $2(x^2-8)\sin\frac{1}{2}x + 8x\cos\frac{1}{2}x + k$

6 $1-3e^{-2}$, $\frac{1}{4}\pi\left(1-13e^{-4}\right)$

7 $\frac{1}{9}\pi$, $\frac{1}{324}\pi^2\left(2\pi^2-3\right)$

8 (a) $-\frac{1}{2}\left(1+e^\pi\right)$

(b) $\frac{1}{10}e^{4\pi} - \frac{1}{10}e^{-4\pi}$

(c) $\dfrac{a-e^{-2a\pi}(a\cos 2b\pi - b\sin 2b\pi)}{a^2+b^2}$

9 (a) $\frac{1}{2}$ (b) $\frac{4}{3}$

Miscellaneous exercise 18 (page 271)

1 $\frac{1}{2}\ln|2x-1| - \dfrac{1}{2(2x-1)} + k$

2 $2\ln 2 - \frac{3}{4}$

3 $\frac{1}{6}\tan^{-1}\frac{2}{3}x + k$

4 $\frac{1}{110}$

5 $\frac{1}{4} - \frac{1}{12}e^{\frac{2}{3}}$

6 (a) $x\cos^{-1}x - \sqrt{1-x^2} + k$

(b) $x\tan^{-1}x - \frac{1}{2}\ln\left(1+x^2\right) + k$

(c) $x\left((\ln x)^2 - 2\ln x + 2\right) + k$

7 $\frac{1}{2}\pi - 1$

8 $\frac{1}{2}\tan^{-1}\frac{1}{2}e^x + k$

9 $\ln\left(1+3x^2\right) + k$

10 $\frac{1}{2}\ln 10$

11 $\frac{2}{5}\sin^5 x + k$

12 $2\sqrt{x} - 2\ln\left(1+\sqrt{x}\right) + k$

13 (a) $\frac{1}{6}x(4x-1)^{\frac{3}{2}} - \frac{1}{60}(4x-1)^{\frac{5}{2}} + k$

(b) $-\frac{2}{3}x(2-x)^{\frac{3}{2}} - \frac{4}{15}(2-x)^{\frac{5}{2}} + k$

(c) $\frac{1}{3}x(2x+3)^{\frac{3}{2}} - \frac{1}{15}(2x+3)^{\frac{5}{2}} + k$

16 $\frac{1}{54}(3x-1)^6 + \frac{1}{45}(3x-1)^5 + k$

17 $\frac{1}{8}\pi + \frac{1}{4}$

18 (a) $\frac{1}{7}x(1+x)^7 - \frac{1}{56}(1+x)^8 + k$

(b) $\frac{1}{15}x(3x-1)^5 - \frac{1}{270}(3x-1)^6 + k$

(c) $\frac{1}{13}\dfrac{x(ax+b)^{13}}{a} - \frac{1}{182}\dfrac{(ax+b)^{14}}{a^2} + k$

19 $1-2e^{-1}$

20 16

21 $\pi(8\ln 2 - 3)$

22 $2x(x-1)^{\frac{3}{2}} - \frac{4}{5}(x-1)^{\frac{5}{2}} + k$

23 1.701; exact value is $\frac{1}{4}\ln 1201$

24 (a) 0.70   (c) 1.69

(d) $\pi\left(\frac{1}{4} + \frac{3}{32}\pi\right) \approx 1.71$

25 $\frac{3}{4}\pi$

26 $3\pi a^2$, $5\pi^2 a^3$

27 $\frac{1}{4}\pi c^2$, $\frac{1}{3}\pi c^3$

28 $\frac{3}{8}\pi a^2$

29 (a) $A=-1, B=1, C=1$ (b) $a=\frac{1}{2}, b=-\frac{1}{2}$

30 $2 + 2\ln 5 - 3\ln 3$

19 Differential equations

Exercise 19A (page 277)

1. (a) $y = x^3 - 5x^2 + 3x + k$
   (b) $x = \frac{1}{2}t - \frac{1}{12}\sin 6t + k$
   (c) $P = 500e^{0.01t} + k$
   (d) $u = k - 50e^{-2t}$
   (e) $y = \frac{2}{3}\sqrt{x}(x+3) + k$
   (f) $x = \ln\sin t + 2\sin t + k$

2. (a) $x = 5e^{0.4t} - 4$
   (b) $v = 3 - 3\cos 2t - 2\sin 3t$
   (c) $y = -\ln(1 - t^2)$

3. (a) $y = \ln x + \frac{1}{x} - 1$ (b) $y = 2\sqrt{x} - 4$
   (c) $y = x - 2\ln(x+1)$

4. 14.1 s

5. 2

6. $-6 - \frac{5}{2}\pi, 2.237$

7. $y = k - \frac{1}{2}x^2$

8. 0.483 m; $-0.1h + 0.0483$

9. (a) $A + kt$ (b) $A + k\left(t - \frac{1}{4\pi}\sin 2\pi t\right)$
   (c) $A + k\left(t + \frac{5}{4\pi}\sin \frac{1}{5}\pi t\right)$
   (d) $A + k\Big(t - \frac{1}{4\pi}\sin 2\pi t + \frac{5}{4\pi}\sin \frac{1}{5}\pi t$
   $\qquad - \frac{5}{176\pi}\sin \frac{11}{5}\pi t - \frac{5}{144\pi}\sin \frac{9}{5}\pi t\Big)$

   All give $A + 10k$

Exercise 19B (page 284)

1. (a) $y = \dfrac{1}{k - x}$ (b) $y = \sin^{-1} e^{x-k}$
   (c) $x = Ae^{4t}$ (d) $z = \sqrt{2t + c}$
   (e) $x = \cos^{-1}(k - t)$ (f) $u = \sqrt[3]{3ax + c}$

2. (a) $x = 3e^{-2t}$ (b) $u = \dfrac{1}{\sqrt{1 - 2t}}$

3. (a) $y = \dfrac{e^x}{2 - e^x}, x < \ln 2$
   (b) $y = -\ln(3 - x), x < 3$

4. About $6\frac{1}{2}$ minutes

5. 3.6 minutes

6. 443 s

7. (a) $\dfrac{dh}{dt} = 0$ when $h = 25$, so the tree stops growing.
   (b) $\dfrac{dh}{dt} = 0.2\sqrt{25 - h}$, $t = -10\sqrt{25 - h} + 50$
   (c) (i) 1.0 years   (ii) 10 years
   (d) $h = t - 0.01t^2$ for $0 \leqslant t \leqslant 50$

8. $m = \frac{1}{3}$; enlargement factors over the 4 months for $m = \frac{1}{3}$ and $m = \frac{1}{2}$ are 25.6 and 35.3 respectively.

Exercise 19C (page 289)

1. (a) $y^3 = x^3 + k$ (b) $y^2 = x^2 + k$
   (c) $y = ce^{\frac{1}{2}x^2}$ (d) $y^2 = \sqrt{2\ln|x| + k}$

2. $y = \dfrac{4x}{x + 1}$

3. $x^2 + y^2 = k$; a set of circles with centre $(0,0)$, $x^2 + y^2 = 25$

4. $(x+1)^2 + (y-2)^2 = k$, circles with centre $(-1,2)$

5. (a) $y = \sqrt{2}x^{\frac{3}{2}}$ (b) $y = 8\sqrt{2}x^{-\frac{3}{2}}$
   (c) $\sin y = \frac{1}{2} - \cos x$ (d) $\cos y = 2\cos x$

6. $v^2 = k - \omega^2 x^2$; $v^2 = \omega^2(a^2 - x^2)$

7. (a) $y^2 + 1 = k(x^2 + 1)^2$, $k > 0$
   (b) $\sec y = c \sec x, c \neq 0$

8. (a) $y = \dfrac{2}{2 - x}$ (b) $2\sin x \cos y = 1$
   (c) $y = \sqrt{\dfrac{2}{x - 1}}$ (d) $y = 2\sec x$

9. (a) $y = \dfrac{2(1 - kx^4)}{1 + kx^4}$ (b) $y = \ln(k + x\ln x)$
   (c) $y^2 = 4\left(k + \tan \frac{1}{2}x\right)$

10. $y = kx^n$

11. (a) $n = 5000e^{0.01(0.05t - 50\sin 0.02t)}$
    (b) 3150

12. $v^2 = \dfrac{20R^2}{x} + V^2 - 20R$

Miscellaneous exercise 19 (page 290)

1. $y = x^2 + 7x + 3\ln x + 2$

2. $t = \dfrac{10z}{20 - z}$

3. $y = x^3 - 4x^2 + 5x + 3$;
   $(1,5)$ maximum, $\left(\frac{5}{3}, 4\frac{23}{27}\right)$ minimum

4 (a) $2\pi r$, $\dfrac{1}{2\pi r}$ (b) $\dfrac{1}{\pi r(t+1)^3}$

(c) $A = 1 - \dfrac{1}{(t+1)^2}$

5 $\dfrac{dx}{dt} = -kx$

6 (a) The number of people served in each minute; $x = 0.7t^2 - 4t + 8$

7 (a) $400x^2 + 25y^2 = 4$ (b) $(\pm 0.1, 0)$, $(0, \pm 0.4)$

8 $N = 3000e^{0.02t} + 5000$, 35 days; $\dfrac{dN}{dt} = \dfrac{1}{50}N - F$;

since N decreases, $\dfrac{dN}{dt} < 0$ but $\left|\dfrac{dN}{dt}\right|$ gets larger,

so N decreases with increasing rapidity.

9 (a) $T = 470 - 6x$, 360 °C

(b) $\dfrac{dT}{dx} = -kx$, $T = 380 - \frac{1}{10}x^2$

10 (a) $q = Q\left(1 - (1-\lambda)e^{-kt}\right)$

11 $y = \pm\sqrt{k(x-1)^2 e^{2x} + 1}$

12 $\ln y = \frac{1}{4} + \left(-\frac{1}{4} + \frac{1}{2}x\right)e^{2x}$

13 2 minutes; $\frac{1}{4}\pi$

14 (a) 12, $4\frac{1}{2}$, rate of increase $\to 0$, N will never exceed 500

(c) $t = 10\ln\dfrac{N}{500 - N} + k$, $t = 10\ln\dfrac{4N}{500 - N}$, 18 years

15 $p = \dfrac{mAe^{kt}}{1 + Ae^{kt}}$; 54 000;

$\displaystyle\int_{30\,000}^{54\,000} \dfrac{5}{p\left(1 - \left(\frac{1}{100\,000}p\right)^5\right)}\, dp$, 3.0 years

16 $y^2 = 1 + c\operatorname{cosec} 2x$

17 $y = \dfrac{3}{\ln\dfrac{x+1}{x-2} - k}$; $y = \dfrac{3}{\ln\dfrac{e^3(x+1)}{2(x-2)}}$

18 $y = \sqrt[3]{\frac{1}{4}(4 + 6x - 3\sin 2x)}$

19 $(1 + y)e^{-y} = (1 - x)e^x$

20 $P = 600e^{kt}$, $P = 600e^{0.005t - 0.4\sin 0.02t}$; the model $P = 600e^{kt}$ is not consistent with the data; the model $P = 600e^{0.005t - 0.4\sin 0.02t}$ is consistent with the data, given to the nearest 100. Smallest number is 549.

21 (a) According to the model, £745.

(b) £$\frac{400}{3}\sqrt{5}$ per hour, which is approximately £298 per hour.

22 (a) $P = P_0 e^{\sin kt}$ (b) e^2

23 0.185; minimum

Revision exercise 3

(page 296)

1 $\dfrac{\frac{1}{2}}{r} - \dfrac{\frac{1}{2}}{r+2}$

2 $(x+2)(x^2 - 2x + 4)$, $\dfrac{2x-2}{x^2 - 2x + 4} - \dfrac{2}{x+2}$; $a = -1, b = 3$

3 (a) $\frac{8}{27}\sqrt{3} - \frac{2}{3} + \frac{1}{12}\pi$ (b) $2(\sqrt{2} - 1)$

(c) $\frac{1}{4}\pi - \frac{1}{32}\pi^2 - \frac{1}{2}\ln 2$ (d) $\frac{3}{25}\left(e^{\frac{3}{2}\pi} - 1\right)$

4 (a) $2 + i$ (b) $-2 \pm 3i$ (c) $i, 2 + i$

(d) $z = 2 + i, w = i$

5 (a) (i) $1, 0$ (ii) $1, \pi$ (iii) $1, 0$

(iv) $1, \pi$ (v) $1, 2\theta$ (vi) $1, 2\theta - 2\pi$

(vii) $1, 2\theta$ (viii) $1, 2\theta + 2\pi$

(b) (i) $2, 0$ (ii) 0, undefined

(iii) $2, 0$ (iv) 0, undefined

(v) $2\cos\theta, \theta$ (vi) $-2\cos\theta, \theta - \pi$

(vii) $2\cos\theta, \theta$ (viii) $-2\cos\theta, \theta + \pi$

7 (a) $(x-1)(x^2 + x + 1)$, $(x+1)(x^2 - x + 1)$

(b) $(z-1)\left(z + \frac{1}{2} + \frac{1}{2}\sqrt{3}i\right)\left(z + \frac{1}{2} - \frac{1}{2}\sqrt{3}i\right)$,

$(z+1)\left(z - \frac{1}{2} + \frac{1}{2}\sqrt{3}i\right)\left(z - \frac{1}{2} - \frac{1}{2}\sqrt{3}i\right)$

(c) $-1, 1$; no roots; $-1, 1$

(d) ± 1, $\pm\frac{1}{2} \pm \frac{1}{2}\sqrt{3}i$; $\pm i$, $\pm\frac{1}{2}\sqrt{3} \pm \frac{1}{2}i$;

± 1, $\pm i$, $\pm\frac{1}{2} \pm \frac{1}{2}\sqrt{3}i$, $\pm\frac{1}{2}\sqrt{3} \pm \frac{1}{2}i$ (where the $\pm$ signs are independent of each other)

8 $1 + \frac{1}{2}x - \frac{1}{2}x^2 + \frac{1}{2}x^3 - \frac{5}{8}x^4$

9 The point does lie on the line.

10 (b) $\cos 2\beta + i\sin 2\beta$

(e) P lies on a circle through A and B with centre C.

11 (a) $-\frac{1}{2}\cos\left(2x+\frac{1}{6}\pi\right)+k$

(b) $\frac{1}{2}x-\frac{1}{12}\sin 6x+k$ (c) $\frac{1}{6}\sin^3 2x+k$

12 $\mathbf{r}=\begin{pmatrix}1\\4\\2\end{pmatrix}+t\begin{pmatrix}-3\\-1\\1\end{pmatrix}$, $(-5,2,4)$

13 $1-\frac{1}{2}x+\frac{5}{8}x^2-\frac{5}{16}x^3$

14 $\sqrt{\frac{2}{3}}$

15 $\mathbf{r}=\begin{pmatrix}2\\-1\\4\end{pmatrix}+s\begin{pmatrix}1\\-1\\1\end{pmatrix}$

16 $3y=4x+2$

17 $a=nc$, $b=\frac{1}{2}n(n-1)c^2$;

$c=\dfrac{a^2-2b}{a}$, $n=\dfrac{a^2}{a^2-2b}$

18 $1-2x+3x^2-4x^3+\ldots$,

$1-3x+6x^2-10x^3+\ldots$

19 (a) $\ln 2$ (b) $\sqrt{3}\pi-\frac{1}{3}\pi^2$

20 $\dfrac{\frac{1}{4}}{1+x}+\dfrac{\frac{1}{4}}{3-x}$, $\dfrac{\frac{1}{16}}{(1+x)^2}+\dfrac{\frac{1}{16}}{(3-x)^2}+\dfrac{\frac{1}{32}}{1+x}+\dfrac{\frac{1}{32}}{3-x}$;

$\frac{1}{2}\ln 3$, $\left(\frac{1}{12}+\frac{1}{16}\ln 3\right)\pi$

22 (b) $P=108\,000\mathrm{e}^{0.001(10-m)t}$

(c) E.g., no immigration or emigration.

(d) $m=10(1-\ln 2)$

(e) If the death rate were zero, i.e., the lowest possible, the population would take $100\ln 2\approx 69$ years to double.

23 1.97; 500, 3550

24 (a) 4 metres   (c) $-\mathbf{k}$, $2\mathbf{i}$

(d) $\left(\frac{1}{5}t\right)\mathbf{i}-6\mathbf{j}+\left(\frac{1}{10}t-2\right)\mathbf{k}$, $\frac{1}{20}t^2-\frac{2}{5}t+40$,

6.26 m

25 (a) $\dfrac{4-3x}{2\sqrt{2-x}}$ (b) $-\frac{2}{15}(4+3x)(2-x)^{\frac{3}{2}}+k$

26 $2x-2y+z=3$ (a) 4   (b) 7.76

27 (a) 22 (b) $12\mathbf{i}-15\mathbf{j}-9\mathbf{k}$;   2.27

28 $3\mathbf{i}+4\mathbf{j}+\mathbf{k}$, $3x-2y-2z=-1$

29 $\frac{1}{4}\pi^2+3\pi\ln 2-2\pi\ln 3$

Practice examinations for P3

Practice examination 1 for P3 (page 300)

2 $-\dfrac{\ln 3}{\ln 2}$, $\dfrac{\ln 3}{\ln 2}$

3 6°

4 (i) $\ln n=1+\ln 5000-\mathrm{e}^{-0.01t}$

(ii) $n\to 5000\mathrm{e}\approx 13\,600$

5 (i) 56.8°

(ii) E.g., $(-1,1,-2)$, $(-3,0,-6)$

(iii) $\mathbf{r}=\begin{pmatrix}-1\\1\\-2\end{pmatrix}+t\begin{pmatrix}2\\1\\4\end{pmatrix}$

6 (ii) 111.5

7 (i) $\dfrac{1}{x-2}-\dfrac{x}{x^2+1}$

8 (a) $\frac{1}{2}x^2\ln x-\frac{1}{4}x^2+k$ (b) $\ln 3$

9 (i) $y-2x=1$ (iii) $0.77, 0.4$

10 (i) 1

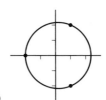

(ii), (iii)

Practice examination 2 for P3 (page 302)

1 $x>-\dfrac{\ln 0.025}{\ln 3}\approx 3.36$

2 (i) 0.551

(ii) Overestimate, since graph of $y=\sec x$ bends upwards

3 (ii) 1.26

4 $\dfrac{\cos 2x-2x\sin 2x}{\cos y}$, $-\frac{1}{2}\pi$

5 (a) $2x-1,1$ (b) $k=\frac{3}{2}$

6 (i) $R=\sqrt{10}$, $\alpha=18.4$

(ii) $69.2, 327.7$

7 (ii) $\ln\frac{3}{2}$

8 (i) $4,\frac{1}{3}\pi$ (ii) $\pm\left(\sqrt{3}+\mathrm{i}\right)$

(iii) $1-\mathrm{i}\left(\sqrt{3}+\sqrt{2}\right)$, $-1+\mathrm{i}\left(\sqrt{3}-\sqrt{2}\right)$

9 (i) $\dfrac{\mathrm{d}V}{\mathrm{d}t}=Cx^2\dfrac{\mathrm{d}x}{\mathrm{d}t}$ (iii) $\frac{2}{5}x^{\frac{5}{2}}=-At+k$

(iv) $\frac{32}{31}$

10 (i) $\begin{pmatrix}1\\1\\9\end{pmatrix}$ (ii) $x+2y-4z=9$

(iii) $\begin{pmatrix}-1\\3\\-1\end{pmatrix}$ (iv) 61.9°

Index

The page numbers refer to the first mention of each term, or the box if there is one.